Galois' theory of algebraic equations

Written, translated and revised by

Jean-Pierre Tignol

Université Catholique de Louvain

Edited by T S Blyth, University of St Andrews

Originally published in French by arrangement with the Institut de Mathématique Pure et Appliquée, Université Catholique de Louvain, Louvain-la-Neuve, Belgium

Copublished in the United States with
John Wiley & Sons, Inc., New York

QA
211
.T5413

Longman Scientific & Technical
Longman Group UK Limited
Longman House, Burnt Mill, Harlow
Essex CM20 2JE, England
and Associated Companies throughout the world.

*Copublished in the United States of America with
John Wiley & Sons, Inc., 605 Third Avenue, New York, NY 10158*

Originally published in French as *Leçons sur la théorie
des équations* (Monographies de Mathématique No. 1)

© 1980 Institut de Mathématique Pure et Appliquée, UCL, Louvain-la-Neuve, Belgium

English translation © Institut de Mathématique Pure et Appliquée, UCL, Louvain-la-Neuve, Belgium, 1988

All rights reserved; no part of this publication
may be reproduced, stored in a retrieval system,
or transmitted in any form or by any means, electronic,
mechanical, photocopying, recording, or otherwise,
without either the prior written permission of the Publishers
or a licence permitting restricted copying issued by the Copyright
Licensing Agency Ltd, 33-34 Alfred Place, London, WC1E 7DP.

English language edition first published by
Longman Group UK Limited, 1988

British Library Cataloguing in Publication Data
Tignol, J.-P.
  Galois' theory of algebraic equations.
  1. Equations
  I. Title II. Blyth, T.S.
  515'.25 QA211
ISBN 0-582-00290-7

Library of Congress Cataloging-in-Publication Data
Tignol, J.-P.
  Galois' theory of algebraic equations.
  Translation of: Leçons sur la théorie des équations.
  Bibliography: p.
  Includes index.
  1. Equations, Theory of. 2. Galois theory.
3. Galois, Evariste, 1811–1832. I. Blyth, T. S.
(Thomas Scott) II. Title.
QA211.T5413   1987   512.9'4   87-16795
ISBN 0-470-20919-4 (USA only)

Printed and Bound in Great Britain at The Bath Press, Avon

à Paul

For inquire, I pray thee, of the former age,
and prepare thyself to the search of their fathers :
For we are but of yesterday, and know nothing,
because our days upon earth are a shadow.

Job 8, 8 - 9.
Authorized King James Version

Contents

Introduction   1

1. Quadratic equations   6
   1. Introduction   6
   2. Babylonian algebra   7
   3. Greek algebra   11
   4. Arabic algebra   16

2. Cubic equations   21
   1. Priority disputes on the solution of cubic equations   21
   2. Cardano's formula   24
   3. Developments arising from Cardano's formula   25

3. Quartic equations   31
   1. The unnaturalness of quartic equations   31
   2. Ferrari's method   32

4. The creation of polynomials in one indeterminate   36
   1. The rise of symbolic algebra   36
   2. Relations between roots and coefficients   42

5. A modern approach to polynomials in one indeterminate   57
   1. Definitions   57
   2. Euclidean division   60
   3. Irreducible polynomials   65
   4. Roots   69
   5. Multiple roots and derivatives   72
   6. Common roots of two polynomials   77

Appendix to chapter 5 : Decomposition of rational fractions in sums of partial fractions   80

6. Alternative methods for cubic and quartic equations   83
   1. Viète on cubic equations   83
   2. Descartes on quartic equations   86
   3. Rational solutions for equations with rational coefficients   88
   4. Tschirnhaus' method   90

7. Roots of unity   97
   1. Introduction   97
   2. The origins of de Moivre's formula   98
   3. The roots of unity   107
   4. Primitive roots and cyclotomic polynomials   113
   Appendix to chapter 7 : Leibniz and Newton on the summation of series   122
   Exercises for chapter 7   124

8. Symmetric functions   129
   1. Introduction   129
   2. Waring's method   133
   3. The discriminant   141
   Appendix to chapter 8 : Euler's summation of the series of reciprocals of perfect squares   146
   Exercices for chapter 8   149

9. The fundamental theorem of algebra   152
   1. Introduction   152
   2. Girard's theorem   154
   3. Proof of the fundamental theorem   157

10. Lagrange   163
    1. Introduction   163
    2. Lagrange's observations on previously known methods   168
    3. First results of group theory and Galois theory   183
    Exercises for chapter 10   201

11. Vandermonde — 203
    1. Introduction — 203
    2. The solution of general equations — 204
    3. Cyclotomic equations — 209
    Exercises for chapter 11 — 217

12. Gauss on cyclotomic equations — 219
    1. Introduction — 219
    2. Number-theoretic preliminaries — 220
    3. Irreducibility of the cyclotomic polynomials of prime index — 230
    4. The periods of cyclotomic equations — 239
    5. Solvability by radicals — 252
    6. Irreducibility of the cyclotomic polynomials — 257
    Appendix to chapter 12 : Ruler and compass construction of regular polygons — 263
    Exercises for chapter 12 — 271

13. Ruffini and Abel on general equations — 273
    1. Introduction — 273
    2. Radical extensions — 277
    3. Abel's theorem on natural irrationalities — 286
    4. Proof of the unsolvability of general equations of degree higher than 4 — 295
    Exercises for chapter 13 — 298

14. Galois — 301
    1. Introduction — 301
    2. The Galois group of an equation — 306
    3. The Galois group under field extension — 332
    4. Solvability by radicals — 345
    5. Applications — 368
    Appendix to chapter 14 : Galois' description of groups of permutations — 387
    Exercises for chapter 14 — 395

Epilogue — 396
    The fundamental theorem of Galois theory — 401

Exercises for the appendix	412
Selected solutions	414
References	423
Index	427

Note to the reader

Definitions, lemmas, propositions, theorems, etc. are numbered consecutively within each chapter. References to such an item are usually made by numbers of the form m.n where m is the chapter number and n indicates the position within chapter m. However, if within chapter m a reference is made to item n of the same chapter, only n appears. References to the bibliography at the end of this volume are indicated in the text by strings of letters (in some cases including numbers) between square brackets, such as [Caj], [Gau2], ... Exercises are given at the end of chapters 7, 8, 10, 11, 12, 13, 14 and the epilogue. Solutions to some of the exercises are given at the end of the book, mostly to those which are harder or for which no hint is given.

Introduction

In spite of the title, the main subject of these lectures is not algebra, even less history, as one could conclude from a glance over the table of contents, but methodology. Their aim is to convey to the audience, which originally consisted of undergraduate students in mathematics, an idea of how mathematics is made. For such an ambitious project, the individual experience of any but the greatest mathematicians seems of little value, so I thought it appropriate to rely instead on the collective experience of generations of mathematicians, on the premise that there is a close analogy between collective and individual experience : the problems over which past mathematicians have stumbled are most likely to cause confusion to modern learners, and the methods which have been tried in the past are those which should come to mind naturally to the (gifted) students of today. The way in which mathematics is made is best learnt from the way mathematics has been made, and that premise accounts for the historical perspective on which this work is based.

The theme used as an illustration for general methodology is the theory of equations. The main stages of its evolution, from its origins in ancient times to its completion by Galois around 1830 will be reviewed and discussed. For the purpose of these lectures, the theory of equations seemed like an ideal topic in several respects : first, it is completely elementary, requiring virtually no mathematical background for the statement of its problems, and yet it leads to profound ideas and to fundamental concepts of modern algebra. Secondly, it underwent a very long and eventful evolution, and several gems lie along the road, like Lagrange's 1770 paper, which brought order and method to the theory in a masterly way, and Vandermonde's visionary glimpse of the solution of certain equations of high degree, which hardly unveiled the principles of Galois theory sixty years before Galois' memoir. Also instructive from a methodological point of view is the relationship between the general theory, as developed by Cardano, Tschirnhaus, Lagrange and Abel, and the attempts by Viète, de Moivre, Vandermonde and Gauss at significant examples, namely the so-called cyclotomic equations, which arise from the division of the circle into equal parts. Works in these two directions are closely intertwined like themes in a counterpoint, until their resolution in Galois' memoir. Finally, the algebraic theory of equations is now a closed subject, which reached complete maturity a long time ago; it is therefore possible to give a fair assessment of its various aspects. This is of course not true of Galois theory, which still provides inspiration for original research in numerous directions, but these lectures are concerned with the theory of equations and not with Galois theory of fields. The evolution from Galois' theory to modern Galois theory falls beyond the scope of this work; it would certainly fill another book like this one.

   As a consequence of emphasis on historical evolution, the exposition of mathematical facts in these lectures is genetic rather than systematic, which means that it aims to retrace the concatenation of ideas by following (roughly) their chronological order of occurrence. Therefore, results which are logically close to each other may be scattered in different chapters, and some topics are discussed several times, by little touches, instead of being given a unique definitive account. The expected reward for these circumlocutions is that the reader could hopefully gain a better insight into the inner workings of the theory, which prompted it to evolve the way it did.

Of course, in order to avoid discussions that are too circuitous, the works of mathematicians of the past -especially the distant past- have been somewhat modernized as regards notation and terminology. Although considering sets of numbers and properties of such sets was clearly alien to the patterns of thinking until the nineteenth century, it would be futile to ignore the fact that (naive) set theory has now pervaded all levels of mathematical education. Therefore, free use will be made of the definitions of some basic algebraic structures such as field and group, at the expense of devaluing some of the most original discoveries of Gauss, Abel and Galois. Except for those definitions and some elementary facts of linear algebra which are needed to clarify some proofs, the exposition is completely self-contained, as can be expected from a genetic treatment of an elementary topic.

It is fortunate to those who want to study the theory of equations that its long evolution is well documented : original works by Cardano, Viète, Descartes, Newton, Lagrange, Gauss, Ruffini, Abel, Galois are readily available through modern publications, some even in English translations. Besides these original works and those of Girard, Cotes, Tschirnhaus and Vandermonde, I have relied on several sources, mainly on Bourbaki's Note historique [Bou1] for the general outline, on Van der Waerden's "Science Awakening" [VW2] for the ancient times and on Edwards' "Galois theory" [E] for the proofs of some propositions in Galois' memoir. For systematic expositions of Galois theory, with applications to the solution of algebraic equations by radicals, the reader can be referred to any of the fine existing accounts, such as Artin's classical booklet [Ar1], Kaplansky's monograph [K], Stewart's book [St] or the relevant chapters of algebra books by Cohn [Coh], Jacobson [J1], [J2] or Van der Waerden [VW1], and presumably to many others I am not aware of. In the present lectures, however, the reader will find a thorough treatment of cyclotomic equations after Gauss and of the solvability of algebraic equations after Galois, with complete proofs. The point of view differs from the one in the quoted references in that it is strictly utilitarian, focusing (albeit to a lesser extent than the original papers) on the concrete problem at hand, which is to solve equations. Incidentally, it is striking to observe, in comparison, what kind of acrobatic tricks are needed to apply modern Galois theory to the solution of algebraic equations.

This book is based on a course taught at the Université Catholique de Louvain since 1978; it is an expanded and completely revised version of my "Leçons sur la théorie des équations" (Cabay, Louvain-la-Neuve, 1980). The historical discussions of quadratic equations, of the origin of polynomials and of de Moivre's formula have been significantly enlarged. A complete proof of Abel's theorem on the impossibility of solving the general equation of degree 5 by radicals has been added and the review of Galois' memoir has been considerably detailed. Moreover, exercises have been appended at the end of some chapters. These exercises point to some extensions of the theory and occasionally provide the proof of some technical fact which is alluded to in the text. They are never indispensable for a good understanding of the text.

I am greatly indebted to the many students who endured my original lectures and made valuable criticism, in particular to Pasquale Mammone and Nicole Vast, who read parts of the manuscript. I wish to express my thanks also to Murray Schacher, whose sabbatical leave at Louvain-la-Neuve was spoiled by numerous questions on English grammar and semantics, to T.S. Blyth, who edited the manuscript, to the staff of the centre général de documentation of the U.C.L. and of the Bibliothèque Royale Albert Ier for their helpfulness and for allowing me to reproduce parts of their books, and to Nicolas Rouche, who gave me access to the riches of his private library.

It is a pleasure to thank once again Suzanne D'Addato, who typed the whole manuscript with such superb results. Finally, my warmest thanks to Céline, Paul and Eve for their infectious joy of living and to Astrid for her patience and constant encouragement. The preparation of this work for publication spanned the whole life of our little Paul. I wish to dedicate this book to his memory.

1 Quadratic equations

§ 1. Introduction

Since the solution of a linear equation $aX = b$ does not use anything more than a division, it hardly belongs to the algebraic theory of equations; it is therefore appropriate to begin these lectures with quadratic equations :

$$aX^2 + bX + c = 0. \qquad (a \neq 0)$$

Dividing both sides by a, we may reduce this to :

$$X^2 + pX + q = 0.$$

The solution of this equation is well-known : when $(p/2)^2$ is added to each side, the square of $X + (p/2)$ appears and the equation can be written :

$$[X + (p/2)]^2 + q = (p/2)^2.$$

(This procedure is called 'completion of the square'). The values of X easily follow :

$$X = -(p/2) \pm \sqrt{(p/2)^2 - q}.$$

This formula is so well-known that it may be rather surprising to note that the solution of quadratic equations could not have been written in this form before the seventeenth century.* Nevertheless, mathematicians had been solving

---

(*) : The first uniform solution for quadratic equations (regardless of the signs of coefficients) is due to Simon

quadratic equations for about 40 centuries before. The
purpose of this first chapter is to give a brief outline of
this 'prehistory' of the theory of quadratic equations.

§ 2. Babylonian algebra

The first known solution of a quadratic equation dates from
about 2000 B.C.; on a Babylonian tablet, one reads :

> "I have subtracted from the area the side of my square :
> 14.30. Take 1, the coefficient. Divide 1 into two
> parts : 30. Multiply 30 and 30 : 15. You add to 14.30,
> and 14.30.15 has the root 29.30. You add to 29.30 the
> 30 which you have multiplied by itself : 30, and this is
> the side of the square" (see [VW2, p. 69]).

This text obviously provides a procedure for finding the
side of a square (say : x) when the difference between the
area and the side (i.e. $x^2 - x$) is given; in other words,
it gives the solution of $x^2 - x = b$.

However, one may be puzzled by the strange arithmetic
used by Babylonians. It can be explained by the fact that
their base for numeration is 60; therefore 14.30 really
means 14.60 + 30, i.e. 870. Moreover, they had no symbol
to indicate the absence of a number or to indicate that
certain numbers are intended as fractions. For instance,
when 1 is divided by 2, the result which is indicated as 30
really means 30.60^{-1}, i.e. 0.5. The square of this 30 is
then 15 which means 0.25, and this explains why the sum of
14.30 and 15 is written as 14.30.15 : in modern notations,
the operation is 870 + 0.25 = 870.25.

After clearing the notational ambiguities, it appears
that the author correctly solves the equation $x^2 - x = 870$,

---

Stevin in "L'Arithmetique" [S, p. 595], published in 1585.
However, Stevin does not use litteral coefficients, which
were introduced some years later by François Viète : see
chapter 4, § 1.

and gets x = 30. The other solution : x = -29 is neglected, since the Babylonians had no negative numbers.

This lack of negative numbers prompted Babylonians to consider various types of quadratic equations, depending on the signs of coefficients. There are three types in all :

$$X^2 + aX = b, \qquad X^2 - aX = b, \qquad \text{and } X^2 + b = aX,$$

where a, b stand for positive numbers. (The fourth type : $X^2 + aX + b = 0$ obviously has no (positive) solution).

Babylonians could not have written these various types in this form, since they did not use letters in place of numbers, but from the example above and from other numerical examples contained on the same tablet, it clearly appears that the Babylonians knew the solution of

$$X^2 + aX = b \qquad \text{as} \qquad X = \sqrt{(a/2)^2 + b} - (a/2)$$

and of

$$X^2 - aX = b \qquad \text{as} \qquad X = \sqrt{(a/2)^2 + b} + (a/2).$$

How they argued to get these solutions is not known, since in every extant example, only the procedure to find the solution is described, as in the example above. It is very likely that they had previously found the solution of geometric problems, such as to find the length and the breadth of a rectangle, when the excess of the length on the breadth and the area are given. Letting x and y respectively denote the length and the breadth of the rectangle, this problem amounts to solving the system :

$$\begin{cases} x - y = a \\ xy = b. \end{cases} \qquad (1)$$

By elimination of y, this system yields the following equation for x :

$$x^2 - ax = b. \tag{2}$$

If x is eliminated instead of y, we get :

$$y^2 + ay = b. \tag{3}$$

Conversely, equations (2) and (3) are equivalent to system (1) after setting y = x - a or x = y + a.

They probably deduced their solution for quadratic equations (2) and (3) from their solution of the corresponding system (1), which could be obtained as follows : let z be the arithmetic mean of x and y.

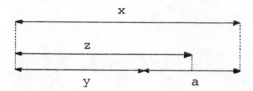

In other words, z is the side of the square which has the same perimeter as the given rectangle :

$$z = x - (a/2) = y + (a/2).$$

Compare then the area of the square (i.e. z^2) to the area of the rectangle (xy = b). We have :

$$xy = (z + (a/2))(z - (a/2))$$

whence

$$b = z^2 - (a/2)^2.$$

Therefore,

$$z = \sqrt{(a/2)^2 + b}$$

and it follows that

$$x = \sqrt{(a/2)^2 + b} + (a/2) \text{ and } y = \sqrt{(a/2)^2 + b} - (a/2).$$

This solves at once the quadratic equations $x^2 - ax = b$ and $y^2 + ay = b$.

Looking at the various examples of quadratic equations solved by Babylonians, one notices a curious fact : the third type $x^2 + b = ax$ does not explicitly appear. This is even more puzzling in view of the frequent occurrence in Babylonian tablets of problems such as to find the length and the breadth of a rectangle when the perimeter and the area of the rectangle are given; which amounts to the solution of :

$$\begin{cases} x + y = a \\ xy = b. \end{cases} \qquad (4)$$

By elimination of y, this system leads to : $x^2 + b = ax$. So, why did Babylonians solve the system (4) and never consider equations like $x^2 + b = ax$?

A clue can be discovered in their solution of system (4), which is probably obtained by comparing the rectangle with sides x, y to the square with perimeter $(a/2)$:

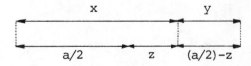

One then sets $x = (a/2) + z$, whence $y = (a/2) - z$, and finishes as before.

Whatever their method, the solution they get is :

$$x = (a/2) + \sqrt{(a/2)^2 - b}$$

$$y = (a/2) - \sqrt{(a/2)^2 - b},$$

thus assigning one value for x and one value for y, while it is clear to us that x and y are interchangeable in the system (4) : we would have given two values for each one of the unknown quantities, and found :

$$x = (a/2) \pm \sqrt{(a/2)^2 - b}, \quad y = (a/2) \mp \sqrt{(a/2)^2 - b}.$$

In the Babylonian phrasing, however, x and y are not interchangeable : they are the length and the breadth of a rectangle, so there is an implicit condition that $x \geqslant y$. According to S. Gandz [Gz, § 9], the type $X^2 + b = aX$ was systematically and purposely avoided by Babylonians because, unlike the two other types, it has <u>two</u> positive solutions (which are the length x and the breadth y of the rectangle). The idea of two values for one quantity was probably very embarrassing to them : it would have struck Babylonians as an illogical absurdity, as sheer nonsense.

However, this observation that algebraic equations of degree higher than 1 have several interchangeable solutions is of fundamental importance : it is the corner-stone of Galois theory, and we shall have the opportunity to see to what clever use it will be put by Lagrange and later mathematicians.

As André Weil commented in relation to another topic [W2, p. 104] :

"This is very characteristic in the history of mathematics. When there is something that is really puzzling and cannot be understood, it usually deserves the closest attention because some time or other some big theory will emerge from it".

§ 3. Greek algebra

The Greeks deserve a prominent place in the history of mathematics, for being the first to perceive the usefulness of proofs. Before them, mathematics were rather empirical. Using deductive reasoning, they built a huge mathematical monument, which is remarkably illustrated by Euclid's celebrated masterwork : "The Elements" (c. 300 B.C.). This book contains a synthesis of the mathematical achievements of the classical period (V^{th} and IVth century B.C.).

The Greeks' major contribution to algebra during this classical period is foundational. They discovered that the naive idea of number (i.e. : integer or rational number) is not sufficient to account for geometric magnitudes. For instance, there is no line segment which could be used as a length unit to measure the diagonal and the side of a square by integers : the ratio of the diagonal to the side (i.e. $\sqrt{2}$) is not a rational number, or in other words : the diagonal and the side are incommensurable.

The discovery of irrational numbers was first made among followers of Pythagoras, probably between 430 and 410 B.C. (see [Kn, p. 49]). It is often credited to Hippasus of Metapontum, who was reportedly drowned at sea for producing a downright counterexample to the Pythagoreans' doctrine that "all things are numbers". However, no direct account is extant, and how the discovery was made is still a matter of conjecture. It is widely believed that the first magnitudes which were shown to be incommensurable are the diagonal and the side of a square, and the following reconstruction of the proof has been proposed by Knorr [Kn, p. 27] :

Assume the side AB and the diagonal AC of the square ABCD are both measured by a common segment : then AB and AC both represent numbers (= integers) and the squares on them, which are ABCD and EFGH, represent square numbers. From the figure, it is clear (by counting triangles) that EFGH is the double of ABCD, so EFGH is an even square number and its side EF is therefore even. It follows that EB also represents a number, whence EBKA is a square number.

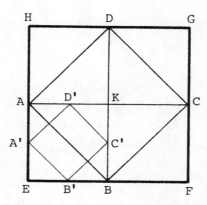

Since the square ABCD clearly is the double of the square EBKA, the same arguments show that AB is even, whence A'B' represents a number.
We now see that A'B' and A'C' (= EB), which are the halves of AB and AC, both represent numbers; but A'B' and A'C' are the side and the diagonal of a new (smaller) square, so we may repeat the same arguments as above.
Iterating this process, we see that the numbers represented by AB and AC are indefinitely divisible by 2 : this is obviously impossible, and this contradiction proves that AB and AC are incommensurable.

This result obviously shows that integers are not sufficient to measure lengths of segments. The right level of generality is that of ratios of lengths. Prompted by this discovery, the Greeks developed new techniques to operate with ratios of geometric magnitudes in a logically coherent way, avoiding the problem of assigning numerical values to these magnitudes. They thus created a "geometric algebra", which is methodically taught by Euclid in "The Elements".

By contrast, Babylonians seem not to have been aware of the theoretical difficulties arising from irrational numbers, although these numbers were of course unavoidable in the treatment of geometric problems : they simply replaced them by rational approximations. For instance, the following approximation of $\sqrt{2}$ has been found on some Babylonian tablet : 1.24.51.10, i.e. $1 + 24/60 + 51/60^2 + 10/60^3$ or 1.41421296296296..., which is accurate up to the fifth place.

Although Euclid does not explicitly deal with quadratic equations, the solution of these equations can be detected under a geometrical garb in some propositions of the Elements. For instance, Proposition 5 of Book II states : [He, v. I, p. 382] :

"If a straight line be cut into equal and unequal segments, the rectangle contained by the unequal segments of the whole together with the square on the straight line between the points of section is equal to the square on the half".

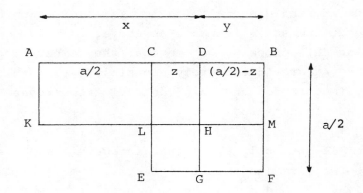

On the figure above, the straight line AB has been cut into equal segments at C and unequal segments at D, and the proposition asserts that the rectangle AH together with the square LG (which is equal to the square on CD) is equal to the square CF. (This is clear from the figure, since the rectangle AL is equal to the rectangle DF).

If we understand that the unequal segments in which the given straight line AB = a is cut are unknown, it appears that this proposition provides us with the core of the solution of the system

$$\begin{cases} x + y = a \\ xy = b. \end{cases}$$

Indeed, setting $z = x - (a/2)$ "the straight line between the points of section", it states that $b + z^2 = (a/2)^2$. It then readily follows that

$$z = \sqrt{(a/2)^2 - b}$$

whence

$$x = (a/2) + \sqrt{(a/2)^2 - b} \quad \text{and} \quad y = (a/2) - \sqrt{(a/2)^2 - b},$$

as in Babylonian algebra. In subsequent propositions, Euclid also teaches the solution of

$$\begin{cases} x - y = a \\ xy = b \end{cases}$$

which amounts to $x^2 - ax = b$ or $y^2 + ay = b$. He returns to the same type of problems, but in a more elaborate form, in propositions 28 and 29 of book VI (Compare [K1, pp. 76-77] and [VW2, p. 121]).

The Greek mathematicians of the classical period thus reached a very high level of generality in the solution of quadratic equations, since they considered equations with (positive) <u>real</u> coefficients. However, geometric algebra, which was the only rigorous method of operating with real numbers before the XIXth century, is very difficult. It imposes tight limitations which are not natural from the point of view of algebra; for instance, a great skill in the handling of proportions is required to go beyond degree three.

To progress in the theory of equations, it was necessary

to think more about formalism and less about the nature of coefficients. Although later Greek mathematicians such as Hero and Diophantus took some steps in that direction, the really new advances were brought by other civilizations. Hindus, and Arabs later, developed techniques of calculation with irrational numbers, which they treated unconcernedly, without worrying about their irrationality. For instance, they were familiar with formulas like :

$$\sqrt{a} + \sqrt{b} = \sqrt{a + b + 2\sqrt{ab}}$$

which they obtained from $(u+v)^2 = u^2 + v^2 + 2uv$ by extracting roots of both sides and replacing u and v by $\sqrt{a}$ and $\sqrt{b}$ respectively. Their notion of mathematical rigor was rather more relaxed than that of Greek mathematicians, but they paved the way to a more formal (or indeed : algebraic) approach to quadratic equations [Kl, ch. 9, § 2].

§ 4. Arabic algebra

The next landmark in the theory of equations is the book : "Al-jabr w' al muqabala" (c. 830 A.D.), due to Mohammed ibn Musa al-Khowarizmi.

The title refers to two basic operations on equations. The first is al-jabr (from which the word "algebra" is derived) which means "the restoration" or "making whole". In this context, it stands for the restoration of equality in an equation by adding to one side a negative term which is removed from the other. For instance, the equation

$$x^2 = 40x - 4x^2$$

is converted into

$$5x^2 = 40x$$

by al-jabr [Ka, p. 105]. The second basic operation al muqabala means : "the opposition" or "balancing"; it is a simplification procedure by which like terms are removed from both sides of an equation. For instance, al muqabala changes

$$50 + x^2 = 29 + 10x$$

into

$$21 + x^2 = 10x$$

[Ka, p. 109].

In this work, al Khowarizmi initiates what might be called the classical period in the theory of equations, by reducing the old methods for solving equations to a few standardized procedures. For instance, in problems involving several unknowns, he systematically sets up an equation for one of the unknowns, and he solves the three types of quadratic equations :

$$X^2 + aX = b, \qquad X^2 + b = aX, \qquad X^2 = aX + b$$

by completion of the square, giving the two (positive) solutions for the type $X^2 + b = aX$.

Al-Khowarizmi first explains the procedure, as a Babylonian would have done :

"The following is an example of squares and roots equal to numbers : a square and 10 roots are equal to 39 units. The question therefore in this type of equation is about as follows : what is the square which combined with ten of its roots will give a sum total of 39 ? The manner of solving this type of equation is to take one-half of the roots just mentioned. Now the roots in the problem before us are 10. Therefore take 5, which multiplied by itself gives 25, an amount which you add to 39, giving 64.

Having taken then the square root of this which is 8, subtract from it the half of the roots, 5, leaving 3. The number three therefore represents one root of this square, which itself, of course, is 9. Nine therefore gives that square" [Ka, pp. 71-73].

However, after explaining the procedure for solving each of the six types : $mX^2 = aX$, $mX^2 = b$, $aX = b$, $mX^2 + aX = b$, $mX^2 + b = aX$ and $mX^2 = aX + b$, he adds :

"We have said enough, says Al-Khowarizmi, so far as numbers are concerned, about the six types of equations. Now, however, it is necessary that we should demonstrate geometrically the truth of the same problems which we have explained in numbers" [Ka, p. 77].

He then gives geometric justifications for his rules for the last three types, using completion of the square as in the following example for $x^2 + 10x = 39$:

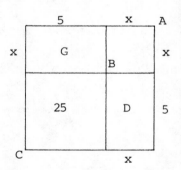

Let x^2 be the square AB. Then 10x is divided into two rectangles G and D, each being 5x and being applied to the side x of the square AB. By hypothesis, the value of the shape thus produced is $x^2 + 10x = 39$. There remains an empty corner of value $5^2 = 25$ to complete the square AC. Therefore, if 25 is added, the square $(x+5)^2$ is completed, and its value is $39 + 25 = 64$. It then follows that $(x+5)^2 = 64$, whence $x + 5 = 8$ and $x = 3$ (see [Ka, p. 81]).

It should be observed that the geometry behind this construction is much more elementary than in Euclid's Elements, since it is not logically connected by deductive reasoning to a small number of axioms, but relies instead on intuitive geometrical evidence. From the point of view of algebra, on the other hand, al-Khowarizmi's work is incommensurately ahead of Euclid's, and it set the stage for the later development of algebra as an independent discipline.

Another remarkable achievement of the Arabs in the theory of equations is a geometrical solution of cubic equations due to Omar Khayyam (c. 1079). For instance, the solution of $x^3 + b^2x = b^2c$ is obtained by intersecting the parabola $x^2 = by$ with the circle of diameter c which is tangent to the axis of the parabola at its vertex:

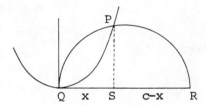

To prove that the segment x as shown on the figure above satisfies $x^3 + b^2x = b^2c$, we start from the relation $x^2 = b \cdot PS$, which yields:

$$\frac{b}{x} = \frac{x}{PS}.$$

On the other hand, since the triangles QSP and PSR are similar, we have:

$$\frac{x}{PS} = \frac{PS}{c-x},$$

whence

$$\frac{b}{x} = \frac{PS}{c-x}.$$

19

As PS = $\frac{x^2}{b}$, this equation yields :

$$\frac{b}{x} = \frac{x^2}{b(c-x)},$$

whence $x^3 = b^2c - b^2x$, as required.

Omar Khayyam also gives geometric solutions for the other types of cubic equations by intersection of conics, but these brilliant solutions are of little use for practical purposes, and an algebraic solution was still longed for.

In 1494, Luca Pacioli closes his book : "Summa de Arithmetica, Geometria, Proportione et Proportionalita" (one of the first printed books in mathematics) with the remark that the solutions of $x^3 + mx = n$ and $x^3 + n = mx$ (in modern notations) are as impossible as the quadrature of the circle. [K1, p. 237], [C, p. 8]. However, unexpected developments were soon to take place.

2 Cubic equations

§ 1. <u>Priority disputes on the solution of cubic equations</u>

The algebraic solution of $X^3 + mX = n$ was first obtained around 1515 by Scipione del Ferro, professor of mathematics in Bologna. Not much is known about him nor about his solution as, for some reason, he decided not to publicize his result. After his death in 1526, his method passed to some of his pupils.

The second discovery of the solution is much better known, through the accounts of its author himself, Niccolo Fontana (c. 1500-1557), nicknamed "Tartaglia" ("Stammerer"), from Brescia (see [Ha, pp. 360 ff]). In 1535, Tartaglia, who had dealt with some very particular cases of cubic equations, was challenged to a public problem-solving contest by Antonio Maria Fior, a former pupil of Scipione del Ferro. When he heard that Fiore had received the solution of cubic equations from his master, Tartaglia threw all his energy and his skill into the struggle. He succeeded in finding the solution just in time to inflict upon Fiore a humiliating defeat.

The news that Tartaglia had found the solution of cubic equations reached Girolamo Cardano (1501-1576), a very versatile scientist, who wrote a number of books on a wide variety of subjects, including medicine, astrology, astronomy, philosophy and mathematics. Cardano then asked Tartaglia to give him his solution, so that he could include it in a treatise on arithmetic, but Tartaglia flatly refused, since he was himself planning to write a book on this topic. It turns out that Tartaglia later changed his mind, at least partially, since in 1539 he handed on to Cardano the solution of $X^3 + mX = n$, $X^3 = mX + n$ and a very brief indication on $X^3 + n = mX$ in verses :

"Quando che'l cubo con le cose appresso
    Se agguaglia a qualche numero discreto :
    Trovan dui altri, differenti in esso.
Dapoi terrai, questo per consueto,
    Che'l lor produtto, sempre sia eguale
    Al terzo cubo delle cose neto;
El residuo poi suo generale,
    Delli lor lati cubi, bene sottratti
    Varra la tua cosa principale.
        ... "

(see [Ha, pp. 364-365]).

This excerpt gives the formula for $X^3 + mX = n$. The equation is indicated in the first two verses : the cube and the things equal to a number. Cosa (= thing) is the word for the unknown. To express the fact that the unknown is multiplied by a coefficient, Tartaglia simply uses the plural form : le cose. He then gives the following procedure : find two numbers which differ by the given number and such that their product is equal to the cube of the third of the number of things. Then the difference between the cube roots of these numbers is the unknown.

With modern notations, we would write that, to find the solution of

$$X^3 + mX = n,$$

we only need to find t, u such that

$$t - u = n$$

and $$tu = (m/3)^3;$$

then

$$x = \sqrt[3]{t} - \sqrt[3]{u}.$$

The values of t and u are easily found (see the system (1) in § 1.2) :

$$t = \sqrt{(n/2)^2 + (m/3)^3} + (n/2)$$

$$u = \sqrt{(n/2)^2 + (m/3)^3} - (n/2).$$

Therefore, a solution x of $x^3 + mX = n$ is given by the following formula :

$$x = \sqrt[3]{\sqrt{(n/2)^2 + (m/3)^3} + (n/2)} - \sqrt[3]{\sqrt{(n/2)^2 + (m/3)^3} - (n/2)}.$$

However, the poem does not provide any justification for this formula. Of course, it "suffices" to check that the value of x given above satisfies the equation $x^3 + mx = n$, but this was far from obvious to a sixteenth century mathematician. The major difficulty was to figure out that

$$(a - b)^3 = a^3 - 3a^2b + 3ab^2 - b^3,$$

a formula which could be properly proved only by dissection of a cube in three-dimensional space.

Having received Tartaglia's poem, Cardano set to work; he not only found justifications for the formulas but he also solved all the other types of cubics. He then published his results, giving due credit to Tartaglia and to del Ferro, in the epoch-making book : "Ars Magna, sive de regulis algebraicis" (The Great Art, or the Rules of Algebra [C]). A bitter quarrel then erupted between Tartaglia and Cardano, the former claiming that Cardano had solemnly sworn never to publish Tartaglia's solution, while the latter countered that there had never been any question of secrecy.

§ 2. Cardano's formula

Although Cardano lists 13 types of cubic equations and gives a detailed solution for each of them, we shall use modern notations in this section, and explain Cardano's method for the general cubic equation :

$$X^3 + aX^2 + bX + c = 0.$$

First, the change of variable : $Y = X + (a/3)$ converts the equation into one which lacks the second degree term :

$$Y^3 + pY + q = 0 \qquad (1)$$

where $p = b - (a^2/3)$ and $q = c - (a/3)b + 2(a/3)^3$.
If $Y = \sqrt[3]{t} + \sqrt[3]{u}$, then

$$Y^3 = t + u + 3(\sqrt[3]{t} + \sqrt[3]{u})(\sqrt[3]{tu})$$

and equation (1) becomes :

$$(t+u+q) + (\sqrt[3]{t} + \sqrt[3]{u})(3\sqrt[3]{tu} + p) = 0.$$

This equation clearly holds if the rational part $t+u+q$ and the irrational part $(\sqrt[3]{t} + \sqrt[3]{u})(3\sqrt[3]{tu} + p)$ both vanish or, in other words, if

$$\begin{cases} t + u = -q \\ tu = -(p/3)^3. \end{cases}$$

This system has the solution :

$$t, u = -(q/2) \pm \sqrt{(p/3)^3 + (q/2)^2}$$

(see § 1); hence a solution for equation (1) is :

$$Y = \sqrt[3]{-(q/2) + \sqrt{(p/3)^3 + (q/2)^2}} + \sqrt[3]{-(q/2) - \sqrt{(p/3)^3 + (q/2)^2}}$$

and a solution for the initial equation

$$X^3 + aX^2 + bX + c = 0$$

easily follows :

$$X = \sqrt[3]{(\tfrac{a}{3})(\tfrac{b}{2}) - (\tfrac{a}{3})^3 - (\tfrac{c}{2}) + \sqrt{(\tfrac{b}{3})^3 - \tfrac{1}{3}(\tfrac{a}{3})^2 (\tfrac{b}{2})^2 + (\tfrac{c}{2})^2 + 2(\tfrac{a}{3})^3(\tfrac{c}{2}) - 2(\tfrac{a}{3})(\tfrac{b}{2})(\tfrac{c}{2})}}$$

$$+ \sqrt[3]{(\tfrac{a}{3})(\tfrac{b}{2}) - (\tfrac{a}{3})^3 - (\tfrac{c}{2}) - \sqrt{(\tfrac{b}{3})^3 - \tfrac{1}{3}(\tfrac{a}{3})^2 (\tfrac{b}{2})^2 + (\tfrac{c}{2})^2 + 2(\tfrac{a}{3})^3(\tfrac{c}{2}) - 2(\tfrac{a}{3})(\tfrac{b}{2})(\tfrac{c}{2})}}$$

$$- (\tfrac{a}{3}).$$

Several terms vanish when $a = 0$, and the simplified formula in this special case (to which the general case is reduced by an appropriate change of variables, as we have seen above) is known as <u>Cardano's formula</u> :

$$X = \sqrt[3]{-\tfrac{q}{2} + \sqrt{(\tfrac{p}{3})^3 + (\tfrac{q}{2})^2}} + \sqrt[3]{-\tfrac{q}{2} - \sqrt{(\tfrac{p}{3})^3 + (\tfrac{q}{2})^2}}$$

is a solution of

$$X^3 + pX + q = 0.$$

§ 3. <u>Developments arising from Cardano's formula</u>

The solution of cubic equations was a remarkable achievement, but Cardano's formula is far less convenient than the corresponding formula for quadratic equations since it has some drawbacks which undoubtedly baffled XVIth century mathematicians (to begin with, its discoverers) :

a) First, when some solution is expected, it is not always

yielded by Cardano's formula. This could have struck Cardano when he was devising examples for illustrating his rules, such as :

$$x^3 + 16 = 12x$$

[C, p. 12] which is constructed to give 2 as an answer. Cardano's formula yields :

$$x = \sqrt[3]{-8} + \sqrt[3]{-8} = -4.$$

Why does it yield -4 and not 2 ?
It is likely that the above observation had first prompted Cardano to investigate a question much more interesting to him : How many solutions does a cubic equation have ?
He was thus led to observe that cubic equations may have three solutions (including the negative ones, which Cardano terms "false" or "fictitious" but not the imaginary ones) and to investigate the relations between these solutions (see [C, Chapter I]).

b) Next, when there is a rational solution, its expression according to Cardano's formula can be rather awkward. For instance, it is easily seen that 1 is solution of

$$x^3 + x = 2,$$

but Cardano's formula yields :

$$x = \sqrt[3]{1 + \frac{2}{3}\sqrt{\frac{7}{3}}} + \sqrt[3]{1 - \frac{2}{3}\sqrt{\frac{7}{3}}}.$$

Now, the equation above has only one real root, since the function $f(x) = x^3 + x$ is monotonically increasing (as it is the sum of two monotonically increasing functions) and, therefore, takes the value 2 only once. We are thus compelled to conclude :

$$1 = \sqrt[3]{1 + \frac{2}{3}\sqrt{\frac{7}{3}}} + \sqrt[3]{1 - \frac{2}{3}\sqrt{\frac{7}{3}}}, \qquad (2)$$

a rather surprising result.

Already in 1540, Tartaglia tried to simplify the irrational expressions arising in his solution of cubic equations [Ha, p. 373]. More precisely, he tried to determine under which condition an irrational expression like $\sqrt[3]{a + \sqrt{b}}$ could be simplified to $u + \sqrt{v}$. This problem can be solved as follows (in modern notations) : starting with

$$\sqrt[3]{a + \sqrt{b}} = u + \sqrt{v} \qquad (3)$$

and taking the cube of both sides, we obtain

$$a + \sqrt{b} = u^3 + 3uv + (3u^2+v)\sqrt{v},$$

whence, equating separately the rational and the irrational parts (this is licit if a, b, u and v are rational numbers),

$$\begin{cases} a = u^3 + 3uv \\ \sqrt{b} = (3u^2+v)\sqrt{v}. \end{cases} \qquad (4)$$

Subtracting the second equation from the first, we then obtain :

$$a - \sqrt{b} = (u-\sqrt{v})^3$$

whence

$$\sqrt[3]{a-\sqrt{b}} = u - \sqrt{v}. \qquad (5)$$

Multiplying (3) and (5), we obtain :

$$\sqrt[3]{a^2-b} = u^2 - v \qquad (6)$$

which can be used to eliminate v from the first equation of system (4). We thus get :

$$a = 4u^3 - 3(\sqrt[3]{a^2-b})u.$$

Therefore, if a and b are rational numbers such that $\sqrt[3]{a^2-b}$ is rational and if the equation

$$4u^3 - 3(\sqrt[3]{a^2-b})u = a \qquad (7)$$

has a rational solution u, then

$$\sqrt[3]{a+\sqrt{b}} = u + \sqrt{v} \quad \text{and} \quad \sqrt[3]{a-\sqrt{b}} = u - \sqrt{v},$$

where v is given by equation (6) :

$$v = u^2 - \sqrt[3]{a^2-b}.$$

This effectively provides a simplification in the irrational expressions

$$\sqrt[3]{-(q/2) + \sqrt{(p/3)^3 + (q/2)^2}} \quad \text{and} \quad \sqrt[3]{-(q/2) - \sqrt{(p/3)^3 + (q/2)^2}}$$

which appear in Cardano's formula for the solution of $X^3 + pX + q = 0$, but this simplification is useless as far as the solution of cubic equations is concerned. Indeed, if $a = -(q/2)$ and $b = (p/3)^3 + (q/2)^2$, one has to find a rational solution of equation (7) :

$$4u^3 + pu = -(q/2),$$

and this exactly amounts to finding a rational solution of the initial equation :

$$X^3 + pX + q = 0,$$

since these equations are related by the change of variable $X = 2u$. However, this process can be used to show, for instance, that

$$\sqrt[3]{1 + \frac{2}{3}\sqrt{\frac{7}{3}}} = \frac{1}{2} + \frac{1}{2}\sqrt{\frac{7}{3}} \text{ and } \sqrt[3]{1 - \frac{2}{3}\sqrt{\frac{7}{3}}} = \frac{1}{2} - \frac{1}{2}\sqrt{\frac{7}{3}},$$

from which formula (2) follows.

c) The most serious drawback of Cardano's formula appears when one tries to solve an equation like :

$$X^3 = 15X + 4.$$

It is easily seen that $X = 4$ is a solution, but Cardano's formula yields a very embarrassing expression :

$$X = \sqrt[3]{2 + \sqrt{-121}} + \sqrt[3]{2 - \sqrt{-121}},$$

in which square roots of negative numbers are extracted. The case where $(p/3)^3 + (q/2)^2 < 0$ is known as the "casus irreducibilis" of cubic equations. For a long time, the validity of Cardano's formula in this case had been a matter of debate, but the discussion of this case had a very important by-product : it prompted the use of complex numbers.

Complex numbers had been, up to then, brushed aside as absurd, nonsensical expressions. A remarkably explicit example of this attitude appears in the following excerpt from chapter 37 of the Ars Magna [C, p. 219] :

> "If it should be said, Divide 10 into two parts the product of which is 30 or 40, it is clear that this case is impossible. Nevertheless, we will work thus : (...)".

Cardano then applies the usual procedure with the given data, which amounts to solving $X^2 - 10X + 40 = 0$, and comes

up with the solution : these parts are $5 + \sqrt{-15}$ and $5 - \sqrt{-15}$.
He then justifies his result :

> "Putting aside the mental tortures involved*, multiply $5 + \sqrt{-15}$ by $5 - \sqrt{-15}$, making $25 - (-15)$ which is $+15$. Hence this product is 40. (...) So progresses arithmetic subtlety the end of which, as is said, is as refined as it is useless".

However, with the "casus irreducibilis" of cubic equations, complex numbers were imposed upon mathematicians. The operations on these numbers are clearly taught, in a nearly modern way, by Rafaele Bombelli (c. 1526-1573), in his influential treatise : "Algebra" (1572). In this book, Bombelli boldly applies to cube roots of complex numbers the same simplification procedure as in (b) above, and he obtains, for instance :

$$\sqrt[3]{2 + \sqrt{-121}} = 2 + \sqrt{-1} \quad \text{and} \quad \sqrt[3]{2 - \sqrt{-121}} = 2 - \sqrt{-1},$$

from which it follows that Cardano's formula gives indeed 4 for a solution of $X^3 = 15X + 4$.

Complex numbers thus appeared, not to solve quadratic equations which lack solutions (and do not need any), but to explain why Cardano's formula, efficient as it may seem, fails in certain cases to provide expected solutions to cubic equations.

(*) In the original text : "dismissis incruciationibus". Perhaps Cardano played on words here, since another translation for this passage is : "the cross-multiples having canceled out", referring to the fact that in the product $(5 + \sqrt{-15})(5 - \sqrt{-15})$, the terms $5\sqrt{-15}$ and $-5\sqrt{-15}$ cancel out.

3 Quartic equations

§ 1. <u>The unnaturalness of quartic equations</u>

The solution of quartic equations was found soon after that of cubic equations. It is due to Ludovico Ferrari (1522-1565), a pupil of Cardano, and it first appeared in the "Ars Magna".

Ferrari's method is very ingenious, relying mainly on transformation of equations, but it aroused less interest than the solution of cubic equations. This is clearly shown by its place in the "Ars Magna" : while Cardano spends thirteen chapters to discuss the various cases of cubic equations, Ferrari's method is briefly sketched in the penultimate chapter.

The reason for this relative disregard may be found in the introduction of the "Ars Magna" : [C, p. 9] :

"Although a long series of rules might be added and a long discourse given about them, we conclude our detailed consideration with the cubic, others being merely mentioned, even if generally, in passing. For as positio [the first power] refers to a line, quadratum [the square] to a surface, and cubum [the cube] to a solid body, it would be very foolish for us to go beyond this point. Nature does not permit it".

This passage shows the equivocal status of algebra in the sixteenth century. Its logical foundations were still geometric, as in the classical Greek period; in this framework, each quantity has a dimension and only quantities of the same dimension can be added or equated. For instance, an equation like $x^2 + b = ax$ makes sense only if x and a are line segments and b is an area, and equations of degree higher than three don't make any sense at all.*

However, from an arithmetical point of view, quantities are regarded as dimensionless numbers, which can be raised to any power and equated unconcernedly. This way of thought was clearly prevalent among Babylonians, since the very statement of the problem : "I have subtracted from the area the side of my square : 14.30" is utter nonsense from a geometrical point of view. The Arabic algebra also stresses arithmetic, although Al-Khowarizmi provides geometric proofs of his rules (see § 1.4).

In the Ars Magna, both the geometric and the arithmetic approaches to equations are present. On one hand, Cardano tries to base his results on Euclid's elements, and on the other hand, he gives the solution of equations of degree 4. He also solves some equations of higher degree, such as $x^9 + 3x^6 + 10 = 15x^3$ [C, p. 159], in spite of his initial statement that it would be "foolish" to go beyond degree 3. However, the arithmetic approach, which would eventually predominate, still suffered from its lack of a logical base until the early seventeenth century (see § 4.1).

§ 2. Ferrari's method

In this section, we use modern notations to discuss Ferrari's solution of quartic equations. Let

$$x^4 + ax^3 + bx^2 + cx + d = 0$$

be an arbitrary quartic equation. By the change of variable

(*) A way out of this difficulty was eventually found by Descartes. In "La Geometrie" [D, p. 5], published in 1637, he introduces the following convention : if a unit line segment e is chosen, then the square x^2 of a line segment x is the side of the rectangle constructed on e which has the same area as the square with side x. Thus, x^2 is a li-

$$Y = X + (a/4)$$

the cubic term cancels out, and the equation becomes:

$$Y^4 + pY^2 + qY + r = 0 \qquad (1)$$

with $p = b - 6(a/4)^2$

$q = c - (a/2)b + (a/2)^3$

$r = d - (a/4)c + (a/4)^2 b - 3(a/4)^4.$

Moving the linear terms to the right-hand side and completing the square on the left-hand side, we obtain:

$$[Y^2 + (p/2)]^2 = -qY - r + (p/2)^2.$$

If we add a quantity u to the expression squared in the left-hand side, we get:

$$[Y^2 + (p/2) + u]^2 = -qY - r + (p/2)^2 + 2uY^2 + pu + u^2.$$
$$(2)$$

The idea is to determine u in such a way that the right-hand side also becomes a square. Looking at the terms in Y^2 and in Y, it is easily seen that if the right-hand side is a square, then it is the square of $\sqrt{2u}\, Y - (q/2\sqrt{2u})$; therefore, we should have:

$$-qY - r + (p/2)^2 + 2uY^2 + pu + u^2 = [\sqrt{2u}\, Y - (q/2\sqrt{2u})]^2$$
$$(3)$$

and, equating the independent terms, we see that this equation holds if and only if:

$$-r + (p/2)^2 + pu + u^2 = q^2/8u$$

---

ne segment, and arbitrary powers of x can be interpreted as line segments in a similar way.

or equivalently, after clearing the denominator and rearranging terms :

$$8u^3 + 8pu^2 + (2p^2 - 8r)u - q^2 = 0. \tag{4}$$

Therefore, by solving this cubic equation, we can find a quantity u for which equation (3) holds. Returning to equation (2), we then have :

$$[Y^2 + (p/2) + u]^2 = [\sqrt{2u}\, Y - (q/2\sqrt{2u})]^2,$$

whence

$$Y^2 + (p/2) + u = \pm\, [\sqrt{2u}\, Y - (q/2\sqrt{2u})].$$

The values of Y are then obtained by solving the two quadratic equations above (one corresponding to the sign + for the right-hand side, the other to the sign -).

To complete the discussion, it remains to consider the case where u = 0 is a root of equation (4), since the calculations above implicitly assume u ≠ 0. But this case occurs only if q = 0 and in this case the initial equation (1) is

$$Y^4 + pY^2 + r = 0,$$

which is easily solved, since it is a quadratic equation in Y^2.

In summary, the solutions of

$$X^4 + aX^3 + bX^2 + cX + d = 0$$

are obtained as follows : let p, q and r be defined as before (see equation (1)) and let u be a solution of

$$8u^3 + 8pu^2 + (2p^2 - 8r)u - q^2 = 0. \tag{4}$$

If $q \neq 0$, the solutions of the initial quartic equation are:

$$X = \varepsilon \sqrt{\frac{u}{2}} + \varepsilon' \sqrt{-\frac{u}{2} - \frac{p}{2} - \frac{\varepsilon q}{2\sqrt{2u}}} + \frac{a}{4}$$

where ε and ε' can be independently $+1$ or -1.
If $q = 0$, the solutions are :

$$X = \varepsilon \sqrt{-\frac{p}{2} + \varepsilon' \sqrt{(\frac{p}{2})^2 - r}} + \frac{a}{4}$$

where ε and ε' can be independently $+1$ or -1.

<u>Remark</u> : Equation (4), on which the solution of the quartic equation depends, is called the <u>resolvent cubic</u> equation (relative to the given quartic equation).
Depending on the way equations are set up, one may come up with other resolvent cubic equations. For instance, from equation (1) one could pass to :

$$(Y^2 + v)^2 = (-pY^2 - qY - r) + 2vY^2 + v^2$$

where v is an arbitrary quantity (which plays the same role as $(p/2) + u$ in the preceding discussion). The condition on v for the right-hand side to be a perfect square is then:

$$8v^3 - 4pv^2 - 8rv + 4pr - q^2 = 0. \qquad (5)$$

After having determined v such that this condition holds, one finishes as before.
    This second method is clearly equivalent to the previous one, by the change of variable $v = (p/2) + u$. Therefore, equation (5), which is obtained from (4) by this change of variable, is also entitled to be called the resolvent cubic.

4 The creation of polynomials in one indeterminate

§ 1. <u>The rise of symbolic algebra</u>

In comparison to the rapid development of the theory of equations around the middle of the sixteenth century, progress during the next two centuries was rather slow. The solution of cubic and quartic equations was a very important breakthrough, and it took some time before the circle of new ideas arising from these solutions was fully explored and understood, and new advances were possible.

First of all, it was necessary to devise appropriate notations for handling equations. In the solution of cubic and quartic equations, Cardano was straining to the utmost the capabilities of the algebraic system available to him. Indeed, his notations were rudimentary : the only symbolism he uses consists in abreviations such as p : for "plus", m : for "minus" and ℞ for "radix" (= root). For instance, the equation $X^2 + 2X = 48$ is written as :

    1. quad. p : 2 pos. aeq. 48

(quad. is for "quadratum", pos. for "positiones" and aeq. for "aequatur"), and

$$(5 + \sqrt{-15})(5 - \sqrt{-15}) = 25 - (-15) = 40$$

is written :

    5p : ℞ m : 15
    5m : ℞ m : 15
    ―――――――――――
    25m : m : 15 q̃d est 40   (see [Caj. § 140]).

Using these embryonic notations, transformation of equations was clearly a tour de force, and more efficient notations had to develop in order to enlighten this new part of algebra.

This development was rather erratic. Advances made by some authors were not immediately taken up by others, and the process of normalization of notations took a long time. For example, the symbols + and - were already used in Germany since the end of the fifteenth century [Caj. § 201], but they were not widely accepted before the early seventeenth century, and the sign = for equality, first proposed by R. Recorde in 1557, had to struggle with Descartes' symbol ∞ for nearly two centuries [Caj. § 267].

These are relatively minor points, since it may be assumed that p :, m : and aeq. were as convenient to Cardano as +, - and = are to us. There is one point, however, where new notations were nearly vital. In effect, they helped create a new mathematical object : polynomials. There is indeed a significant step from :

1. quad. p : 2 pos. aeq. 48

which is the mere statement of a problem, to the calculation with polynomials like $X^2 + 2X - 48$, and this step was considerably facilitated by suitable notations.

Significant as it may be, the evolution from equations to polynomials is rather subtle, and leading mathematicians of this period rarely took the time to clarify their views on the subject; the rise of the concept of polynomial was most often overshadowed by its application to the theory of equations, and it can only be gathered from indirect indications.

Two milestones in this evolution are "L'Arithmetique" (1585) of Simon Stevin (1548-1620) and "In Artem Analyticem Isagoge" (= Introduction to the analytic art) (1591) of François Viète (1540-1603).

a) L'Arithmetique

This book combines notational advances made by Bombelli and earlier authors (see [Caj. § 296]) with theoretical advances made by Pedro Nunes (1502-1578) (see [Bos, p. 165]) to present a comprehensive treatment of polynomials. Stevin's notation for polynomials, which he terms "multinomials" [S, p. 521] or "integral algebraic numbers" [S, p. 518] (see also pp. 570 ff) has a surprising touch of modernity : the indeterminate is denoted by ①, its square by ②, its cube by ③, etc., and the independent term is indicated by ⓪ (sometimes omitted), so that a "multinomial" appears as an expression like :

$$3 \,③ + 5 \,② - 4 \,① + 6 \,⓪ \quad (\text{or } 3 \,③ + 5 \,② - 4 \,① + 6).$$

Such an expression could be regarded (from a modern point of view) as a finite sequence of real numbers, or, better, as a sequence of real numbers which are all zero except for a finite number of them, or as a function from $\mathbb{N}$ to $\mathbb{R}$ with finite support (compare § 5.1).

This exponential notation (which was not unprecedented) probably helped to abolish the psychological barrier of the third degree (see § 3.1), by placing all the powers of the unknown on an equal footing. It is however rather unfortunate for equations with several unknowns.

Most important is Stevin's observation that the operations on "integral algebraic numbers" share many features with those on "integral arithmetic numbers" (= integers). In particular, he shows [S, Probleme 53, p. 577] that Euclid's algorithm for determining the greatest common divisor of two integers applies nearly without change to find the greatest common divisor of two polynomials (see § 5.2, and particularly n° 5.7).

Although the concept of polynomial is quite clear, the way equations are set up is rather awkward in "L'Arithmetique", since equations are replaced by proportions and the solution of equations is called by Stevin : "the rule of three of quantities". In modern notations, the idea is to replace the solution of an equation like

$$X^2 - aX - b = 0$$

by the following problem : find the fourth proportional u in :

$$\frac{X^2}{aX + b} = \frac{X}{u}$$

or, more generally, find P(u) in :

$$\frac{X^2}{aX + b} = \frac{P(X)}{P(u)},$$

where P(X) is an arbitrary polynomial in X : see [S, p. 592]. (Of course, the solutions which are equal to zero must then be rejected).

In Stevin's words : "Estant donnez trois termes, desquels le premier ②, le second ① ⓪, le troisiesme nombre algebraique quelconque : Trouver leur quatriesme terme proportionnel" [S, p. 595]. (Given three terms, of which the first is X^2, the second aX + b, the third an arbitrary polynomial : to find their fourth proportional term).

This fancy approach to equations may have been prompted by Stevin's methodical treatment of polynomials : an equality like

$$X^2 = aX + b$$

would mean that the <u>polynomials</u> X^2 and aX + b are equal; but polynomials are equal if and only if the coefficients of similar powers of the unknown are the same in both polynomials, and this is clearly not the case here, since X^2 appears on the left-hand side but not on the right-hand side. Stevin's own explanation, while not quite convincing, at least shows that he was fully aware of this notational difficulty :

> "The reason why we call rule of three, or invention of the fourth proportional of quantities, that which is commonly called equation of quantities : (...)
> Because this word "equation" let the beginners think that it was some singular matter, which however is common in the usual arithmetic, since we seek to three given terms a fourth proportional. As that, which is called equation, does not consist in the equality of absolute quantities, but in equality of their values, so this proportion is concerned with the value of quantities, as the same is usual in everyday life" [S, pp. 581-582]. ("Quantity" is Stevin's word for the unknown or any power of it).

This approach met with little success. Even Albert Girard, the first editor of Stevin's works, did not follow Stevin's set up in his own work, and it was soon abandoned.

Stevin's formal treatment of polynomials is rather isolated too; in later works, polynomials were most often considered as functions, although formal operations like Euclid's algorithm were performed. For instance, here is the definition of an equation according to René Descartes (1596-1650) :

> "An equation consists of several terms, some known and some unknown, some of which are together equal to the rest; or rather, all of which taken together are equal to nothing; for this is often the best form to consider" [D, p. 156].

A polynomial then appears as "the sum of an equation" [D, p. 159].

On the whole, the idea of polynomial in the seventeenth century is not very different from the modern notion, and

the need for a more formal definition was not felt for a
long time, but one can get some feeling of the difference
between Descartes' view and ours from the following excerpt
of "La Geometrie" (1637) : (the emphasis is mine) :

"Multiplying together the two <u>equations</u> $x - 2 = 0$ and
$x - 3 = 0$, we have $x^2 - 5x + 6 = 0$, or $x^2 = 5x - 6$.
This is an equation in which x has the value 2 and <u>at
the same time</u> the value 3" [D, p. 159].

b) <u>In_Artem_Analyticem_Isagoge</u>

A major advance in notation, with far-reaching consequences,
was François Viète's idea, put forward in his "Introduction
to the Analytic Art" (1591), of designating by letters all
the quantities, known or unknown, occuring in a problem.
Although letters were occasionally used for unknowns as
early as the third century A.D. (by Diophantus of Alexandria [Caj, § 101]), the use of letters for known quantities
was very new. It also proved to be very useful, since for
the first time it was possible to replace various numerical
examples by a single 'generic' example, from which all the
others could be deduced by assigning values to the letters.
However, it should be observed that this progress did not
reach its full extent in Viète's works, since Viète completely disregards negative numbers : therefore, his letters
always stand for <u>positive</u> numbers only.

This slight limitation notwithstanding, the idea had
another important consequence : by using symbols as his
primary means of expression and showing how to calculate
with those symbols, Viète initiated a completely formal
treatment of algebraic expressions, which he called "logistice speciosa" [V2, p. 17] (as opposed to the "logistice
numerosa", which deals with numbers). This symbolic logistic gave some substance, some legitimacy to algebraic calculations, which allowed Viète to free himself from the
geometric diagrams used so far as justifications.

However, Viète's calculations are somewhat hindered by his insistence that each coefficient in an equation be endowed with a dimension, in such a way that all the terms have the same dimension : the "prime and perpetual law of equations" is that "homogeneous terms must be compared with homogeneous terms" [V2, p. 15]. Moreover, Viète's notations are not as advanced as they could be, since he does not use numerical exponents. For instance, instead of :

Let $A^3 + 3BA = 2Z$,

Viète writes :

"Proponatur A cubus + B plano 3 in A aequari Z solido 2" [Caj, § 177],

insisting that B and Z have degree 2 and 3 respectively.

These minor flaws were soon corrected. In "La Geometrie" (1637) [D], René Descartes shaped the notations that are still in use today (except for his above-mentioned sign ∞ for equality). Thus, in less than one century, algebraic notations had dramatically improved, reaching the same level of generality and the same versatility as ours. These notational advances fostered a deeper understanding of the nature of equations, and the theory of equations was soon advanced in some important points, such as the number of roots and the relations between roots and coefficients of an equation.

§ 2. Relations between roots and coefficients

Cardano's observations on the number of roots of cubic and quartic equations (see § 2.3) were substantially generalized during the next century.

Progress in plane trigonometry brought rather unexpected insights into this question.

In 1593, at the end of the preface of his book "Ideae

Mathematicae" [Ro] (see also [Go, § 1.6]), Adriaan van Roomen[(*)] (1561-1615) issued the following challenge to "all the mathematicians throughout the whole world" :

PROBLEMA MATHEMATICVM
omnibus totius orbis Mathematicis ad construendū propositum.

SI duorum terminorum prioris ad posteriorem proportio sit, ut 1 (1) ad 45 (1) -- 3795 (3) + 9,5634 (5) -- 113,8500 (7) + 781,1375 (9) -- 3451,2075 (11) + 1,0530,6075 (13) -- 2,3267,6280 (15) + 3,8494,2375 (17) -- 4,8849,4125 (19) + 4,8384,1800 (21) -- 3,7865,8800 (23) + 2,3603,0652 (25) -- 1,1767,9100 (27) + 4695,5700 (29) -- 1494,5040 (31) + 376,4565 (33) -- 74,0259 (35) + 11,1150 (37) -- 1,2300 (39) + 945 (41) -- 45 (43) + 1 (45), deturque terminus posterior, invenire priorem.

Exemplum primum datum.

SIt terminus posterior $r\,bin.\,2 + r\,bin.\,2 + r\,bin\,2 + r\,2$, quæritur terminus prior. SOLVTIO. Dico terminū priorem esse $r\,bin.\,2 - r\,bin.\,2 + r\,bin,\,2 + r\,bin.\,2 + r\,3$.

Exemplum secundum datum.

Sit terminus posterior $r\,bin.\,2 + r\,bin.\,2 - r\,bin.\,2 - r\,bin.\,2 - r\,bin.\,2 - r\,2$. quæritur terminus prior. SOLVTIO. Terminus prior est $r\,bin.\,2 - r\,bin.\,2 + r\,bin.\,2 + r\,bin.\,2 + r\,bin.\,2 + r\,2$.

Exemplum tertium datum.

Sit terminus posterior $r\,bin.\,2 + r\,2$, quæritur terminus prior.
SOL. Terminus prior est $r\,bin.\,2 - r\,quadrin.\,2 + r\,\frac{3}{16} + r\,\frac{15}{16} + r\,bin.\,\frac{5}{8} - r\,\frac{5}{64}$.
Si in numeris absolutis solinomijs id proponere libuerit : Sit posterior terminus $r\,3\,\frac{4142,1356,2373,0950,4880,1688,7242,0969,8078,5696,7187,5375}{10000,0000,0000,0000,0000,0000,0000,0000,0000,0000,0000,0000}$.
Quæritur terminus prior. SOLVTIO. Terminus prior erit
$r\,\frac{27,4093,0490,8522,5243,1015,8831,2112,6838,8180}{10000,0000,0000,0000,0000,0000,0000,0800,0000,0000}$.

EXEMPLVM QVÆSITVM.

SIt posterior terminus $r\,trinomia\,1\,\frac{3}{4} -- r\,\frac{5}{16} -- r\,bin.\,1\,\frac{7}{8} -- r\,\frac{45}{64}$.
quæritur terminus prior. Hoc exemplum omnibus Mathematicis ad construendum sit propositum. Non dubito quin *Ludolf van Cöllen* ejus solutionem, saltem in numeris solinomijs sit inventurus.

M E-

[Ro, p. **iijv°], Copyright Bibliothèque Albert 1er, Bruxelles, Réserve précieuse, cote VB4973ALP.

[(*)] then professor at the University of Louvain.

Find the solution of the equation :

$$45X - 3795X^3 + 95634X^5 - 1138500X^7 + 7811375X^9 - 34512075X^{11}$$
$$+ 105306075X^{13} - 232676280X^{15} + 384942375X^{17} - 488494125X^{19}$$
$$+ 483841800X^{21} - 378658800X^{23} + 236030652X^{25} - 117679100X^{27}$$
$$+ 46955700X^{29} - 14945040X^{31} + 3764565X^{33} - 740259X^{35}$$
$$+ 111150X^{37} - 12300X^{39} + 945X^{41} - 45X^{43} + X^{45} = A.$$

He gave the following examples, the second of which is erroneous :
a) if $A = \sqrt{(2+\sqrt{(2+\sqrt{(2+\sqrt{2})})})}$, then $X = \sqrt{(2-\sqrt{(2+\sqrt{(2+\sqrt{(2+\sqrt{3})})})})}$.
b) if $A = \sqrt{(2+\sqrt{(2-\sqrt{(2-\sqrt{(2-\sqrt{2})})})})}$ [it should be :
$A = \sqrt{(2-\sqrt{(2-\sqrt{(2+\sqrt{(2+\sqrt{2})})})})}$], then $X = \sqrt{(2-\sqrt{(2+\sqrt{(2+\sqrt{(2+\sqrt{(2+\sqrt{2})})})})})}$,
[it should be : $X = \sqrt{(2-\sqrt{(2+\sqrt{(2+\sqrt{(2+\sqrt{(2+\sqrt{3})})})})})}$].
c) if $A = \sqrt{(2+\sqrt{2})} =$

$$\sqrt{3 + \frac{4142,1356,2373,0950,4880,1688,7242,0969,8078,5696,7187,5375}{10000,0000,0000,0000,0000,0000,0000,0000,0000,0000,0000,0000}}$$

then $X = \sqrt{(2-\sqrt{(2+\sqrt{3/16} + \sqrt{15/16} + \sqrt{(5/8) - \sqrt{5/64}})})}$

$$= \sqrt{\frac{27,4093,0490,8522,5243,1015,8831,2112,6838,8180}{10000,0000,0000,0000,0000,0000,0000,0000,0000,0000}}$$

and he asked for a solution when

$$A = \sqrt{(1 + (3/4) - \sqrt{5/16} - \sqrt{(1 + (7/8) - \sqrt{45/64})})}.$$

Of course, this was not just any 45-th degree equation; its coefficients had been very carefully chosen.
When this problem was submitted to Viète, he recognized that the left-hand side of the equation is the polynomial by which $2\sin(45\alpha)$ is expressed as a function of $2\sin\alpha$ (see equation (2) below). Therefore, it suffices to find

an arc α such that $2\sin(45\alpha) = A$, and the solution of Van Roomen's equation is $X = 2\sin\alpha$. In Van Roomen's examples :

a) $A = 2\sin(15\pi/2^5)$, and $X = 2\sin(\pi/2^5.3)$

b) A should be $2\sin(15\pi/2^6)$, and $X = 2\sin(\pi/2^6.3)$

c) $A = 2\sin(3\pi/2^3)$ and $X = 2\sin(\pi/2^3.3.5)$

and in the proposed problem $A = 2\sin(\pi/3.5)$, whence $X = 2\sin(\pi/3^3.5^2)$. That Van Roomen's examples correspond to these arcs can be verified by the formulas :

$2\sin(\alpha/2) = \pm\sqrt{2-2\cos\alpha}$ $\quad 2\cos(\alpha/2) = \pm\sqrt{2+2\cos\alpha}$

$2\cos(\pi/3) = 1$ $\quad 2\sin(\pi/3) = \sqrt{3}$

$2\cos(2\pi/5) = (\sqrt{5}-1)/2$ $\quad 2\sin(2\pi/5) = \sqrt{(5+\sqrt{5})/2}$

(see also remark 7.9). From these last results, the value of $\sin(\pi/15)$ and of $\cos(\pi/15)$ can be calculated by the addition formulas, since $\pi/15 = (2\pi/5) - (\pi/3)$.

It turns out that the solution $2\sin(\pi/3^3.5^2)$ of Van Roomen's equation could also be expressed by radicals, but this expression, which does not involve square roots only, is of little use for the determination of its numerical value since it requires the extraction of roots of complex numbers : see remark 7.9. Only the numerical value, suitably approximated to the ninth place, is given by Viète. But Viète does not stop there. While Van Roomen asked for "the" solution of his 45-th degree equation, Viète shows that this equation has 23 positive solutions, and, in passing, he points out that it also has 22 negative solutions [V1, Cap. 6]. Indeed, if α is an arc such that $2\sin(45\alpha) = A$, then, letting

$$\alpha_k = \alpha + k(2\pi/45) \qquad \text{for } k=0,1,\ldots,44,$$

one also has $2\sin(45\alpha_k) = A$ for all k, so that $2\sin(\alpha_k)$

is a solution of Van Roomen's equation for k=0,1,...,44. If A ⩾ 0 (and A ⩽ 2), then one can choose α between 0 and π/90, whence

$$2 \sin(\alpha_k) \geq 0 \quad \text{for } k=0,\ldots,22$$

and $\quad 2 \sin(\alpha_k) < 0 \quad$ for $k=23,\ldots,44$.

Another interesting feature of Viète's brilliant solution [V1] is that, instead of solving directly Van Roomen's equation, which amounts, as we have seen, to the division of an arc into 45 parts, Viète decomposes the problem : since $45 = 3^2.5$, the problem can be solved by the trisection of the arc, followed by the trisection of the resulting arc and the division into 5 parts of the arc thus obtained. As Viète shows, $2 \sin(n\alpha)$ is given as a function of $2 \sin \alpha$ by an equation of degree n, for n odd (see equation (2) below), whence the solutions of Van Roomen's equation of degree 45 can be obtained by solving successively two equations of degree 3 and one equation of degree 5. This idea of solving an equation step by step was to play a central role in Lagrange's and Gauss' investigations, two hundred years later (see chapters 10 and 12).

In modern language, Viète's results on the division of arcs can be stated as follows : for any integer n ⩾ 1, let [n/2] be the greatest integer which is less than (or equal to) n/2, and define :

$$f_n(X) = \sum_{i=0}^{[n/2]} (-1)^i \frac{n}{n-i} \binom{n-i}{i} X^{n-2i}$$

where $\binom{n-i}{i} = \frac{(n-i)!}{i!(n-2i)!}$ is the binomial coefficient. Then,

for all n ⩾ 1 $\quad 2 \cos(n\alpha) = f_n(2\cos\alpha)$ \hfill (1)

and for all <u>odd</u> n ⩾ 1 $\quad 2 \sin(n\alpha) = (-1)^{(n-1)/2} f_n(2\sin\alpha)$. \hfill (2)

Formula (1) can be proved by induction on n, using :

$$2 \cos(n+1)\alpha = (2\cos\alpha)(2\cos n\alpha) - 2\cos(n-1)\alpha$$

and (2) is easily deduced from (1), by applying (1) to
$\beta = (\pi/2) - \alpha$, which is such that $\cos\beta = \sin\alpha$. (The original formulation is not quite so general, but Viète shows how to compute recursively the coefficients of f_n : see [V1, Cap. 9], [V2, pp. 432 ff] or [Go, § 1.6]).

For each integer $n \geq 1$, the equation

$$f_n(X) = A$$

has degree n, and the same arguments as for Van Roomen's equation show that this equation has n solutions (at least when $|A| \leq 2$). These examples, which are quite explicit for n = 3,5,7 in Viète's works [V1, Cap. 9], [V2, pp. 445 ff], may have been influential in the progressive emergence of the idea that equations of degree n have n roots, although this idea was still somewhat obscured by Viète's insistence on considering positive roots only (see also § 6.1).

In later works, such as : "De Recognitione Aequationum" (On Understanding Equations), published posthumously in 1615, Viète also stressed the importance of understanding the structure of equations, meaning by this the relations between roots and coefficients. However, the theoretical tools at his disposal were not sufficiently developed, and he failed to grasp these relations in their full generality. For example, he shows [V2, pp. 210-211] that if an equation

$$B^p A - A^3 = Z^s \qquad (p = \text{plano};\ s = \text{solido})$$

(in the indeterminate A) has two roots A and E, then assuming A > E, one has :

$$B^P = A^2 + E^2 + AE$$

$$Z^S = A^2E + E^2A.$$

The proof is as follows : since

$$B^PA - A^3 = Z^S \quad \text{and} \quad B^PE - E^3 = Z^S,$$

one has

$$B^PA - A^3 = B^PE - E^3,$$

whence

$$B^P(A-E) = A^3 - E^3$$

and, dividing both sides by $A - E$:

$$B^P = A^2 + E^2 + AE.$$

The formula for Z^S is then obtained by replacing B^P in the initial equation $B^PA - A^3 = Z^S$.

The structure of equations was eventually discovered in its proper generality and its simplest form by Albert Girard (1595-1632), and published in "Invention nouvelle en l'algebre" (1629) [Gi] :

> "Le Theoreme qui doit suivre ayant besoin de nouveaux termes, les definitions s'ensuivront premierement" [Gi, p. E2v°] "As the next theorem needs new terms, the definitions will be given first".

Girard calls an equation <u>incomplete</u> if it lacks at least one term (i.e. if at least one of the coefficients is zero); the various terms are called <u>minglings</u> ("meslés") and the last is called the <u>closure</u>. The <u>first faction</u> of the solutions is their sum, the <u>second faction</u> is the sum of their products two by two, the <u>third</u> is the sum of their products three by three and the <u>last</u> is their product. Finally, an equation is in the <u>alternative order</u> when the odd powers of the unknown are on one side of the equality and the even powers on the other side, and when moreover the coefficient of the highest power is 1.

Girard's main theorem is then :

"Toutes les equations d'algebre reçoivent autant de solutions, que la denomination de la plus haute quantité le demonstre, excepté les incomplettes : & la premiere faction des solutions est esgale au nombre du premier meslé, la seconde faction des mesmes, est esgale au nombre du deuxiesme meslé; la troisiesme, au troisiesme, & tousjours ainsi, tellement que la derniere faction est esgale à la fermeture, & ce selon les signes qui se peuvent remarquer en l'ordre alternatif" [Gi, p. E4]. "All the equations of algebra receive as many solutions as the exponent of the highest quantity shows, except the incomplete ones : and the first faction of the solutions is equal to the number of the first mingling, the second faction of the same is equal to the number of the second mingling; the third to the third, and so on, so that the last faction is equal to the closure, and this according to the signs that can be noticed in the alternative order".

The restriction to complete equations is not easy to explain. Half a page later, Girard points out that incomplete equations have not always as many solutions, and that in this case some solutions are imaginary ("impossible" is Girard's own word). However, it is clear that even complete equations may have imaginary solutions (consider for instance $x^2 + x + 1 = 0$), and this fact could not have escaped Girard.

At any rate, Girard claims that the relations between roots and coefficients also hold in this case, provided that the equation be completed by adding powers of the unknown with coefficient 0.

Therefore, the theorem asserts that each equation :

$$x^n + s_2 x^{n-2} + s_4 x^{n-4} + \ldots = s_1 x^{n-1} + s_3 x^{n-3} + s_5 x^{n-5} + \ldots$$

or :

$$x^n - s_1 x^{n-1} + s_2 x^{n-2} - s_3 x^{n-3} + \ldots + (-1)^n s_n = 0$$

has n roots $x_1, \ldots, x_n$ such that

$$s_1 = \Sigma_{i=1}^n x_i$$

$$s_2 = \Sigma_{i<j} x_i x_j$$

$$s_3 = \Sigma_{i<j<k} x_i x_j x_k \tag{3}$$

$$\ldots\ldots$$

$$s_n = x_1 x_2 \ldots x_n.$$

It should be observed that this 'theorem' is not nearly as precise as the modern formulation of the fundamental theorem of algebra, since Girard does not explicitly assert that the roots are of the form $a + b\sqrt{-1}$. It is therefore more a postulatum than a theorem : it claims the existence of "impossible" roots of polynomials, but it is essentially unprovable[*] since nothing else is said about these roots,

---

[*] at least, the proof and, for that matter, even a correct statement of the theorem were very far beyond the reach of seventeenth century mathematicians : see § 9.2.

except (implicitly) that one can calculate with them as if they were numbers. Of course, in all the examples, it turns out that the impossible roots are of the form $a + b\sqrt{-1}$, but Girard nowhere explains what he has in mind.

> "On pourroit dire à quoy sert ces solutions qui sont impossibles, je respond pour trois choses, pour la certitude de la reigle generale, & qu'il ny a point d'autre solutions, & pour son utilité" [Gi, p. Fr°]. "One could say of what use are these solutions which are impossible, I answer for three things, for the certitude of the general rule, and because there is no other solution, and for its utility".

Girard elaborates as follows on the utility of impossible roots : if one seeks the (positive) values of $(x+1)^2 + 2$, where x is such that $x^4 = 4x - 3$, then since the solutions of this last equation are 1, 1, $-1 + \sqrt{-2}$ and $-1 - \sqrt{-2}$, one gets : $(x+1)^2 + 2 = 6, 6, 0$ or 0, so that 6 is the unique result. One would never have been so sure of that whithout the impossible roots.

Of course, Girard does not provide the faintest hint of a proof of his theorem. It would have been very interesting to see at least how he found the relations (3) between the roots and the coefficients of an equation. These relations readily follow from the identification of coefficients in :

$$x^n - s_1 x^{n-1} + s_2 x^{n-2} - \ldots + (-1)^n s_n = (X-x_1)(X-x_2)\ldots(X-x_n)$$

but this equality was probably not known to Girard. Indeed, Girard does not seem to have been aware of the fact that a number a is a root of a polynomial P(X) if and only if X - a divides P(X) : this observation is usually credited to Descartes [D, p. 159], although Nunes may have been aware of it as early as 1567, and perhaps earlier (see

[Bos, pp. 163ff]).

On the subject of impossible roots, Descartes first seems more cautious than Girard :

> "Every equation <u>can</u> have as many distinct roots as the number of dimensions of the unknown quantity in the equation" [D, p. 159], (the emphasis is mine).

This at least can be proved by Descartes' preceding observation (see theorem 5.20). However, Descartes further states :

> "Neither the true nor the false roots are always real; sometimes they are imaginary; that is, while we can always conceive of as many roots for each equation as I have already assigned, yet there is not always a definite quantity corresponding to each root so conceived of" [D, p. 175].

Around the middle of the seventeenth century, the fact that the number of solutions of an equation is equal to the degree was becoming common knowledge, like a piece of mathematical "folklore", accepted without proof and never questioned. At least, it was a good working hypothesis, and mathematicians started to calculate formally with roots of equations without worrying about their nature, pushing further Girard's results to discover what kind of information can be obtained rationally from the coefficients of an equation.

Girard himself had shown (omitting the details of his calculations) that the sum of the squares of the solutions, the sum of their cubes, and the sum of their fourth powers, can be calculated from the coefficients [Gi, p. F2r°] : if $x_1,\ldots,x_n$ are the solutions of

$$x^n - s_1 x^{n-1} + s_2 x^{n-2} - s_3 x^{n-3} + \ldots + (-1)^n s_n = 0,$$

let $\sigma_k = \Sigma_{i=1}^{n} x_i^k$ for any integer k;

then
$$\sigma_1 = s_1$$
$$\sigma_2 = s_1^2 - 2s_2$$
$$\sigma_3 = s_1^3 - 3s_1s_2 + 3s_3$$
$$\sigma_4 = s_1^4 - 4s_1^2s_2 + 4s_1s_3 + 2s_2^2 - 4s_4.$$

Around 1666, general formulas for the sum of any power of the solutions were found by Isaac Newton (1642-1727) [N1,v.I, p. 519] (who was probably unaware of Girard's work : see footnote (12) in [N1, v.I, p. 518]). Newton's clever observation is that, while the formulas for $\sigma_1, \sigma_2, \ldots$ in terms of $s_1, \ldots, s_n$ do not seem to follow a simple pattern, yet there are simple formulas expressing σ_k in terms of $s_1, \ldots, s_n$ and $\sigma_1, \ldots, \sigma_{k-1}$. These formulas can be used to calculate recursively the various σ_k in terms of $s_1, \ldots, s_n$.
Newton's formulas are :

$$\sigma_1 = s_1$$
$$\sigma_2 = s_1\sigma_1 - 2s_2$$
$$\sigma_3 = s_1\sigma_2 - s_2\sigma_1 + 3s_3$$
$$\sigma_4 = s_1\sigma_3 - s_2\sigma_2 + s_3\sigma_1 - 4s_4$$
$$\sigma_5 = s_1\sigma_4 - s_2\sigma_3 + s_3\sigma_2 - s_4\sigma_1 + 5s_5$$

and, generally :

for $k \leq n$: $\sigma_k = \Sigma_{i=1}^{k-1} (-1)^{i+1} s_i\sigma_{k-i} + (-1)^{k+1} k s_k$

for $k > n$: $\sigma_k = \Sigma_{i=1}^{n} (-1)^{i+1} s_i\sigma_{k-i}.$

These formulas (for $k \leq n$) were published without proof in

"Arithmetica Universalis" (1707) [N2, p. 107] (see also [N1, vol. I, p. 519; vol. V, p. 361]). Various ingenious proofs have since been proposed (see for instance [Bou1, App. 1 n° 3]); the following elementary proof is perhaps not very different from Newton's own calculations :

For any integers a,b with $1 \leq a \leq n$ and $b \geq 1$, let $\tau(a,b)$ be the sum of the various different terms obtained from $x_1^b x_2 \ldots x_a$ by permutation of $x_1, \ldots, x_n$: thus

$$\tau(a,1) = s_a, \qquad \tau(1,b) = \sigma_b$$

and, for $a,b \geq 2$:

$$\tau(a,b) = \sum_{i_1=1}^{n} \sum_{\substack{i_2 < i_3 < \ldots < i_a \\ i_k \neq i_1}} x_{i_1}^b x_{i_2} \ldots x_{i_a}.$$

Since, for $b \geq 2$, each term $x_{i_1}^b x_{i_2} \ldots x_{i_a}$ can be obtained as a product :

$$x_{i_1}^b x_{i_2} \ldots x_{i_a} = (x_{i_1} x_{i_2} \ldots x_{i_a})(x_{i_1}^{b-1}),$$

the product $s_a \cdot \sigma_{b-1}$ yields all the terms of $\tau(a,b)$; moreover, this product also yields terms like $x_{i_1}^{b-1} x_{i_2} \ldots x_{i_{a+1}}$, if $a \leq n-1$. Furthermore, if $b = 2$, then each of these latter terms is obtained $(a+1)$ times, while they are obtained only once if $b > 2$. Therefore, we have the following results :

for $a < n$ and $b > 2$: $\tau(a,b) = s_a \sigma_{b-1} - \tau(a+1,b-1)$ \hfill (4)

for $a < n$: $\tau(a,2) = s_a \sigma_1 - (a+1)s_{a+1}$ \hfill (5)

for $b \geq 2$: $\tau(n,b) = s_n \sigma_{b-1}$. \hfill (6)

Since $\tau(1,k) = \sigma_k$, equation (4) with $a = 1$ and $b = k$ yields:

$$\sigma_k = s_1\sigma_{k-1} - \tau(2,k-1).$$

Equation (4) with $a = 2$ and $b = k - 1$ can then be used to eliminate $\tau(2,k-1)$:

$$\sigma_k = s_1\sigma_{k-1} - s_2\sigma_{k-2} + \tau(3,k-2).$$

Next, we use (4) with $a = 3$ and $b = k - 2$ to eliminate $\tau(3,k-2)$, and so on. After a certain number of steps, we obtain :

if $k \leq n$: $\sigma_k = s_1\sigma_{k-1} - s_2\sigma_{k-2} + \ldots + (-1)^k \tau(k-1,2)$

whence, using (5) :

$$\sigma_k = s_1\sigma_{k-1} - s_2\sigma_{k-2} + \ldots + (-1)^k s_{k-1}\sigma_1 + (-1)^{k+1} k s_k$$

and, if $k > n$, we obtain :

$$\sigma_k = s_1\sigma_{k-1} - s_2\sigma_{k-2} + \ldots + (-1)^{n+1} \tau(n,k+1-n)$$

whence, by (6) :

$$\sigma_k = s_1\sigma_{k-1} - s_2\sigma_{k-2} + \ldots + (-1)^{n+1} s_n\sigma_{k-n}. \qquad \square$$

This, of course, was not Newton's most prominent achievement, even if we consider only his contributions to the theory of equations. Indeed, Newton was much more interested in numerical aspects (see for instance [Go, chapter 2]).

Numerical methods to find the roots of polynomial equations were at first one of the several aims of the theory of equations, developed by some of the same authors who developed other aspects : see for instance Cardano's "golden rule" [C, ch. 30], Stevin's "Appendice algebraique"

[S, pp. 740-745] or Viète's "De numerosa Potestatum ad Exegesim Resolutione" (On the numerical resolution of powers by exegetics) [V2, pp. 311-370]. These numerical methods were much more successful for the solution of explicit numerical equations than the algebraic formulas "by radicals"; indeed, algebraic formulas are available only for degrees up to 4 and they are by no means more accurate than the numerical methods (see also § 2.3). Therefore, the numerical solution of equations soon developed into a new branch of mathematics, growing more accurate and powerful while the algebraic theory of equations was progressively stalled.

Since the discussion of numerical methods falls beyond the scope of this book, we take the occasion of this period of relatively low activity in algebra to justify a pause in the historical exposition. We turn in the next chapter to a modern exposition of the above-mentioned results on polynomials in one indeterminate, in order to show on which mathematical base later results were grounded.

5 A modern approach to polynomials in one indeterminate

§ 1. Definitions

<u>1</u>. In modern terminology, a polynomial in one indeterminate with coefficients in a ring A can be defined as a map

$$P : \mathbb{N} \to A$$

such that the set supp $P = \{n \in \mathbb{N} \mid P_n \neq 0\}$, called the <u>support</u> of P, is finite. The addition of polynomials is the usual addition of maps :

$$(P+Q)_n = P_n + Q_n$$

and the product is the convolution product :

$$(PQ)_n = \Sigma_{i+j=n} \, P_i \cdot Q_j.$$

Every element $a \in A$ is identified with the polynomial $a : \mathbb{N} \to A$ which maps 0 to a and n to 0 for $n \neq 0$. Denoting by

$$X : \mathbb{N} \to A$$

the polynomial which maps 1 onto the unit element $1 \in A$ and the other integers to 0, it is then easily seen that every polynomial P can be uniquely written as :

$$P = \Sigma_{i \in \mathbb{N}} \, P_i \cdot X^i.$$

Therefore, we shall henceforth denote by

$$a_0 + a_1 X + \ldots + a_n X^n \tag{1}$$

(as is usual) the polynomial which maps $i \in \mathbb{N}$ to a_i for $i = 0, \ldots, n$ and to 0 for $i > n$. Accordingly, the set of all polynomials with coefficients in A (or : polynomials over A) is denoted $A[X]$. Straightforward calculations show that $A[X]$ is a ring, which is commutative if and only if A is commutative.

The ring of polynomials in any number m of indeterminates over A can be similarly defined as the ring of maps from $\mathbb{N}^m$ to A with finite support, with the convolution product.

<u>2</u>. Of course, the definition above is not quite natural. The naive approach to polynomials is to consider expressions like (1), where X is an undefined object, called an <u>indeterminate</u>, or a <u>variable</u>. While this terminology will be retained in the sequel, it should be observed that, without any other proper definition, to say something is an indeterminate or a variable is hardly a definition. Moreover, it fosters confusion between the polynomial

$$P(X) = a_0 + a_1 X + \ldots + a_n X^n$$

and the associated polynomial function

$$P(.) : A \to A$$

which maps $x \in A$ onto $P(x) = a_0 + a_1 x + \ldots + a_n x^n$. This same confusion has prompted the use of the term : <u>constant</u> polynomials for the elements of A, considered as polynomials. While this confusion is not so serious when A is a field with infinitely many elements (see corollary 21 below), it could be harmful when A is finite : for instance, if $A = \{a_1, \ldots, a_n\}$ (with $n \geq 2$), then the polynomial $(X - a_1) \ldots (X - a_n)$ is not the zero polynomial since the coefficient of X^n is 1, but its associated polynomial function maps every element of A to 0.

<u>3</u>. The <u>degree</u> of a non-zero polynomial P is the greatest integer n for which the coefficient of X^n in the expression of P is not zero; this coefficient is called the <u>leading coefficient</u> of P, and P is said to be <u>monic</u> if its leading coefficient is 1.

The degree of P is denoted by deg P. One also sets : deg $0 = -\infty$, so that the following relations hold without restriction if A is a domain (i.e. a ring in which $ab = 0 \Rightarrow a = 0$ or $b = 0$) :

$$\deg(P+Q) \leq \max(\deg P, \deg Q)$$
$$\deg(PQ) = \deg P + \deg Q.$$

<u>4</u>. When A is a (commutative) field, the ring A[X] has a field of fractions A(X), constructed as follows : the elements of A(X) are the equivalence classes of couples f/g, where $f, g \in A[X]$ and $g \neq 0$, under the equivalence relation :

$$f/g = f'/g' \quad \text{if} \quad g'f = f'g.$$

The addition of equivalence classes is defined by :

$$(f_1/g_1) + (f_2/g_2) = (f_1 g_2 + f_2 g_1)/g_1 g_2$$

and the multiplication by :

$$(f_1/g_1) \cdot (f_2/g_2) = (f_1 f_2)/(g_1 g_2).$$

It is then easily verified that A(X) is a field, called the field of <u>rational fractions</u> in one indeterminate X over the field A. The same construction can be applied to the ring of polynomials in m indeterminates and yields the field of rational fractions in m indeterminates over the field A.

However, the ring $A[X]$ of polynomials in one indeterminate over a field A has particularly nice properties, which follow from the Euclidean division algorithm. These properties are reviewed in the next section.

§ 2. Euclidean division

From now on, we only consider polynomials over a field F.

5. THEOREM (Euclidean Division Property) : Let $P_1, P_2 \in F[X]$. If $P_2 \neq 0$, then there exist polynomials $Q, R \in F[X]$ such that

$$P_1 = P_2 Q + R$$

and $\deg R < \deg P_2$.

Moreover, the polynomials Q and R are uniquely determined by these properties.

They are called respectively the <u>quotient</u> and the <u>remainder</u> of the division of P_1 by P_2.

<u>Proof</u> : The existence of Q and R is proved by induction on $\deg P_1$. If $\deg P_1 < \deg P_2$, we set $Q = 0$ and $R = P_1$. If $\deg P_1 - \deg P_2 = d \geq 0$, then letting $c \in F$ be the quotient of the leading coefficient of P_1 by that of P_2, we have :

$$\deg(P_1 - c X^d P_2) < \deg P_1.$$

Therefore, by the induction hypothesis, there exist Q and $R \in F[X]$ such that :

$$P_1 - c X^d P_2 = P_2 Q + R \quad \text{and} \quad \deg R < \deg P_2.$$

This equation yields :

$$P_1 = P_2(Q + c\, x^d) + R$$

whence $Q + c\, x^d$ and R satisfy the required properties.

To prove the uniqueness of Q and R, assume:

$$P_1 = P_2 Q_1 + R_1 = P_2 Q_2 + R_2$$

with $\deg R_1 < \deg P_2$ and $\deg R_2 < \deg P_2$.

Then $R_1 - R_2 = P_2(Q_2 - Q_1)$

and this equality is impossible if both sides are non-zero, since the degree of the right-hand side is then at least equal to $\deg P_2$, while the degree of the left-hand side is strictly less than $\deg P_2$. □

6. DEFINITIONS : Let $P_1, P_2 \in F[X]$. We say P_2 <u>divides</u> P_1 if

$$P_1 = P_2 Q \quad \text{for some} \quad Q \in F[X]$$

or, equivalently when $P_2 \neq 0$, if the remainder of the division of P_1 by P_2 is 0.

A <u>greatest common divisor</u> (G.C.D.) of P_1 and P_2 is a polynomial $D \in F[X]$ which has the following two properties:
a) D divides P_1 and P_2
b) If S is a polynomial which divides P_1 and P_2, then S divides D.

If 1 is a G.C.D. of P_1 and P_2, then P_1 and P_2 are said to be <u>relatively prime</u>.

7. Since it is by no means obvious that any two polynomials P_1, P_2 have a G.C.D., our first objective is to devise a method of finding such a G.C.D., thereby proving its existence. We shall closely follow Euclid's algorithm for finding the G.C.D. of two integers or the greatest common meas-

ure of two line segments, assuming (without loss of generality) that deg $P_1 \geq$ deg P_2.

If $P_2 = 0$, then P_1 is a G.C.D. of P_1 and P_2. Otherwise, we divide P_1 by P_2 :

$$P_1 = P_2 Q_1 + R_1. \qquad (2.1)$$

Next, we divide P_2 by the remainder R_1, provided that it is not zero :

$$P_2 = R_1 Q_2 + R_2 \qquad (2.2)$$

then we divide the first remainder by the second, and so on as long as the remainders are not zero :

$$R_1 = R_2 Q_3 + R_3 \qquad (2.3)$$

$$R_2 = R_3 Q_4 + R_4 \qquad (2.4)$$

...   ...

$$R_{n-1} = R_n Q_{n+1} + R_{n+1}. \qquad (2.n+1)$$

...   ...

Since deg $P_2 >$ deg $R_1 >$ deg $R_2 > \ldots$, this sequence of integers cannot extend indefinitely. Therefore, $R_{n+1} = 0$ for some n.

<u>Claim</u> : R_n is then a G.C.D. of P_1 and P_2. (If $n = 0$, then set $R_n = P_2$).

To see that R_n divides P_1 and P_2, observe that equation (2.n+1), together with $R_{n+1} = 0$, implies that R_n divides R_{n-1}. It then follows from equation (2.n) that R_n divides also R_{n-2}. Going up in the sequence of equations (2.n), (2.n-1), (2,n-2),..., we conclude recursively that R_n divides $R_{n-3},...,R_2,R_1,P_2$ and P_1.

Assume next that P_1 and P_2 are both divisible by some polynomial S. Then equation (2.1) shows that S also divides R_1. Since it divides P_2 and R_1, S divides also R_2, by equation (2.2). Going down in the above sequence of equations (2.2), (2.3),..., we finally see that S divides R_n.

This completes the proof that R_n is a G.C.D. of P_1 and P_2.

However, the G.C.D. of two polynomials is not unique (except over the field with two elements), as the following theorem shows :

8. THEOREM : Any two polynomials $P_1, P_2 \in F[X]$ which are not both zero have a unique monic greatest common divisor D_1, and a polynomial $D \in F[X]$ is a greatest common divisor of P_1 and P_2 if and only if

$$D = c D_1 \quad \text{for some} \quad c \in F^\times (= F - \{0\}).$$

Moreover, if D is a greatest common divisor of P_1 and P_2, then

$$D = P_1 U_1 + P_2 U_2 \quad \text{for some} \quad U_1, U_2 \in F[X].$$

Proof : Euclid's algorithm already yields a greatest common divisor R_n of P_1 and P_2. Dividing R_n by its leading coefficient, we get a monic G.C.D. of P_1 and P_2. Now, assume D and D' are G.C.D. of P_1 and P_2. Then D divides D' since D satisfies condition (a) and D' satisfies condition (b). The same argument, with D and D' exchanged, shows that D' divides D. Let then

$$D' = DQ \quad \text{and} \quad D = D'Q';$$

it follows that

$$QQ' = 1,$$

so that Q and Q' are constants, which are inverse of each other. This proves at once the second statement and the uniqueness of the monic G.C.D. of P_1 and P_2, since D and D' cannot be both monic unless $Q = Q' = 1$.

Moreover, if D is any G.C.D. of P_1 and P_2, then D and the greatest common divisor R_n found by Euclid's algorithm are related by :

$$D = c R_n \quad \text{for some} \quad c \in F^\times.$$

Therefore, it suffices to prove the last statement for R_n. To this end, we consider again the sequence of equations (2.1),...,(2.n) : from equation (2.n), we get :

$$R_n = R_{n-2} - R_{n-1}Q_n.$$

We then use equation (2.n-1) to eliminate R_{n-1} in this expression of R_n, and we obtain :

$$R_n = -R_{n-3}Q_n + R_{n-2}(1 + Q_{n-1}Q_n).$$

This shows that R_n is a sum of multiples of R_{n-2} and R_{n-3}. Now, R_{n-2} can be eliminated using (2.n-2) : we thus obtain an expression of R_n as a sum of multiples of R_{n-3} and R_{n-4}. Going up in the sequence of equations (2.n-3), (2.n-4),... we end up with an expression of R_n as a sum of multiples of P_1 and P_2 :

$$R_n = P_1 U_1 + P_2 U_2 \quad \text{for some} \quad U_1, U_2 \in F[X]. \qquad \square$$

Because of its repeated use in the sequel, the following special case seems to be worth pointing out explicitly :

9. __COROLLARY__ : If P_1, P_2 are relatively prime polynomials in $F[X]$, then there exist polynomials $U_1, U_2 \in F[X]$ such that

$$P_1 U_1 + P_2 U_2 = 1.$$

10. __REMARKS__ :

a) The proof given above is effective : Euclid's algorithm yields a procedure for constructing the polynomials U_1, U_2 in theorem 8 and corollary 9.

b) Since the G.C.D. of two polynomials $P_1, P_2 \in F[X]$ can be found by rational calculations (i.e. calculations involving only the four basic operations of arithmetic), it does not depend on particular properties of the field F. The point of this observation is that if the field F is embedded in a larger field K, then the polynomials $P_1, P_2 \in F[X]$ can be regarded as polynomials over K, but their monic G.C.D. in $K[X]$ is the same as their monic G.C.D. in $F[X]$. This is noteworthy in view of the fact that the __irreducible__ factors of P_1 and P_2 depend on the base field F, as is clear from example 12.b below.

§ 3. Irreducible polynomials

11. __DEFINITION__ : A polynomial $P \in F[X]$ is said to be __irreducible in__ $F[X]$ (or : __over__ F) if deg P > 0 and P is not divisible by any polynomial $Q \in F[X]$ such that

$$0 < \deg Q < \deg P.$$

From this definition, it follows that if a polynomial D divides an irreducible polynomial P, then either D is a constant or deg D = deg P. In the latter case, the quotient of P by D is a constant, whence D is the product of P by a non-zero constant. In particular, for any polynomial $S \in F[X]$, either 1 or P is a G.C.D. of P and S. Consequently, either P divides S or P is relatively prime to S.

12. EXAMPLES :

a) By definition, it is clear that every polynomial of degree 1 is irreducible. It will be proved later that over the field of complex numbers, only these polynomials are irreducible.

b) Theorem 17 below will show that if a polynomial of degree at least 2 has a root in the base field, then it is not irreducible. The converse is true for polynomials of degree 2 and 3; namely : if a polynomial of degree 2 or 3 has no root in the base field, then it is irreducible over this field : see corollary 18. It follows that, for instance, the polynomial $X^2 - 2$ is irreducible over $\mathbb{Q}$, but not over $\mathbb{R}$. Thus, the irreducibility of polynomials of degree at least 2 depends on the base field (compare remark 10(b)).

c) A polynomial (of degree at least 4) may be reducible over a field without having any root in this field. For instance, in $\mathbb{Q}[X]$, the polynomial $X^4 + 4$ is reducible since

$$X^4 + 4 = (X^2+2X+2)(X^2-2X+2).$$

<u>Remark</u> : To determine whether a given polynomial with rational coefficients is irreducible or not over $\mathbb{Q}$ may be difficult, although a systematic procedure has been devised by Kronecker : see [VW1, § 32]. This procedure is not unlike that which is used to find the rational roots of polynomials with rational coefficients : see § 6.3.

13. THEOREM : Every non constant polynomial $P \in F[X]$ is a finite product :

$$P = c.P_1. \ldots .P_n$$

where $c \in F^\times$ and $P_1,\ldots,P_n$ are monic irreducible polynomials (not necessarily distinct). Moreover, this factorization is unique, except for the order of the factors.

Proof : The existence of the above factorization is easily proved by induction on deg P : if deg P = 1 or, more generally, if P is irreducible, then

$$P = c \cdot P_1$$

where c is the leading coefficient of P and $P_1 = c^{-1}P$ is irreducible and monic. If P is reducible, then it can be written as a product of two polynomials of degree strictly less than deg P. By induction, each of these two polynomials has a finite factorization as above, and these factorizations multiply up to a factorization of P.

To prove the uniqueness of the factorization, we shall use the following lemma :

14. LEMMA : If a polynomial divides a product of r factors and is relatively prime to the first r - 1 factors, then it divides the last one.

Proof : It suffices to consider the case r = 2, since the general case then easily follows by induction.

Assume that a polynomial S divides a product T.U and is relatively prime to T. By corollary 9, we can find polynomials V, W such that

$$SV + TW = 1.$$

Multiplying both sides of this equality by U, we obtain :

$$S(UV) + (TU)W = U.$$

Now, S divides the left-hand side, since it divides TU by hypothesis; it then follows that S divides U. □

Completion of the proof of theorem 13 : It remains to prove the uniqueness (up to the order of factors) of factori-

zations. Assume :

$$P = c P_1 \ldots P_n = d Q_1 \ldots Q_m \qquad (3)$$

where $c, d \in F^\times$ and $P_1, \ldots, P_n$, $Q_1, \ldots, Q_m$ are monic irreducible polynomials.

First, $c = d$ since c and d are both equal to the leading coefficient of P. Hence, (3) yields :

$$P_1 \ldots P_n = Q_1 \ldots Q_m. \qquad (4)$$

Next, since P_1 divides the product $Q_1 \ldots Q_m$, it follows from lemma 14 that it cannot be relatively prime to all the factors. Changing the numbering of $Q_1, \ldots, Q_m$ if necessary, we may assume that P_1 is not relatively prime to Q_1, so that their monic G.C.D., which we denote by D, is not a constant polynomial. Since D divides P_1, which is irreducible, it is equal to P_1 up to a constant factor. Since moreover D and P_1 are both monic, the constant factor is 1, whence

$$D = P_1.$$

We may argue similarly with Q_1, since Q_1 is also monic and irreducible, and we thus obtain $D = Q_1$, whence

$$P_1 = Q_1.$$

Cancelling P_1 ($= Q_1$) in equation (4), we get a similar equality, with one factor less on each side :

$$P_2 \ldots P_n = Q_2 \ldots Q_m.$$

Using inductively the same argument, we get (changing the numbering of $Q_2, \ldots, Q_m$ if necessary) :

$$P_2 = Q_2, P_3 = Q_3, \ldots, P_n = Q_n,$$

and it follows that $n \leq m$. If $n < m$, then comparing the degrees of both sides of (4) we get :

$$\deg Q_{n+1} = \ldots = \deg Q_m = 0$$

and this is absurd since Q_i is irreducible for $i=1,\ldots,m$. Therefore, $n = m$ and the proof is complete. □

Lemma 14 has another consequence which is worth noting in view of its repeated use in the sequel :

<u>15. PROPOSITION</u> : If a polynomial is divisible by pairwise relatively prime polynomials, then it is divisible by their product.

<u>Proof</u> : Let $P_1,\ldots,P_r$ be pairwise relatively prime polynomials which divide a polynomial P. We argue by induction on r, the case $r = 1$ being trivial. By the induction hypothesis,

$$P = P_1 \ldots P_{r-1} Q$$

for some polynomial Q. Since P_r divides P, lemma 14 shows that P_r divides Q, so that $P_1 \ldots P_r$ divides P. □

§ 4. <u>Roots</u>

As in the preceding sections, F denotes a field. For any polynomial

$$P = a_0 + a_1 X + \ldots + a_n X^n \in F[X]$$

we denote by $P(.)$ the associated polynomial function :

$$P(.) : F \to F$$

which maps any $x \in F$ onto $P(x) = a_0 + a_1 x + \ldots + a_n x^n$. It is readily verified that for any two polynomials $P, Q \in F[X]$ and any $x \in F$,

$$(P+Q)(x) = P(x) + Q(x) \text{ and } (P.Q)(x) = P(x).Q(x);$$

hence the map $P \to P(.)$ is a homomorphism from the ring $F[X]$ to the ring of functions from F to F.

16. DEFINITION : An element $a \in F$ is a <u>root</u> of a polynomial $P \in F[X]$ if $P(a) = 0$.

17. THEOREM : An element $a \in F$ is a root of a polynomial $P \in F[X]$ if and only if $(X-a)$ divides P.

<u>Proof</u> : Since $\deg(X-a) = 1$, the remainder R of the division of P by $(X-a)$ is a constant polynomial. Evaluating at a the polynomial functions associated with each side of the equation

$$P = (X-a)Q + R$$

we get :

$$P(a) = (a-a)Q(a) + R,$$

whence

$$P(a) = R.$$

This shows that $P(a) = 0$ if and only if the remainder of the division of P by $X-a$ is 0. The theorem follows, since the last condition means that $(X-a)$ divides P. □

18. COROLLARY : Let $P \in F[X]$ be a polynomial of degree 2 or 3. Then P is irreducible over F if and only if it has no root in F.

Proof : This readily follows from the theorem, since the hypothesis on deg P implies that if P is not irreducible, then it has a factor of degree 1, whence a factor of the form X-a. □

19. DEFINITIONS : The <u>multiplicity</u> of a root a of a non-zero polynomial P is the exponent of the highest power of X-a which divides P. Thus, the multiplicity is m if $(X-a)^m$ divides P but $(X-a)^{m+1}$ does not divide P. A root is called <u>simple</u> when its multiplicity is 1; otherwise it is called <u>multiple</u>.

When the multiplicity of a as a root of P is considered as a function of P, it is also called the <u>valuation</u> of P at a, and denoted : $v_a(P)$; we then set $v_a(P) = 0$ if a is not a root of P. By convention, we also set $v_a(0) = \infty$, so that the following relations hold for any $P,Q \in F[X]$ and any $a \in F$:

$$v_a(P+Q) \geq \min(v_a(P), v_a(Q)).$$

$$v_a(PQ) = v_a(P) + v_a(Q).$$

These properties are exactly the same as those of the function

$$- \deg : F[X] \to \mathbb{Z} \cup \{\infty\}$$

which maps any polynomial to the <u>opposite</u> of its degree. Accordingly, (- deg) is sometimes considered as the valuation at infinity.

20. THEOREM : Every non-zero polynomial $P \in F[X]$ has a finite number of roots. If $a_1, \ldots, a_r$ are the various roots of P in F, with respective multiplicities $m_1, \ldots, m_r$, then

$$\deg P \geq m_1 + \ldots + m_r$$

and $P = (X-a_1)^{m_1}\ldots(X-a_r)^{m_r} Q$

for some polynomial $Q \in F[X]$ which has no root in F.

<u>Proof</u> : If $a_1,\ldots,a_r$ are distinct roots of P in F, with respective multiplicities $m_1,\ldots,m_r$, then the polynomials $(X-a_1)^{m_1},\ldots,(X-a_r)^{m_r}$ are relatively prime and divide P, whence proposition 15 shows that

$$P = (X-a_1)^{m_1}\ldots(X-a_r)^{m_r} Q,$$

for some polynomial Q.
It readily follows that

$$\deg P \geq m_1 + \ldots + m_r;$$

hence P cannot have infinitely many roots. The rest follows from the observation that if $a_1,\ldots,a_r$ are <u>all</u> the roots of P in F, then Q has no root. □

<u>21. COROLLARY</u> : Let $P,Q \in F[X]$. If F is infinite, then $P = Q$ if and only if the associated polynomial functions $P(.)$ and $Q(.)$ are equal.

<u>Proof</u> : If $P(.) = Q(.)$, then all the elements in F are roots of $P-Q$, whence $P-Q = 0$, since F is infinite. The converse is clear. □

§ 5. Multiple roots and derivatives

The aim of this section is to derive a method of determining whether a polynomial has multiple roots without actually finding the roots, and to reduce to 1 the multiplicity of the roots. (More precisely, the method yields, for any given polynomial P, a polynomial P_s which has the same roots as P, each with multiplicity 1).

This method is due to Johann Hudde (1633-1704). It uses the derivative of polynomials, which was introduced purely algebraically by Hudde in his letter : "De Reductione Aequationum" (On the reduction of equations) (1657) [H1] and subsequently applied to find maxima and minima of polynomials and rational fractions in his letter : "De Maximis et Minimis" (1658) [H2].

22. DEFINITION : The <u>derivative</u> ∂P of a polynomial $P = a_0 + a_1 X + \ldots + a_n X^n$ with coefficients in a field F is the polynomial

$$\partial P = a_1 + 2a_2 X + 3a_3 X^2 + \ldots + na_n X^{n-1}.$$

Straightforward calculations show that the following (familiar) relations hold :

$$\partial(P+Q) = \partial P + \partial Q$$
$$\partial(PQ) = (\partial P)Q + P(\partial Q).$$

<u>Remark</u> : The integers which appear in the coefficients of ∂P are regarded as elements in F : thus, n stands for $1 + 1 + \ldots + 1$ (n terms), where 1 is the unit element of F. This requires some caution, since it could happen that $n = 0$ in F even if $n \neq 0$ (as an integer). For instance, if F is the field with two elements $\{0,1\}$, then $2 = 1 + 1 = 0$ in F, whence non-constant polynomials like $X^2 + 1$ have derivative zero over F.

23. LEMMA : Let $a \in F$ be a root of a polynomial $P \in F[X]$. Then a is a multiple root of P (i.e. $v_a(P) \geq 2$) if and only if a is a root of ∂P.

<u>Proof</u> : Since a is a root of P, one has :

$$P = (X-a)Q \quad \text{for some} \quad Q \in F[X],$$

whence

$$\partial P = Q + (X-a)\partial Q.$$

This equality shows that X-a divides ∂P if and only if it divides Q. Since this last condition amounts to : $(X-a)^2$ divides P, i.e. $v_a(P) \geq 2$, the lemma follows. □

24. PROPOSITION : Let $P \in F[X]$ be a polynomial which splits into a product of linear factors(*) over some field K containing F. A necessary and sufficient condition for the roots of P in K to be all simple is that P and ∂P be relatively prime.

<u>Proof</u> : If some root a of P in K is not simple, then, by the preceding lemma, P and ∂P have the common factor X-a in K[X]; therefore, they are not relatively prime in K[X], whence also in F[X] (see remark 10(b)).

Conversely, if P and ∂P are not relatively prime, they have in K[X] a common irreducible factor, which has degree 1 since all the irreducible factors of P in K[X] are linear. We can thus find a polynomial $X-a \in K[X]$ which divides both P and ∂P, and it follows from the preceding lemma that a is a multiple root of P in K. □

To improve lemma 23, we now assume that the characteristic of F is zero, which means that every non-zero integer is non-zero in F. (The characteristic of a field F is either 0 or the smallest integer n > 0 such that n = 0 in F).

25. PROPOSITION : Assume that char.F = 0 and let $a \in F$ be a root of a non-zero polynomial $P \in F[X]$. Then

$$v_a(\partial P) = v_a(P) - 1.$$

(*) In § 9.2, it is proved that this condition holds for every (non constant) polynomial.

Proof : Let $m = v_a(P)$ and let $P = (X-a)^m Q$ where $Q \in F[X]$ is not divisible by $X-a$. Then

$$\partial P = (X-a)^{m-1}[mQ + (X-a)\partial Q],$$

and the hypothesis on the characteristic of F ensures that $mQ \neq 0$. Then, since $X-a$ does not divide Q, it does not divide $mQ + (X-a)\partial Q$ either, whence $v_a(\partial P) = m - 1$. □

As an application of this result, we now derive Hudde's method to reduce to 1 the multiplicity of the roots of a non-zero polynomial P over a field F of characteristic zero : let D be a G.C.D. of P and ∂P, and let $P_s = P/D$.

26. THEOREM : Let K be an arbitrary field containing F. Then the roots of P_s in K are the same as those of P, and every root of P_s in K is simple, i.e. has multiplicity 1.

Proof : Since P_s is a quotient of P, it is clear that every root of P_s is a root of P. Conversely, if $a \in K$ is a root of P, of multiplicity m, then the preceding proposition shows that $v_a(\partial P) = m - 1$, and it follows that $(X-a)^{m-1}$ is the highest power of $X-a$ which divides both P and ∂P. It is therefore the highest power of $X-a$ which divides D. Consequently, P_s is divisible by $X-a$ but not by $(X-a)^2$, which means that a is a simple root of P_s. □

27. COROLLARY : If $P \in F[X]$ is irreducible, then its roots in every field K containing F are simple.

Proof : Since P is irreducible and does not divide ∂P, since deg ∂P = deg P - 1, the constant polynomial 1 is a G.C.D. of P and ∂P. Therefore, $P_s = P$ and the preceding theorem shows that every root of P in any field K containing F is simple. □

This corollary does not hold if char.$F \neq 0$. For instance, if char.$F = 2$ and $a \in F$ is not a square in F, then X^2-a is irreducible in $F[X]$, but it has $\sqrt{a}$ as a double root in $F(\sqrt{a})$. (Observe that $\sqrt{a} = -\sqrt{a}$ since the characteristic is 2).

28. REMARKS :

a) Since a G.C.D. of P and ∂P can be calculated by Euclid's algorithm (see n° 7 above), it is not necessary to find the roots of P in order to construct the polynomial P_s. Thus, there is no serious restriction if we henceforth assume, when trying to solve an equation $P = 0$, that all the roots of P in F or in any field containing F are simple.

b) In his work, Hudde does not explicitly introduce the derivative polynomial ∂P, but indirectly he uses it. His formulation [H1, Reg. 10, pp 433 ff] is as follows : to reduce to 1 the multiplicity of the roots of a polynomial

$$P = a_0 + a_1 X + a_2 X^2 + \ldots + a_n X^n,$$

form a new polynomial P_1 by multiplying the coefficients of P by the terms of an arbitrary arithmetical progression: m, m+r, m+2r, ..., m+nr :

$$P_1 = a_0 m + a_1(m+r)X + a_2(m+2r)X^2 + \ldots + a_n(m+nr)X^n.$$

Then, the quotient of P by a greatest common divisor D_1 of P and P_1 is the required polynomial.

The relation between this rule and its modern translation is easy to see, since

$$P_1 = mP + rX \partial P.$$

Therefore, $D_1 = D$ up to a non-zero constant factor, and possibly to a factor X in D_1, if 0 is a root of P. So,

Hudde's method yields the same equation with simple roots as its modern equivalent, except that Hudde's equation lacks the root 0 whenever 0 is a root of the initial equation.

§ 6. Common roots of two polynomials

As the preceding discussion shows, it is sometimes useful to determine whether two polynomials P,Q are relatively prime or not. The most straightforward method is of course to calculate a G.C.D. of P and Q, but there is another construction, due to L. Euler (1707-1783) (with different notations) : the <u>resultant</u> of P and Q. This construction is also basic for elimination theory; it will be used in §§ 6.4 and 10.1.

<u>29</u>. Let $\quad P = a_n X^n + a_{n-1} X^{n-1} + \ldots + a_1 X + a_0 \quad (a_n \neq 0)$

and $\quad Q = b_m X^m + b_{m-1} X^{m-1} + \ldots + b_1 X + b_0 \quad (b_m \neq 0)$

be polynomials over a field F. The <u>resultant</u> of P and Q is the following $(m+n) \times (m+n)$ determinant :

$$R = \det \begin{pmatrix} a_n & a_{n-1} & \cdots & & a_1 & a_0 & & & \\ & a_n & & \cdots & & a_1 & a_0 & 0 & \\ 0 & & \cdots\cdots\cdots & & & & & & \\ & & & a_n & a_{n-1} & \cdots & & a_1 & a_0 \\ b_m & b_{m-1} & \cdots & b_1 & b_0 & & & & \\ & b_m & \cdots & & b_1 & b_0 & 0 & & \\ & & & \cdots\cdots & & & & & \\ 0 & & & & & & & & \\ & & & b_m & b_{m-1} & \cdots & b_1 & b_0 & \end{pmatrix} \begin{matrix} \left.\vphantom{\begin{matrix}1\\1\\1\\1\end{matrix}}\right\}m \\ \\ \left.\vphantom{\begin{matrix}1\\1\\1\\1\end{matrix}}\right\}n \end{matrix}$$

30. **THEOREM** : Assume(*) P and Q split into products of linear factors over some field K containing F, and let R denote the resultant of P and Q. The following conditions are then equivalent :
a) P and Q are not relatively prime
b) P and Q have a common root in K
c) R = 0.

Proof : (a) ⇒ (b) : Let D be a G.C.D. of P and Q. By hypothesis, D is not constant. Moreover, since D divides P (and Q) which splits into linear factors over K, the irreducible factors of D in K[X] have degree 1. Therefore, D has at least one root in K, which is also a common root of P and Q since D divides P and Q.

(b) ⇒ (c) : Let $u \in K$ be a common root of P and Q. Let

$$P = (X-u)P_1 \text{ and } Q = (X-u)Q_1$$

where $P_1, Q_1 \in K[X]$. Then

$$PQ_1 = QP_1 \ (= PQ/(X-u)).$$

From this equality, we obtain a system of $m+n$ equations by equating the coefficients of like terms. More precisely, if

$$P = a_n X^n + a_{n-1} X^{n-1} + \ldots + a_1 X + a_0$$

$$Q = b_m X^m + b_{m-1} X^{m-1} + \ldots + b_1 X + b_0$$

as above, and if

$$-P_1 = z_1 X^{n-1} + z_2 X^{n-2} + \ldots + z_{n-1} X + z_n$$

and $\quad Q_1 = y_1 X^{m-1} + y_2 X^{m-2} + \ldots + y_{m-1} X + y_m$

(*) Theorem 9.3 will show that this hypothesis always holds.

then by equating to zero the coefficient of x^k in $PQ_1 - QP_1$ we get :

$$\sum_{i+j=k} a_i y_{m-j} + b_i z_{n-j} = 0 \qquad (5.k)$$

(where we set $a_i = 0$ (resp. $b_i = 0$) if $i > n$ (resp. $i > m$) and $y_{m-j} = 0$ (resp. $z_{n-j} = 0$) if $j > m$ (resp. $j > n$)).
More explicitly,

$$a_n y_1 \qquad\qquad\qquad + b_m z_1 \qquad\qquad\qquad\qquad = 0 \quad (5.m+n-1)$$

$$a_{n-1} y_1 + a_n y_2 \qquad\qquad + b_{m-1} z_1 + b_m z_2 \qquad\qquad = 0 \quad (5.m+n-2)$$

$$a_{n-2} y_1 + a_{n-1} y_2 + a_n y_3 \quad + b_{m-2} z_1 + b_{m-1} z_2 + b_m z_3 = 0 \quad (5.m+n-3)$$

$$\cdots\cdots\cdots\cdots\cdots\cdots\cdots\cdots\cdots\cdots\cdots$$

$$a_0 y_{m-1} + a_1 y_m \qquad\qquad + b_0 z_{m-1} + b_1 z_m \qquad = 0 \quad (5.1)$$

$$a_0 y_m \qquad\qquad\qquad\qquad + b_0 z_m \qquad = 0 \quad (5.0)$$

Equations $(5.m+n-1)$, $(5.m+n-2),\ldots,(5.0)$ can be regarded as a system of $m+n$ homogeneous linear equations in the indeterminates $y_1,\ldots,y_m,z_1,\ldots,z_n$. It is easily verified that the coefficient matrix of this system is the transpose of the matrix which appears in the definition of R. It follows that $R = 0$, since $y_1,\ldots,y_m,z_1,\ldots,z_n$ is a non-trivial solution of this system.

(c) ⇒ (a) : Assume now that $R = 0$; then, reversing the steps in the proof of (b) ⇒ (c), we observe that the system $(5.m+n-1),\ldots,(5.0)$ has a non-trivial solution and conclude that there exist non-zero polynomials P_1 and Q_1 such that

$$PQ_1 = QP_1, \qquad \deg P_1 \leq n-1, \qquad \deg Q_1 \leq m-1.$$

(Note that inequalities $\deg P_1 < n-1$ and $\deg Q_1 < m-1$ occur when $y_1 = z_1 = 0$). The preceding equality shows that P divides QP_1. If P is relatively prime to Q, then it divides P_1, by lemma 14. This is impossible since $\deg P_1 < \deg P$; therefore, P and Q are not relatively prime. □

Appendix to chapter 5 : Decomposition of rational fractions in sums of partial fractions

To find a primitive of a rational fraction P/Q (where $P, Q \in \mathbb{R}[x]$ and $Q \neq 0$), it is customary to decompose it as a sum of partial fractions in the following way : factor Q into irreducible factors :

$$Q = Q_1^{m_1} \ldots Q_r^{m_r}$$

where $Q_1, \ldots, Q_r$ are distinct irreducible polynomials. Then there exist polynomials $P_0, P_1, \ldots, P_r$ such that :

$$P/Q = P_0 + (P_1/Q_1^{m_1}) + \ldots + (P_r/Q_r^{m_r})$$

and $\deg P_i < \deg Q_i^{m_i}$ for $i=1,\ldots,r$.

To prove the existence of the polynomials $P_0, \ldots, P_r$, it suffices to prove that if $Q = S_1 S_2$ where S_1 and S_2 are relatively prime polynomials, then there exist polynomials P_0, P_1, P_2 such that

$$P/Q = P_0 + (P_1/S_1) + (P_2/S_2)$$

and $\deg P_i < \deg S_i$ for $i=1,2$.

The property then easily follows by induction on the number of distinct irreducible factors of Q.

<u>Proof</u> : Since S_1 and S_2 are relatively prime, corollary 9 yields :

$$1 = S_1T_1 + S_2T_2$$

for some polynomials T_1, T_2. Multiplying both sides by P/Q, we get:

$$P/Q = (PT_1/S_2) + (PT_2/S_1).$$

By Euclidean division of PT_1 by S_2 and of PT_2 by S_1, we have:

$$PT_1 = S_2U_2 + P_2 \quad \text{and} \quad PT_2 = S_1U_1 + P_1$$

for some polynomials U_1, U_2, P_1, P_2 with $\deg P_i < \deg S_i$ for $i=1,2$. Replacing PT_1 and PT_2 in the preceding equation, we obtain:

$$P/Q = (U_1+U_2) + (P_2/S_2) + (P_1/S_1). \qquad \square$$

To facilitate integration, each partial fraction P/Q^m (where Q is irreducible and $\deg P < \deg Q^m$) can be decomposed further as:

$$P/Q^m = (P_1/Q) + (P_2/Q^2) + \ldots + (P_m/Q^m)$$

with $\deg P_i < \deg Q$ for $i=1,\ldots,m$.

To obtain this decomposition, let P_1 be the quotient of the Euclidean division of P by Q^{m-1}:

$$P = P_1 Q^{m-1} + R_1 \quad \text{with} \quad \deg R_1 < \deg Q^{m-1}.$$

Since $\deg P < \deg Q^m$, it follows that $\deg P_1 < \deg Q$. Let then P_2 be the quotient of the Euclidean division of R_1 by Q^{m-2}, and so on. Then:

$$P = P_1 Q^{m-1} + P_2 Q^{m-2} + \ldots + P_{m-1} Q + P_m, \tag{6}$$

and the required decomposition follows after the division of both sides by Q^m.

Remark : The right-hand side of (6) is the 'Q-adic' expansion of P. When P and Q are replaced by integers, equation (6) shows that the integer P is written as $P_1P_2...P_m$ in the base Q.

6 Alternative methods for cubic and quartic equations

With their improved notations, mathematicians of the seventeenth century devised new methods for solving cubic and quartic equations. The aim of this chapter is to review some of these advances, in particular the important method proposed by Ehrenfried Walter Tschirnhaus in 1683.

§ 1. Viète on cubic equations

Viète's contribution to the theory of cubic equations is twofold : in "De Recognitione Aequationum" he gave a trigonometric solution for the irreducible case and in "De Emendatione Aequationum" a solution for the general case which requires the extraction of only one cube root. These methods were both posthumously published in "De Aequationum Recognitione et Emendatione Tractatus Duo" (1615) ("Two Treatises on the Understanding and Amendment of Equations"; see [V2]).

a) Trigonometric solution for the irreducible case [V2, p. 174].

The irreducible case of cubic equations

$$X^3 + pX + q = 0$$

occurs when $(p/3)^3 + (q/2)^2 < 0$ (see § 2.3(c)). This inequality of course implies $p < 0$, hence there is no loss of generality if the above equation is written as :

$$X^3 - 3a^2 X = a^2 b, \tag{1}$$

and the condition $(p/3)^3 + (q/2)^2 < 0$ becomes $a > |b/2|$. (Note that one can obviously assume $a > 0$, since only a^2 occurs in the given equation).

From the formula for the cosine of a sum of arcs (or from the general formula (1) in chapter 4), it follows that for all α :

$$(2a \cos \alpha)^3 - 3a^2(2a \cos \alpha) = 2a^3 \cos 3\alpha.$$

Comparing with equation (1), we see that if α is an arc such that

$$\cos 3\alpha = b/2a,$$

then $2a \cos \alpha$ is a solution of equation (1). The two other solutions are easily derived from this one; if $\alpha_k = \alpha + k(2\pi/3)$, with $k = 1, 2$, then we also have

$$\cos 3\alpha_k = b/2a,$$

whence the solutions of equation (1) are :

and
$$\begin{aligned}&2a \cos \alpha \\ &2a \cos(\alpha+(2\pi/3)) = -a \cos \alpha - a\sqrt{3} \sin \alpha \\ &2a \cos(\alpha+(4\pi/3)) = -a \cos \alpha + a\sqrt{3} \sin \alpha.\end{aligned}$$

Since Viète systematically avoids negative numbers, he gives only the first solution (which is positive if $b > 0$). However, he points out immediately afterwards that $a \cos \alpha + a\sqrt{3} \sin \alpha$ and $a \cos \alpha - a\sqrt{3} \sin \alpha$ are solutions of the equation

$$3a^2Y - Y^3 = a^2b.$$

This shows how clear was Viète's notion of the number of roots of cubic equations.

b) Algebraic solution for the general case [V2, p. 287].

Viète suggested an ingenious change of variable to solve the equation

$$X^3 + pX + q = 0. \qquad (2)$$

Setting $X = (p/3Y) - Y$ and substituting in the equation above, he gets for Y the equation :

$$Y^6 - qY^3 - (p/3)^3 = 0,$$

whence Y^3 can be found by solving a quadratic equation :

$$Y^3 = -(q/2) \pm \sqrt{(p/3)^3 + (q/2)^2}.$$

Thus, a solution of the cubic equation (2) is given by

$$X = (p/3Y) - Y,$$

where

$$Y^3 = -(q/2) + \sqrt{(p/3)^3 + (q/2)^2}.$$

<u>Remarks</u> :
a) If the other determination of Y^3 is chosen, namely :

$$Y'^3 = -(q/2) - \sqrt{(p/3)^3 + (q/2)^2},$$

then the value of X does not change : indeed, since $(YY')^3 = -(p/3)^3$, we have

$$(p/3Y) = -Y' \quad \text{and} \quad (p/3Y') = -Y,$$

whence

$$(p/3Y) - Y = (p/3Y') - Y'.$$

Incidentally, this remark also shows that Viète's method yields the same result as Cardano's formula, since after replacing Y and Y' by their values in the formula X = -Y - Y', we obtain

$$X = \sqrt[3]{-(q/2) + \sqrt{(p/3)^3 + (q/2)^2}} + \sqrt[3]{-(q/2) - \sqrt{(p/3)^3 + (q/2)^2}}.$$

b) The case where Y = 0 occurs only if p = 0; this case is therefore readily solved.

c) Viète gives only one root, because in the original formulation the equation is :

$$A^3 + 3B^p A = 2Z^s$$

(p = plano; s = solido; unknowns are always designated by vowels), which has only one real root if B^p is positive, since the function $A^3 + 3B^p A$ is then monotonically increasing and takes therefore the value $2Z^s$ only once.

§ 2. Descartes on quartic equations

New insights into the solution of equations arose from the arithmetic of polynomials. In "La Geometrie", Descartes recommends the following way of attacking equations of any degree : "First, try to put the given equation into the form of an equation of the same degree obtained by multiplying together two others, each of a lower degree" [D, p. 192]. He himself shows how this method can be successfully applied to quartic equations [D, pp. 180 ff] :

After cancelling out the cubic term, as in Ferrari's method (chapter 3), the general quartic equation is set in the form :

$$X^4 + pX^2 + qX + r = 0, \tag{3}$$

and we may assume $q \neq 0$, otherwise the equation is quadratic in X^2 and is therefore easily solved. We then determine a,b,c,d in such a way that

$$X^4 + pX^2 + qX + r = (X^2+aX+b)(X^2+cX+d).$$

Equating the coefficients of similar powers of X, we obtain from this equation :

$$0 = a + c \qquad (4)$$
$$p = b + d + ac \qquad (5)$$
$$q = ad + bc \qquad (6)$$
$$r = bd. \qquad (7)$$

From equations (4), (5), (6), the values of b, c and d are easily derived in terms of a :

$$c = -a$$
$$b = (a^2/2) + (p/2) - (q/2a)$$
$$d = (a^2/2) + (p/2) + (q/2a).$$

(Observe that $a \neq 0$ since $q \neq 0$). Replacing b and d in equation (7), we get the following equation for a :

$$a^6 + 2pa^4 + (p^2-4r)a^2 - q^2 = 0. \qquad (8)$$

This is a cubic equation in a^2, which can therefore be solved. If a is a solution of this equation, then the given equation (3) factorizes into two quadratic equations :

$$X^2 + aX + (a^2/2) + (p/2) - (q/2a) = 0$$

and $$X^2 - aX + (a^2/2) + (p/2) + (q/2a) = 0,$$

whence the solutions are easily found.

§ 3. Rational solutions for equations with rational coefficients

The rational solutions of equations with rational coefficients of arbitrary degree can be found by a finite trial and error process. This seems to have been first observed by Albert Girard [Gi, D.4 v°]; it also appears in "La Geometrie" [D, p. 176].

Let

$$a_n X^n + a_{n-1} X^{n-1} + \ldots + a_1 X + a_0 = 0 \qquad (9)$$

be an equation with rational coefficients : $a_i \in \mathbb{Q}$ for $i = 0,\ldots,n$. Multiplying both sides by a common multiple of the denominators of the coefficients if necessary, we may assume $a_i \in \mathbb{Z}$ for $i = 0,\ldots,n$. Multiplying then both sides by a_n^{n-1}, the equation becomes :

$$(a_n X)^n + a_{n-1}(a_n X)^{n-1} + a_{n-2} a_n (a_n X)^{n-2} + \ldots + a_1 a_n^{n-2}(a_n X) +$$
$$+ a_0 a_n^{n-1} = 0.$$

Letting $Y = a_n X$, we are then reduced to a <u>monic</u> equation with integral coefficients :

$$Y^n + b_{n-1} Y^{n-1} + b_{n-2} Y^{n-2} + \ldots + b_1 Y + b_0 = 0 \qquad (10)$$

$$(b_i \in \mathbb{Z})$$

1. THEOREM : All the rational roots of a monic equation with integral coefficients are integers which divide the independent term.

Proof : Discarding the null roots and dividing the left-hand side of (10) by a suitable power of Y, we may assume $b_0 \neq 0$. Let then $y \in \mathbb{Q}$ be a rational root of (10). Write $y = y_1/y_2$ where y_1, y_2 are relatively prime integers.

From

$$(y_1/y_2)^n + b_{n-1}(y_1/y_2)^{n-1} + \ldots + b_1(y_1/y_2) + b_0 = 0$$

it follows, by multiplication by y_2^n and rearrangement of the terms:

$$y_1^n = -y_2[b_{n-1}y_1^{n-1} + \ldots + b_1 y_1 y_2^{n-2} + b_0 y_2^{n-1}].$$

This equation shows that each prime factor of y_2 divides y_1^n, whence also y_1; since y_1 and y_2 are relatively prime, this is impossible unless y_2 has no prime factor: hence $y_2 = \pm 1$ and $y \in \mathbb{Z}$.

To prove that y divides b_0, consider again:

$$y^n + b_{n-1}y^{n-1} + \ldots + b_1 y + b_0 = 0$$

and separate off on one side the independent term:

$$b_0 = -y[y^{n-1} + b_{n-1}y^{n-2} + \ldots + b_1].$$

This equation shows that y divides b_0, since the factor between brackets is an integer. □

Tracing back through the transformations from equation (9) to equation (10), we get the following result:

2. COROLLARY: Each rational solution of the equation with integral coefficients:

$$a_n X^n + a_{n-1} X^{n-1} + \ldots + a_1 X + a_0 = 0 \qquad (a_i \in \mathbb{Z})$$

has the form (y/a_n), where $y \in \mathbb{Z}$ is a divisor of $a_0 a_n^{n-1}$.

This last condition is very useful, in that it gives a bound on the number of trials which are necessary to find a rational root of the proposed equation, provided that $a_o \neq 0$. Of course, this can always be assumed, after dividing by a suitable power of X.

For example, the theorem (or its corollary) shows that an equation like

$$X^n + a_{n-1}X^{n-1} + \ldots + a_1 X \pm 1 = 0$$

with $a_i \in \mathbb{Z}$ for $i = 1,\ldots,n-1$, has no rational root, except possibly +1 or -1.

"Other example, once very difficult" [Gi, E r°] : the rational solutions of

$$X^3 = 7X - 6$$

are among $\pm 1, \pm 2, \pm 3, \pm 6$. Trying successively the various possibilities, one finds 1, 2 and -3 as solutions.

§ 4. <u>Tschirnhaus' method</u>

Although research on the theory of equations was not quite as active at the end of the seventeenth century, substantial progress arose from a 4-page note by Tschirnhaus [T], in 1683. This note proposes a uniform method to solve equations of any degree.

The basic idea is very simple; it starts from the observation that it is always possible to remove the second term of any equation :

$$X^n + a_{n-1}X^{n-1} + \ldots + a_1 X + a_o = 0$$

by a simple change of variable : $Y = X + (a_{n-1}/n)$. (See e.g. § 2.2 and § 3.2). By allowing more general changes of variable, such as :

$$Y = X^m + b_{m-1}X^{m-1} + \ldots + b_1 X + b_o, \qquad (11)$$

Tschirnhaus aims to cancel out several terms of the proposed equation. More precisely, by a suitable choice of the n parameters $b_0, b_1, \ldots, b_{m-1}$, the above change of variable yields an equation in Y :

$$Y^n + c_{n-1} Y^{n-1} + \ldots + c_1 Y + c_0 = 0$$

in which any m coefficients c_i can be chosen to vanish : roughly speaking, this is because the m parameters $b_0, \ldots, b_{m-1}$ provide m degrees of freedom, which can be used to fulfill m conditions.

In particular, taking m = n-1, all the terms except the first and the last could be removed, hence the equation in Y takes the form :

$$Y^n + c_0 = 0,$$

and is thus readily solved by radicals. Plugging in the solution $Y = \sqrt[n]{-c_0}$ in equation (11), we then obtain a solution of the proposed equation of degree n by solving an equation of degree m = n-1 :

$$X^{n-1} + b_{n-2} X^{n-2} + \ldots + b_1 X + b_0 = \sqrt[n]{-c_0}.$$

Arguing by induction on the degree, it thus follows that equations of any degree can be solved by radicals.

There is however a major obstacle, which was soon noticed by Leibniz [Lz2, p. 449; p. 403] : the conditions which ensure that all the coefficients $c_1, \ldots, c_{n-1}$ vanish yield a system of equations of various degrees in the parameters b_i, and this system is very difficult to solve. Indeed, solving this system actually amounts to solving a single equation of degree (n-1)!; it thus appears that this method does not work for n > 3, unless the resulting equation of degree (n-1)! has some particular features which makes it reducible to equations of degree less than n. This turns

91

out to be the case for n = 4 : the resulting sextic can be seen to factorize into a product of factors of degree 2 whose coefficients are solutions of cubic equations (see [L, Art. 41-45]), but for $n \geq 5$, no such simplification is apparent. (Note that for composite n, Tschirnhaus' method can be applied differently, and possibly more easily : for instance if n = 4, then cancelling out the coefficients of Y and Y^3 reduces the equation in Y to a quadratic equation in Y^2).

To discuss Tschirnhaus' method in some detail, we start by explaining how the equation in Y can be found. This is a special instance of a general type of problem which is dealt with by elimination theory : the problem is to eliminate the indeterminate X between the two equations :

$$X^n + a_{n-1}X^{n-1} + a_{n-2}X^{n-2} + \ldots + a_1X + a_0 = 0 \quad (12)$$

$$X^m + b_{m-1}X^{m-1} + b_{m-2}X^{m-2} + \ldots + b_1X + b_0 = Y \quad (13)$$

$$(m < n),$$

i.e. to find an equation R(Y) = 0, called a <u>resulting equation</u>, which has the following properties : (a) whenever x and y are such that equations (12) and (13) hold, then R(y) = 0; (b) whenever y is such that R(y) = 0, then equations (12) and (13) have a common root x. This last property shows that if R(Y) = 0 can be solved, then one (at least) of the roots of equation (12) is among the roots of equation (13).

The properties of R(Y) can be rephrased as follows, considering (12) and (13) as equations in X with coefficients in the field of rational fractions in Y : R(Y) = 0 if and only if the polynomials

$$P(X) = X^n + a_{n-1}X^{n-1} + \ldots + a_1X + a_0$$

and $$Q(X) = X^m + b_{m-1}X^{m-1} + \ldots + b_1X + (b_0-Y)$$

have a common root. As theorem 5.30 shows, a solution of this problem is the resultant[(*)] of P and Q :

$$R(Y) = \det \begin{pmatrix} 1 & a_{n-1} & \cdots & a_1 & a_0 & & & \\ & 1 & a_{n-1} & \cdots & a_1 & a_0 & & 0 \\ & & & \cdots\cdots & & & & \\ & 0 & & 1 & a_{n-1} & \cdots & a_1 & a_0 \\ 1 & b_{m-1} & \cdots & b_1 & b_0-Y & & & \\ & 1 & b_{m-1} & \cdots & b_1 & b_0-Y & & 0 \\ & & & \cdots\cdots & & & & \\ & 0 & & 1 & b_{m-1} & \cdots & b_1 & b_0-Y \end{pmatrix} \begin{matrix} \left.\vphantom{\begin{matrix}1\\1\\1\\1\end{matrix}}\right\} m \\ \\ \left.\vphantom{\begin{matrix}1\\1\\1\\1\end{matrix}}\right\} n \end{matrix}$$

Since the indeterminate Y appears only in the last n lines, it is easily verified that R is a polynomial of degree n in Y. Moreover, since the determinant is an alternating sum of products of entries from different rows and columns, it follows that products of only k factors b_i occur in the coefficient of Y^{n-k}. Therefore,

$$R(Y) = c_n Y^n + c_{n-1} Y^{n-1} + \ldots + c_1 Y + c_0$$

where c_{n-k} is a polynomial of degree k in $b_0,\ldots,b_{m-1}$. (Actually, $c_n = (-1)^n$).

In order to cancel out $c_{n-1}, c_{n-2}, \ldots, c_1$, consider now $m = n - 1$. The preceding discussion shows that

$$c_{n-1} = c_{n-2} = \ldots = c_1 = 0$$

is a system of n - 1 equations of degrees $1, 2, 3, \ldots, n-1$ in the variables $b_0, \ldots, b_{n-2}$. Between these equations, n-2

[(*)] It is slightly anachronistic to resort to determinants in this context, since they came into use somewhat later, but the actual calculations in elimination theory were equivalent and they in fact motivated the development of variables can be eliminated, and the resulting equation in a single variable has degree $1.2.3.\ldots.(n-1) = (n-1)!$ (see for instance [Web, § 53]). This was proved only much later by Bezout, but by considering some examples one soon realizes that the solution of the above system of equations is far from easy.

Let us consider for instance the cubic equation:

$$X^3 + pX + q = 0 \qquad (p \neq 0) \tag{14}$$

and let

$$Y = X^2 + b_1 X + b_0. \tag{15}$$

Elimination of X between these two equations according to the method explained above yields the following resulting equation in Y:

$$c_3 Y^3 + c_2 Y^2 + c_1 Y + c_0 = 0 \tag{16}$$

where

$$c_3 = -1$$
$$c_2 = 3b_0 - 2p$$
$$c_1 = 4pb_0 - 3qb_1 - 3b_0^2 - pb_1^2 - p^2$$
$$c_0 = q^2 + p^2 b_0 - pqb_1 + 3qb_0 b_1 - 2pb_0^2$$
$$\quad + b_0^3 - qb_1^3 + pb_0 b_1^2.$$

Thus, in order to cancel out c_2 and c_1, it suffices to let $b_0 = 2p/3$ and to choose for b_1 a root of the quadratic equation:

---

determinants.

$$pb_1^2 + 3qb_1 - p^2/3 = 0,$$

for instance

$$b_1 = (3/p)[\sqrt{(p/3)^3 + (q/2)^2} - (q/2)].$$

With the above choice of b_0 and b_1, and letting $A = \sqrt{(p/3)^3 + (q/2)^2}$, we have

$$c_0 = 2^3 A^3 (3/p)^3 [A - (q/2)].$$

Therefore, a root of the resulting equation (16) in Y is :

$$Y = 2A(3/p) \sqrt[3]{A - (q/2)}.$$

A root of the proposed cubic equation (14) is then found by solving the quadratic equation (15), which is now :

$$X^2 + (3/p)[A - (q/2)]X + 2(p/3) = 2A(3/p)\sqrt[3]{A - (q/2)}. \tag{17}$$

However, in general only one of the roots of this quadratic equation is a root of the proposed cubic equation (14). A better way to solve (14) is to find the common roots of (14) and (17), which are the roots of their greatest common divisor.

Letting $B = \sqrt[3]{A - (q/2)}$, one gets by Euclid's algorithm 5.7 the following greatest common divisor, if $A \neq 0$:

$$2A(3/p)^2 (B^2 + (p/3)) [BX + ((p/3) - B^2)].$$

(It is easy to see that $B^2 + (p/3) \neq 0$ if $A \neq 0$ and $p \neq 0$). There is thus only one common root of (14) and (17), which is :

$$X = \frac{B^2 - (p/3)}{B}.$$

Since $B = \sqrt[3]{-(q/2) + \sqrt{(p/3)^3 + (q/2)^2}}$, it is easily verified that

$$-p/3B = \sqrt[3]{-(q/2) - \sqrt{(p/3)^3 + (q/2)^2}};$$

thus the above formula for X is identical to Cardano's formula. (Compare also Viète's method in § 1(b)).

If $A = 0$, then the left-hand side of (17) divides the given cubic polynomial, so both roots of (17) are roots of the proposed equation.

7 Roots of unity

§ 1. Introduction

New branches of mathematics, such as analytic geometry and differential calculus came into being during the seventeenth century, and it is therefore not surprising that investigations in the algebraic theory of equations came to a near standstill at the end of this century, being pursued only occasionally by leading mathematicians such as Tschirnhaus. However, progress in other branches indirectly brought some new advances in algebra. A case in point is the well-known 'de Moivre's formula' : for every integer n and every $\alpha \in \mathbb{R}$:

$$(\cos \alpha + i \sin \alpha)^n = \cos(n\alpha) + i \sin(n\alpha), \qquad (1)$$

which is easily proved by induction on n, since from the addition formulas for sines and cosines it readily follows that

$$(\cos \alpha + i \sin \alpha)(\cos \beta + i \sin \beta) = \cos(\alpha+\beta) + i \sin(\alpha+\beta). \qquad (2)$$

This formula (1), and its proof through (2), were first given by Euler in 1748 [Sm, vol. 2, p. 450] but it was already implicit in earlier works by Cotes and by de Moivre, and it turns out that the above proof, simple as it is, is deceitful, since it does not keep any record of the slow evolution which led to de Moivre's formula. It is the purpose of this chapter to sketch this evolution and to discuss the significance of de Moivre's formula for the algebraic theory of equations.

§ 2. The origins of de Moivre's formula

While the differential calculus was being shaped by Leibniz and Newton, the integration (or primitivation) of rational fractions was unavoidable. Very soon, formulas equivalent to :

$$\int x^n \, dx = x^{n+1}/(n+1) \qquad \text{(for } n \neq -1\text{)}$$

and $\quad \int dx/x = \log x$

became familiar, and the integration of any rational fraction in which the denominator is a power of a linear polynomial easily follows by a change of variable. Moreover, around 1675, Leibniz had also obtained :

$$\int dx/(x^2+1) = \tan^{-1} x,$$

from which the integration of other rational fractions can be derived.

The integration of rational fractions is the main theme of a 1702 paper by Leibniz in the Acta Eruditorum of Leipzig : "Specimen novum Analyseos pro Scientia infiniti circa Summas et Quadraturas". (|| "New specimen of the Analysis for the Science of the infinite about Sums and Quadratures" [Lz1, n° 24]). In this paper, Leibniz points out the usefulness of the decomposition of rational fractions into sums of partial fractions (see the appendix to chapter 5) to reduce the integration of rational fractions to the integration of dx/x and $dx/(x^2+1)$ or, in his words, to the quadrature of the hyperbola or the circle. Since this decomposition requires that the denominator be factored in a product of irreducible polynomials, he is thus led to investigate the factorization of real polynomials, coming close to the "fundamental theorem of algebra" according to which every real polynomial of positive degree is a product of factors of degree 1 or 2 (see chapter 9) :

"Now, this leads us to a question of utmost importance : whether all the rational quadratures may be reduced to the quadrature of the hyperbola and of the circle, which by our analysis above amounts to the following : whether every algebraic equation or real integral formula in which the indeterminate is rational can be decomposed into simple or plane real factors" (= real factors of degree 1 or 2) [Lz1, p. 359].

Leibniz then proposes the following counterexample : since

$$x^4 + a^4 = (x^2 + a^2\sqrt{-1})(x^2 - a^2\sqrt{-1}),$$

it follows that :

$$x^4 + a^4 = (x+a\sqrt{\sqrt{-1}})(x-a\sqrt{\sqrt{-1}})(x+a\sqrt{-\sqrt{-1}})(x-a\sqrt{-\sqrt{-1}});$$

failing to observe that

$$\sqrt{\sqrt{-1}} = (1+\sqrt{-1})/\sqrt{2} \quad \text{and} \quad \sqrt{-\sqrt{-1}} = (1-\sqrt{-1})/\sqrt{2},$$

he draws the erroneous conclusion that no non-trivial combination of the four factors above yields a real divisor of $x^4 + a^4$.

"Therefore, $\int dx/(x^4+a^4)$ cannot be reduced to the squaring of the circle or the hyperbola by our analysis above, but founds a new kind of its own" [Lz1, p. 360].

Even without deeper considerations about complex numbers, Leibniz could have avoided this mistake if he had observed that, by adding and subtracting $2a^2x^2$, one gets (*) :

$$x^4 + a^4 = (x^2 + a^2)^2 - 2a^2x^2$$

$$= (x^2 + a^2 + \sqrt{2}\ ax)(x^2 + a^2 - \sqrt{2}\ ax).$$

(*) This was shown by N. Bernoulli in the Acta Eruditorum of 1719.

As it appears from [N1, v. IV, pp. 205 ff], Newton had also tried his hand in the same questions as early as 1676, and he had obtained this factorization of $x^4 + a^4$, as well as factorizations of $1 \pm x^n$ for various values of the integer n, (see the appendix), but in 1702 he presumably did not care enough about mathematics any more to point out the mistake in Leibniz's paper, had he been aware of it.

Leibniz's argument was definitively refuted by Roger Cotes (1682-1716), who thoroughly investigated the factorization of the binomials $a^n \pm x^n$, obtaining the following formulas :

$$a^{2m} + x^{2m} = \Pi_{k=0}^{m-1} (a^2 - 2a \cos((2k+1)\pi/2m)x + x^2) \qquad (3)$$

$$a^{2m+1} + x^{2m+1} = (a+x) \Pi_{k=0}^{m-1} (a^2 - 2a \cos((2k+1)\pi/2m+1)x + x^2) \qquad (4)$$

$$a^{2m} - x^{2m} = (a-x)(a+x) \Pi_{k=1}^{m-1} (a^2 - 2a \cos(2k\pi/2m)x + x^2) \qquad (5)$$

$$a^{2m+1} - x^{2m+1} = (a-x) \Pi_{k=1}^{m} (a^2 - 2a \cos(2k\pi/2m+1)x + x^2). \qquad (6)$$

These formulas appear in a compilation of Cotes' papers : "Theoremata tum Logometrica tum Trigonometrica Datarum Fluxionum Fluentes exhibentia, per Methodum Mensurarum Ulterius extensam" (1722) "Theorems, some logometric, some trigonometric, which yield the fluents of given fluxions by the method of measures further developed" [Co, pp. 113-114], in a very elegant form : to find the factors of $a^\lambda \pm x^\lambda$, it is prescribed to divide a circle of radius a into 2λ equal parts AB, BC, CD, DE, EF, etc. Let O be the center of the circle and let P be a point on the radius OA, at a distance OP = x (< a) from O.

Then

$$a^\lambda - x^\lambda = OA^\lambda - OP^\lambda = AP.CP.EP.etc$$

and

$$a^\lambda + x^\lambda = OA^\lambda + OP^\lambda = BP.DP.FP.\text{etc.}$$

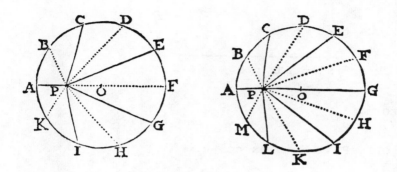

Exempli gratia si λ sit 5, dividatur circumferentia in 10 partes æquales, eritque $AP \times CP \times EP \times GP \times IP = OA^s - OP^s$ existente P intra circulum: & $BP \times DP \times FP \times HP \times KP = OA^s + OP^s$. Similiter si λ sit 6, divisa circumferentia in 12 partes æquales: erit $AP \times CP \times EP \times GP \times IP \times LP = OA^6 - OP^6$, existente P intra circulum; & $BP \times DP \times FP \times HP \times KP \times MP = OA^6 + OP^6$.

[Co, p. 114] (Univ. Cath. Louvain, Centre général de Documentation).

To check that this formulation is equivalent to the previous one, it suffices to observe that, on the figure below :

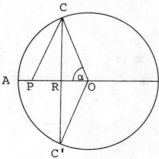

we have, by Pythagoras' theorem :

$$CP^2 = PR^2 + RC^2$$

and since

$$PR = OP - OR = x - a \cos \alpha \text{ and } RC = a \sin \alpha,$$

it follows that

$$CP = \sqrt{x^2 - 2ax \cos \alpha + a^2}.$$

Therefore,

$$CP \cdot C'P = x^2 - 2ax \cos \alpha + a^2.$$

Cotes' formulas were given without justification, but a proof was eventually supplied in 1730 by Abraham de Moivre (1667-1754), who had already obtained some interesting results on the division of the circle. In a 1707 paper entitled "Aequationum quaerundam Potestatis tertiae, quintae, septimae, novae, & c superiorum, ad infinitum usque pergendo, in terminis finitis, ad imstar Regularum pro Cubicis quae vocantur, <u>Cardani</u>, Resolutio Analytica" ("The analytic solution in finite terms of certain equations of the third, fifth, seventh, ninth and other higher powers, by rules similar to those called Cardano's for the cubics") [Sm, vol. 2, pp. 441 ff], he had observed that the equation

$$f_n(X) = 2a \qquad \qquad (n \text{ odd})$$

where f_n is the polynomial by which $2\cos(n\alpha)$ is expressed as a function of $2 \cos \alpha$ (see equation (1) in chapter 4), has the solution :

$$X = \sqrt[n]{a + \sqrt{a^2-1}} + \sqrt[n]{a - \sqrt{a^2-1}},$$

for any value of a, whatsoever.
In particular, if $a = \cos(n\alpha)$, it follows that

$$2 \cos \alpha = \sqrt[n]{\cos(n\alpha) + \sqrt{-1} \sin(n\alpha)} + \sqrt[n]{\cos(n\alpha) - \sqrt{-1} \sin(n\alpha)}, \quad (7)$$

although this formula does not appear explicitly in de Moivre's paper of 1707.

De Moivre's basic observation was that the equation $f_n(X) = 2a$ can be obtained by elimination of z between the two equations :

$$1 - 2az^n + z^{2n} = 0 \quad (8)$$

$$1 - Xz + z^2 = 0. \quad (9)$$

Indeed, equation (9) yields :

$$1 + z^2 = Xz$$

and, squaring both sides of this equation, we get :

$$1 + z^4 = (X^2 - 2)z^2.$$

These last equations show that, for $n = 1, 2$:

$$1 + z^{2n} = f_n(X) z^n; \quad (10)$$

from these initial steps, it is easily verified by induction that equation (10) holds for every integer n, using the recurrence formula :

$$f_{n+1}(X) = X f_n(X) - f_{n-1}(X) \quad (11)$$

(see § 4.2).
Comparing (10) with (8), we obtain :

$$f_n(X) = 2a.$$

Now, dividing by z both sides of (9), it follows that :

$$X = z + z^{-1}$$

while (8) yields :

$$z^n = a \pm \sqrt{a^2-1}.$$

We thus obtain several equivalent expressions for X :

$$X = \sqrt[n]{a + \sqrt{a^2-1}} + (\sqrt[n]{a + \sqrt{a^2-1}})^{-1}$$

or $\quad X = \sqrt[n]{a + \sqrt{a^2-1}} + \sqrt[n]{a - \sqrt{a^2-1}}$

or $\quad X = (\sqrt[n]{a - \sqrt{a^2-1}})^{-1} + \sqrt[n]{a - \sqrt{a^2-1}}$

or $\quad X = (\sqrt[n]{a - \sqrt{a^2-1}})^{-1} + (\sqrt[n]{a + \sqrt{a^2-1}})^{-1}.$

The equivalence of these expressions is easily seen from :

$$(a + \sqrt{a^2-1})(a - \sqrt{a^2-1}) = 1.$$

De Moivre repeatedly returned to these questions in the sequel, displaying formula (7) quite explicitly on p. 1 of his book "Miscellanea Analytica" (1730) [Sm, vol. 2, p. 446]. It is noteworthy that for $X = 2 \cos \alpha$ and $a = \cos(n\alpha)$, the values of z obtained by solving equations (8) and (9) are :

$$\sqrt[n]{a \pm \sqrt{a^2-1}} = \sqrt[n]{\cos(n\alpha) \pm \sqrt{-1} \sin(n\alpha)}$$

and

$$(X/2) \pm \sqrt{(X/2)^2 - 1} = \cos \alpha \pm \sqrt{-1} \sin \alpha$$

so that

$$\sqrt[n]{\cos(n\alpha) \pm \sqrt{-1}\sin(n\alpha)} = \cos\alpha \pm \sqrt{-1}\sin\alpha, \qquad (12)$$

but this was never written out explicitly by de Moivre.

Nevertheless, de Moivre's approach turned out to be quite fruitful, since Cotes' formulas can be easily proved by pushing the preceding calculations a little further (see exercise 1).

In 1739, de Moivre used the trigonometric representation of complex numbers and presumably also his formula, which certainly was thoroughly familiar to him by then, to extract the n-th root of the "impossible binomial $a + \sqrt{-b}$" [Sm, vol. 2, p. 449]. He states the procedure as follows : let φ be an angle such that

$$\cos\varphi = a/\sqrt{a^2+b} \; ;$$

then the n-th roots of $a + \sqrt{-b}$ are :

$$\sqrt[2n]{a^2+b}\;(\cos\psi + \sqrt{\cos^2\psi - 1})$$

where ψ ranges over φ/n, $(2\pi-\varphi)/n$, $(2\pi+\varphi)/n$, $(4\pi-\varphi)/n$, $(4\pi+\varphi)/n$, etc. until the number of them is equal to n. (This result is correct up to the sign of the imaginary part : see proposition 2 below).

As a result of this work, the credibility of the "fundamental theorem of algebra" was significantly enhanced, since the objection that Leibniz had raised was definitely answered : it was clear that extraction of roots of complex numbers does not produce imaginary numbers of a new kind. Moreover, since equations of degree at most 4 can be solved by radicals, it follows from de Moivre's result that polynomials of degree at most 4 split into products of linear factors over the field of complex numbers. It was not long afterwards that the first attempts to prove the fundamental theorem were made (without formulas for the solution of higher degree equations by radicals), and we shall come back to this topic in chapter 9.

Another consequence with far-reaching implications is that the n-th root of any (non-zero) number is ambiguous : it has n different determinations. Therefore, every formula which involves the extraction of a root needs some clarification as to <u>which</u> root should be chosen. This observation, which was conspicuously used as a starting-point in Vandermonde's subsequent investigations, sheds a completely new light on the problem of solving equations by radicals, and even on known solutions : indeed, Cardano's formula, as it appears in § 2.2, involves the extraction of two cube roots; it we consider various determinations of these cube roots, we obtain the three solutions of the cubic equation : this solves the puzzle of § 2.3(a) [*].

Moreover, even de Moivre's formula as it appears in (12) above is ambiguous. To express it properly, one has to raise $\cos \alpha + \sqrt{-1} \sin \alpha$ to the n-th power instead of extracting the n-th root of $\cos(n\alpha) + \sqrt{-1} \sin(n\alpha)$. This viewpoint was adopted by Euler in his "Introductio in Analysin Infinitorum" (1748) [Sm, vol. 2, p. 450], in which he proves de Moivre's formula (1) :

$$(\cos \alpha + \sqrt{-1} \sin \alpha)^n = \cos(n\alpha) + \sqrt{-1} \sin(n\alpha)$$

as in § 1. Later in the same book, comparing the power series expansions of the exponential and of the sine and cosine functions, Euler also states :

$$e^{\alpha \sqrt{-1}} = \cos \alpha + \sqrt{-1} \sin \alpha,$$

a relation from which de Moivre's formula readily follows. Of course, once de Moivre's formula is established, the

---

[*] With the notations of § 2.2, the cube roots should be determined in such a way that their product be $-p/3$, as it appears from the proof of Cardano's formula in § 2.2 or alternatively from Viète's method in § 6.1(b).

other major results of this section, and in particular Cotes' formulas, can be seen as easy applications. We devote the next section to a streamlined exposition of de Moivre's results on the roots of complex numbers along these lines.

§ 3. The roots of unity

1. Let a and b be real numbers, not both zero, and denote by $\sqrt{a^2+b^2}$ the (real, positive) square root of a^2+b^2. Since

$$(a/\sqrt{a^2+b^2})^2 + (b/\sqrt{a^2+b^2})^2 = 1,$$

there is a unique angle φ such that $0 \leq \varphi < 2\pi$,

$$\cos \varphi = a/\sqrt{a^2+b^2} \quad \text{and} \quad \sin \varphi = b/\sqrt{a^2+b^2}.$$

We thus obtain the trigonometric expression of the complex number $a + bi \neq 0$:

$$a + bi = \sqrt{a^2+b^2} \, (\cos \varphi + i \sin \varphi).$$

2. PROPOSITION : For any positive integer n, the n distinct n-th roots of a + bi are :

$$\sqrt[2n]{a^2+b^2} \, (\cos(\varphi+2k\pi)/n + i \sin(\varphi+2k\pi)/n) \tag{13}$$

for $k=0,\ldots,n-1$.
(In this formula, $\sqrt[2n]{a^2+b^2}$ is the unique real positive 2n-th root of a^2+b^2).

Proof : De Moivre's formula (1) yields

$$[\sqrt[2n]{a^2+b^2} \, (\cos(\varphi+2k\pi)/n + i \sin(\varphi+2k\pi)/n)]^n =$$

$$\sqrt{a^2+b^2} \, (\cos \varphi + i \sin \varphi),$$

so that each of the expressions (13) is an n-th root of a + bi. Moreover, these expressions are easily seen to be pairwise distinct for k=0,...,n-1, since for these values of k it is impossible that two among the angles $(\varphi+2k\pi)/n$ differ by a multiple of 2π. □

3. <u>DEFINITION</u> : A complex number ζ is called an n-th root of unity, for some integer n, if

$$\zeta^n = 1.$$

The set of all n-th roots of unity is denoted by μ_n. Thus,

$$X^n - 1 = \Pi_{\zeta \in \mu_n} (X-\zeta).$$

By the preceding proposition, we have :

$$\mu_n = \{e^{2k\pi i/n} = \cos(2k\pi/n) + i\,\sin(2k\pi/n) \,|\, k=0,\ldots,n-1\},$$

hence

$$X^n - 1 = \Pi_{k=0}^{n-1} (X - \cos(2k\pi/n) - i\,\sin(2k\pi/n)). \qquad (14)$$

This formula can be used to produce a factorization of $X^n - 1$ into real factors : indeed, if $k + \ell = n$, then

$$\cos(2k\pi/n) = \cos(2\ell\pi/n) \text{ and } \sin(2k\pi/n) = -\sin(2\ell\pi/n),$$

whence

$$(X-\cos(2k\pi/n) - i\,\sin(2k\pi/n))(X-\cos(2\ell\pi/n) - i\,\sin(2\ell\pi/n)) =$$
$$X^2 - 2\cos(2k\pi/n)X + 1.$$

Therefore, multiplying in the right-hand side of (14) the corresponding pairs of factors, we obtain :

$$X^n - 1 = (X-1) \prod_{k=1}^{(n-1)/2} (X^2 - 2\cos(2k\pi/n)X + 1) \quad \text{if n is odd}$$

and

$$X^n - 1 = (X-1)(X+1) \prod_{k=1}^{(n/2)-1} (X^2 - 2\cos(2k\pi/n)X + 1) \quad \text{if n is even.}$$

Substituting a/x for X in these formulas and multiplying both sides by x^n to clear denominators, we recover Cotes' formulas (5) and (6). Formulas (3) and (4) can be similarly derived by considering n-th roots of (-1) instead of n-th roots of 1.

It should be observed that, in a rectangular coordinate system, the points $(\cos(2k\pi/n), \sin(2k\pi/n))$ for $k=0,1,\ldots,n-1$, which represent the n-th roots of unity in the planar representation of $\mathbb{C}$, are the vertices of a regular polygon with n sides : they divide the unit circle into n equal parts. For this reason, the theory which is concerned with n-th roots of unity or with the values of the cosine and sine functions at $2k\pi/n$ for integers k, n, is called cyclotomy, meaning literally : division of the circle [into equal parts].

Likewise, the n-th roots of any non-zero complex number are represented in the plane of complex numbers by the vertices of a regular n-gon, as proposition 2 shows.

That the roots of 1 deserve special interest comes from the fact that, if an n-th root u of some complex number v has been found, then the various determinations of $\sqrt[n]{v}$ are the products ωu, where ω runs over the set of n-th roots of unity. This is easily seen from :

$$(\omega u)^n = \omega^n u^n = 1 \cdot u^n = v$$

or, equivalently, from proposition 2.

While the n-th roots of unity have been determined above by

a trigonometric expression, yet the problem of deciding whether these roots of unity have an expression by radicals has been untouched. We now turn to this problem, and prove, after some ideas of de Moivre, the following :

4. THEOREM : Let n be a positive integer. If, for each prime factor p of n, the p-th roots of unity can be expressed by radicals, then the n-th roots of unity can be expressed by radicals.

This theorem follows by induction (see n° 6 below) from the following result :

5. LEMMA : Let r and s be positive integers. If $\xi_1,\ldots,\xi_r$ (resp. $\eta_1,\ldots,\eta_s$) are the r-th roots of unity (resp. the s-th roots of unity), then the rs-th roots of unity are of the form :

$$\xi_i \sqrt[r]{\eta_j} \quad \text{for} \quad i=1,\ldots,r \text{ and } j=1,\ldots,s.$$

Proof : From the factorization of $Y^s - 1$, it follows by letting $Y = X^r$ that

$$X^{rs} - 1 = \Pi_{j=1}^{s} (X^r - \eta_j).$$

Therefore, the rs-th roots of unity are the r-th roots of the various η_j, for $j=1,\ldots,s$. □

6. Proof of theorem 4 : We argue by induction on the number of factors of n. If n is a prime number, then there is nothing to prove. Assume then that $n = r.s$ for some positive integers $r,s \neq 1$. Then the number of factors of r (resp. s) is strictly less than the number of factors of n, whence, by induction, the r-th roots of unity $\xi_1,\ldots,\xi_r$ and the s-th roots of unity $\eta_1,\ldots,\eta_s$ can be expressed by radicals. Since the n-th roots of unity are of the form

$\xi_i \sqrt[r]{\eta_j}$, these can also be expressed by radicals. □

7. REMARK : Of course, the expressions thus obtained are not necessarily the simplest ones or the most suitable for the actual calculation of these roots. For instance, since the 4-th roots of unity are 1, -1, $\sqrt{-1}$ and $-\sqrt{-1}$, the 8-th roots of unity are obtained as :

$$1,\ -1,\ \sqrt{-1},\ -\sqrt{-1},\ \sqrt{\sqrt{-1}},\ -\sqrt{\sqrt{-1}},\ \sqrt{-\sqrt{-1}}\ \text{and}\ -\sqrt{-\sqrt{-1}},$$

while they can also be expressed as :

$$\pm 1,\ \pm\sqrt{-1},\ (1 \pm \sqrt{-1})/\sqrt{2}\ \text{and}\ -(1 \pm \sqrt{-1})/\sqrt{2},$$

since $\cos(\pi/4) = \sin(\pi/4) = 1/\sqrt{2}$.
Moreover, the result of lemma 5 can be improved when r and s are relatively prime : in this case, one of the determinations of $\sqrt[r]{\eta_j}$ is a s-th root of unity η_k, so that the rs-th roots of unity are the products of the form $\xi_i \eta_k$ ($i=1,\ldots,r$ and $k=1,\ldots,s$), see remark 16 and exercise 3.

8. Theorem 4 shows that, in order to find expressions by radicals for the n-th roots of unity, for any integer n, it suffices to consider the case where n is prime. Since the equation $X^n - 1 = 0$ has the obvious root $X = 1$, we may divide $X^n - 1$ by $X - 1$, and the question reduces to the following problem : solve by radicals the equation :

$$X^{n-1} + X^{n-2} + \ldots + X + 1 = 0, \text{ for } n \text{ prime.} \qquad (15)$$

This equation is readily solved for $n = 2$ and 3 : the roots are :

for $n = 2$: -1
for $n = 3$: $(-1 \pm \sqrt{-3})/2$.

For $n \geq 5$, the following trick (due to de Moivre) is useful: after division by $X^{(n-1)/2}$, the change of variable $Y = X + X^{-1}$ transforms equation (15) into an equation of degree $(n-1)/2$ in Y. (This trick succeeds because in the polynomial (15), the coefficients of the terms which are symmetric with respect to the middle term are equal). Thus, for $n = 5$, we first divide both sides of

$$X^4 + X^3 + X^2 + X + 1 = 0$$

by X^2, and the change of variable $Y = X + X^{-1}$ transforms the resulting equation

$$X^2 + X + 1 + X^{-1} + X^{-2} = 0$$

into :

$$Y^2 + Y - 1 = 0.$$

We thus find :

$$Y = (-1 \pm \sqrt{5})/2$$

and the values of X are obtained by solving $X + X^{-1} = Y$ for the various values of Y. Thus, the 5-th roots of unity (other than 1) are the roots of the equations :

$$X^2 - X(-1 + \sqrt{5})/2 + 1 = 0 \text{ and } X^2 - X(-1 - \sqrt{5})/2 + 1 = 0$$

which are :

$$(\sqrt{5} - 1 \pm \sqrt{-10-2\sqrt{5}})/4 \text{ and } (-\sqrt{5} - 1 \pm \sqrt{-10+2\sqrt{5}})/4.$$

Similarly, for $n = 7$, de Moivre's trick yields for Y ($= X + X^{-1}$) a cubic equation :

$$Y^3 + Y^2 - 2Y - 1 = 0$$

which can be solved by radicals; the 7-th roots of unity can therefore be expressed by radicals.

However, for the next prime number, which is 11, de Moivre's trick yields an equation of degree 5, for which no general formula by radicals is known. Solving this equation was one of the greatest achievements of Vandermonde (see chapter 11).

9. REMARKS : 1) Since the roots of equation (15) are $e^{2k\pi i/n}$ for $k=1,\ldots,n-1$, the roots of the equation in Y $(= X + X^{-1})$ are :

$$e^{2k\pi i/n} + e^{-2k\pi i/n} = 2\cos(2k\pi/n) \quad \text{for } k=1,\ldots,(n-1)/2.$$

Therefore, the calculations above yield expressions by radicals for $2\cos(2\pi/5)$ and $2\cos(4\pi/5)$ as the positive and the negative root of $Y^2 + Y - 1 = 0$:

$$2\cos(2\pi/5) = (\sqrt{5}-1)/2 \quad \text{and} \quad 2\cos(4\pi/5) = -(\sqrt{5}+1)/2.$$

2) From theorem 4 and the results in n° 8, it follows that the $2.3^3.5^2$-th roots of unity can be expressed by radicals. Hence, $\sin(\pi/3^3.5^2)$ can also be expressed by radicals, since

$$\sin(\pi/3^3.5^2) = (e^{2\pi i/2.3^3.5^2} - e^{-2\pi i/2.3^3.5^2})/2i$$

(see § 4.2).

§ 4. Primitive roots and cyclotomic polynomials

In this section, we complete our discussion of the elementary aspects of the theory of roots of unity by including several results which are more or less straightforward consequences of de Moivre's formula[*]. They are a natural

---

[*] More precisely, these results follow from the fact that the set μ_n of n-th roots of unity is a finite subgroup of the multiplicative group of complex numbers.

outgrowth of the theory developed so far, and became known during the second half of the eighteenth century. The central notion is the following :

10. DEFINITIONS : The <u>exponent</u> of a root of unity ζ is the smallest integer e > 0 such that

$$\zeta^e = 1.$$

For instance, the exponent of 1 is 1 and the exponent of -1 is 2, although 1 is a n-th root of unity for every n and -1 is a n-th root of unity for every even n.
The n-th roots of unity of exponent n are also called : <u>primitive</u> n-th roots of unity.

We aim to give a complete description of the primitive n-th roots of unity and to show that these roots are indeed primitive, in the sense that the other n-th roots of unity can be obtained as powers of any such root. A basic ingredient in the proofs is the following number-theoretic proposition :

11. THEOREM : Let d be the (positive) greatest common divisor of two integers n_1, n_2. Then there exist integers m_1, m_2 such that

$$d = n_1 m_1 + n_2 m_2.$$

In particular, if n_1 and n_2 are relatively prime, then there exist integers m_1, m_2 such that

$$n_1 m_1 + n_2 m_2 = 1.$$

<u>Proof</u> : Duplicate the arguments in the proof of theorem 5.8 with integers instead of polynomials. □

By the way, it is useful to note that most of the arguments

in §§ 5.2 and 5.3 can be carried out with integers instead of polynomials, using the absolute value of integers instead of the degree of polynomials; the point is that we also have a Euclidean division property for integers : if m and n are integers and if $m \neq 0$, then there are integers q, r such that

$$n = mq + r$$
and $\quad 0 \leq r < |m|$.

Moreover, the integers q and r are uniquely determined by these properties.
Thus, mimicking the proofs in §§ 5.2 and 5.3, we get a proof of the unique factorization of integers into prime factors and obtain along the way other useful results like: "If an integer divides a product of r factors and is relatively prime to the (r-1) first factors, then it divides the last one" (compare lemma 5.14) or "If an integer is divisible by pairwise relatively prime integers, then it is divisible by their product" (compare proposition 5.15).

Theorem 11 is sometimes known as "Bezout's theorem". This is clearly a misnomer, since it can be traced back at least to Bachet de Méziriac, in his "Problèmes plaisans et délectables qui se font par les nombres" (1624). However, it could be fair to associate the name of Bezout to a similar statement for polynomials with coefficients in a field of rational fractions in one or several indeterminates (i.e. theorem 5.8 with $F = K(X_1,\ldots,X_n)$ for some field K). This last result implies the existence, for any two polynomials $P_1(X_1,\ldots,X_{n+1})$ and $P_2(X_1,\ldots,X_{n+1})$ in n+1 indeterminates, of polynomials $Q_1(X_1,\ldots,X_{n+1})$, $Q_2(X_1,\ldots,X_{n+1})$ and $D(X_1,\ldots,X_n)$ such that :

$$P_1(X_1,\ldots,X_{n+1})Q_1(X_1,\ldots,X_{n+1}) + P_2(X_1,\ldots,X_{n+1})Q_2(X_1,\ldots,X_{n+1}) = D(X_1,\ldots,X_n).$$

Since the indeterminate X_{n+1} does not appear in D, one says that D is obtained by elimination of X_{n+1} between P_1 and P_2. (The relation with the resultant of P_1 and P_2 (considered as polynomials in X_{n+1}) is pointed out in exercise 7).

We now come back to roots of unity with the following result which characterizes the exponent of roots of unity :

12. LEMMA : Let e be the exponent of a root of unity ζ and let m be an integer. Then

$$\zeta^m = 1$$

if and only if e divides m. In particular, the exponent of an n-th root of unity divides n.

Proof : If e divides m, then m = ef for some integer f, and since $\zeta^e = 1$, it readily follows from $\zeta^m = (\zeta^e)^f$ that $\zeta^m = 1$.
Conversely, assume

$$\zeta^m = 1$$

and let d be the greatest common divisor of m and e. By theorem 11, there are integers r and s such that

$$mr + es = d.$$

Since

$$\zeta^d = (\zeta^m)^r (\zeta^e)^s,$$

we have

$$\zeta^d = 1;$$

but since e is the smallest exponent for which the power of

ζ is 1, it follows that $d \geq e$, whence $d = e$ since d divides e. Therefore, e divides m. □

13. PROPOSITION : Let ζ and η be roots of unity of exponents e and f respectively. If e and f are relatively prime, then $\zeta\eta$ is a root of unity of exponent ef.

<u>Proof</u> : Since $\zeta^e = 1$ and $\eta^f = 1$, we have

$$(\zeta\eta)^{ef} = 1,$$

hence it follows from the preceding lemma that the exponent of $\zeta\eta$, which we denote by k, divides ef.
On the other hand, from

$$(\zeta\eta)^k = 1,$$

it follows that

$$\zeta^k = \eta^{-k}$$

and, raising both sides to the power f :

$$\zeta^{kf} = 1.$$

The preceding lemma then shows that e divides kf. Since e is relatively prime to f, by hypothesis, it follows that e divides k. Likewise, interchanging ζ and η, we see that f divides k. Since e and f are relatively prime and divide k, their product ef divides k; but we have already observed that k divides ef, hence $k = ef$. □

We now show that the primitive n-th roots of unity generate the other n-th roots of unity :

14. PROPOSITION : If ζ is a primitive n-th root of unity, then the n-th roots of unity are of the form :

$$\zeta^0 = 1, \zeta, \zeta^2, \ldots, \zeta^{n-1}.$$

Conversely, if ζ is an n-th root of unity such that every n-th root of unity is a power of ζ, then ζ is primitive.

<u>Proof</u> : Since ζ is an n-th root of unity, all the powers of ζ are n-th roots of unity, whence the set

$$S = \{\zeta^i \mid i=0,1,\ldots,n-1\}$$

consists of n-th roots of unity : $S \subseteq \mu_n$. To prove that all the n-th roots of unity are contained in S, i.e. $S = \mu_n$, it then suffices to prove that S has n distinct elements. This amounts to prove the following claim : <u>the various powers ζ^i for i=0,...,n-1 are pairwise distinct.</u> Assume on the contrary :

$$\zeta^i = \zeta^j \quad \text{for some i,j with} \quad 0 \leq i < j \leq n-1.$$

Then $\zeta^{j-i} = 1$, whence, by lemma 12, the exponent n of ζ divides j-i. This is impossible, since $0 < j-i < n$, and this contradiction proves the claim. Conversely, assume that every n-th root of unity is a power of ζ. If ζ is not primitive, then $\zeta^m = 1$ for some positive integer m < n. Then every power of ζ, hence every n-th root of unity, is an m-th root of unity : $\mu_n \subseteq \mu_m$. This is clearly impossible, since there are n distinct roots of unity while the number of m-th roots of unity is only m. □

<u>Remark</u> : The proposition above shows that the (multiplicative) group μ_n is generated by a single element : one says that μ_n is a <u>cyclic</u> group. The proposition also shows that the generators of μ_n are the primitive n-th roots of unity.

A complete description of the primitive n-th roots of unity will be obtained as a consequence of the following result:

<u>15. PROPOSITION</u> : Let ζ be a primitive n-th root of unity and let k be an integer. Then ζ^k is a primitive n-th root of unity if and only if k is relatively prime to n.

<u>Proof</u> : Assume first that k and n have a common factor $d \neq 1$. Then

$$(\zeta^k)^{n/d} = (\zeta^n)^{k/d},$$

whence $(\zeta^k)^{n/d} = 1$ since $\zeta^n = 1$. Therefore, the exponent of ζ^k divides n/d, hence ζ^k is not a primitive n-th root of unity.

Conversely, assume that ζ^k is not primitive; then there is a positive integer $m < n$ such that

$$(\zeta^k)^m = 1.$$

Since the exponent of ζ is n, it follows from lemma 12 that n divides km. If n is relatively prime to k, then it divides m; but this is impossible since $m < n$. Therefore, n and k are not relatively prime. □

<u>16. REMARK</u> : Let r and s be relatively prime integers and let η be a primitive s-th root of unity. Proposition 15 shows that η^r is a primitive s-th root of unity, hence by proposition 14 the s-th roots of unity are :

$$\mu_s = \{\eta^{ri} \mid i=0,\ldots,s-1\}.$$

Therefore, every s-th root of unity can be written as η^{ri} for some (unique) integer i between 0 and s-1, hence every s-th root of unity has a r-th root in μ_s.

17. COROLLARY : The primitive n-th roots of unity are :

$$e^{2k\pi i/n} = \cos(2k\pi/n) + i\sin(2k\pi/n)$$

where k runs over the positive integers which are less than n and relatively prime to n. In particular, if n is prime, then every n-th root of unity except 1 is primitive.

<u>Proof</u> : Let $\zeta = \cos(2\pi/n) + i\sin(2\pi/n)$. Since

$$\mu_n = \{\cos(2k\pi/n) + i\sin(2k\pi/n) \mid k=0,\ldots,n-1\},$$

we have by de Moivre's formula :

$$\mu_n = \{\zeta^k \mid k=0,\ldots,n-1\}.$$

Therefore, proposition 14 shows that ζ is a primitive n-th root of unity and it follows from proposition 15 that every primitive n-th root of unity is of the form ζ^k where k is a positive integer relatively prime to n between 0 and n-1.□

We now introduce the polynomials which have as roots the primitive n-th roots of unity. Because of their relation with the division of the circle, these polynomials are called : <u>cyclotomic polynomials</u>.

18. DEFINITION : The cyclotomic polynomials Φ_n (n=1,2,3,...) are defined inductively by :

$$\Phi_1(X) = X - 1$$

and :

$$\Phi_n(X) = \frac{X^n - 1}{\Pi_d \Phi_d(X)} \qquad (n \geq 2)$$

where d runs over the set of divisors of n, with $d \neq n$.

In particular, if p is a prime number, we have :

$$\Phi_p(X) = \frac{X^p - 1}{X - 1} = X^{p-1} + X^{p-2} + \ldots + X + 1.$$

However, if n is not prime, it is not clear a priori that Φ_n is a polynomial. We prove this and the fact that the roots of Φ_n are the primitive n-th roots of unity simultaneously :

<u>19. PROPOSITION</u> : For every integer $n \geq 1$, the rational fraction Φ_n is a monic polynomial with integral coefficients, and

$$\Phi_n(X) = \prod_\zeta (X-\zeta)$$

where ζ runs over the set of primitive n-th roots of unity.

<u>Proof</u> : We argue by induction on n and since the proposition is trivial for n = 1, we assume that it holds for all integers up to n - 1. Then, for every divisor d of n, with $d \neq n$, we have :

$$\Phi_d(X) = \prod_\zeta (X-\zeta) \in \mathbb{Z}[X]$$

where ζ runs over the roots of unity of exponent d. Since the roots of exponent d are n-th roots of unity, it follows that Φ_d divides $X^n - 1$. Moreover, if d_1 and d_2 are distinct divisors of n, then Φ_{d_1} and Φ_{d_2} are relatively prime since their roots are pairwise distinct. Therefore, by proposition 5.15, the product $\prod_d \Phi_d(X) \in \mathbb{Z}[X]$ divides $X^n - 1$, and it follows that $\Phi_n(X)$ is a polynomial in $\mathbb{Q}[X]$, which is monic since $X^n - 1$ and Φ_d (for every proper divisor d of n) are monic. Moreover, the proof of the Euclidean division property (Theorem 5.5) shows that the quotient of a polynomial in $\mathbb{Z}[X]$ by a monic polynomial in $\mathbb{Z}[X]$ is in $\mathbb{Z}[X]$. Therefore, $\Phi_n(X) \in \mathbb{Z}[X]$.

Now, the effect of dividing $X^n - 1$ by $\Pi_d \Phi_d(X)$ is to remove from

$$X^n - 1 = \Pi_{\zeta \in \mu_n} (X - \zeta)$$

all the factors $X - \zeta$ where ζ has exponent $d \neq n$. Therefore, in $\Phi_n(X)$ remain all the factors $X - \zeta$ where ζ has exponent n, and only those factors. Thus,

$$\Phi_n(X) = \Pi_\zeta (X - \zeta)$$

where ζ runs over the set of primitive n-th roots of unity. □

20. REMARK : It will be shown in theorem 12.41 that Φ_n is irreducible in $\mathbb{Q}[X]$, for every n. The proof is easier when n is prime : see theorem 12.11.

Appendix to chapter 7 : Leibniz and Newton on the summation of series

Around 1675, Leibniz obtained the following result, of which he was justifiably proud :

$$1 - \frac{1}{3} + \frac{1}{5} - \frac{1}{7} + \ldots = \frac{\pi}{4}.$$

His method was to use the formula

$$\int \frac{dx}{x^2+1} = \tan^{-1} x, \qquad (16)$$

which yields :

$$\int_0^1 \frac{dx}{x^2+1} = \frac{\pi}{4},$$

together with the power series expansion of $(x^2+1)^{-1}$:

$$(x^2+1)^{-1} = 1 - x^2 + x^4 - x^6 + x^8 - \ldots$$

from which it follows that

$$\int_0^1 \frac{dx}{x^2+1} = \int_0^1 dx - \int_0^1 x^2 dx + \int_0^1 x^4 dx - \ldots$$

$$= 1 - \frac{1}{3} + \frac{1}{5} - \ldots$$

Of course, the fact that the integral of the power series expansion of $(x^2+1)^{-1}$ is equal to the series of the integrals of the terms needs some justification, which was supplied much later.

In 1676, Newton sent to Leibniz the following variant:

$$1 + \frac{1}{3} - \frac{1}{5} - \frac{1}{7} + \frac{1}{9} + \frac{1}{11} - \frac{1}{13} - \frac{1}{15} + \ldots = \frac{\pi}{2\sqrt{2}},$$

with the terse hint that it depended on the reduction of an integrand to partial fractions (see [N1, v. IV, p. 212]). It is very likely that Newton's result was obtained from the evaluation of

$$\int_0^1 \frac{(x^2+1)\,dx}{x^4+1}.$$

Indeed, the rational fraction $(x^2+1)(x^4+1)^{-1}$ has the following decomposition into partial fractions :

$$\frac{x^2+1}{x^4+1} = \frac{1}{2}\frac{1}{X^2+\sqrt{2}\,X+1} + \frac{1}{2}\frac{1}{X^2-\sqrt{2}\,X+1}$$

and, by the change of variable $Y = \sqrt{2}\,X+1$ (resp. $Y = \sqrt{2}\,X-1$), it easily follows from (16) that

$$\int \frac{dx}{x^2+\sqrt{2}\,x+1} = \sqrt{2}\,\tan^{-1}(\sqrt{2}\,x+1) \quad (\text{resp.} \int \frac{dx}{x^2-\sqrt{2}\,x+1} = \sqrt{2}\,\tan^{-1}(\sqrt{2}\,x-1)).$$

Therefore,

$$\int_0^1 \frac{(x^2+1)\,dx}{x^4+1} = \frac{1}{\sqrt{2}} [\tan^{-1}(\sqrt{2}+1) + \tan^{-1}(\sqrt{2}-1)],$$

and since $(\sqrt{2}+1)(\sqrt{2}-1) = 1$, we have:

$$\tan^{-1}(\sqrt{2}+1) + \tan^{-1}(\sqrt{2}-1) = \pi/2,$$

whence

$$\int_0^1 \frac{(x^2+1)\,dx}{x^4+1} = \frac{\pi}{2\sqrt{2}}.$$

On the other hand, from the power series expansion

$$(x^4+1)^{-1} = 1 - x^4 + x^8 - x^{12} + x^{16} - \ldots$$

it follows that

$$(x^2+1)(x^4+1)^{-1} = 1 + x^2 - x^4 - x^6 + x^8 + x^{10} - \ldots$$

whence

$$\int_0^1 \frac{(x^2+1)\,dx}{x^4+1} = \int_0^1 dx + \int_0^1 x^2\,dx - \int_0^1 x^4\,dx - \int_0^1 x^6\,dx + \ldots$$

$$= 1 + \frac{1}{3} - \frac{1}{5} - \frac{1}{7} + \ldots\,.$$

This proves Newton's result.

Exercises for chapter 7

The aim of the first exercise is to prove Cotes' formulas along the lines of de Moivre's calculations with polynomials (§ 2).

1. Denote by $f_n(X) \in \mathbb{Q}[X]$ the monic polynomial of degree n such that $f_n(2\cos\alpha) = 2\cos n\alpha$ (see equation (1) in chapter 4) and let

$$P_n(X,z) = 1 - f_n(X)z^n + z^{2n} \in \mathbb{Q}[X,z] \quad \text{for } n=1,2,3,\ldots$$

a) Prove : $P_1(X,z)$ divides $P_n(X,z)$ in $\mathbb{Q}[X,z]$ for every n.
[Hint : use induction on n and the recurrence relation (11)].

b) Prove : $P_n(2\cos\alpha, z) = \prod_{k=0}^{n-1} P_1(2\cos(\alpha + (2k\pi/n)), z)$

$$\text{for } \alpha \in \mathbb{R}.$$

[Hint : prove it first for $\alpha \notin (\pi/n)\mathbb{Z}$, using (a) and observing that for $\alpha \notin (\pi/n)\mathbb{Z}$ the polynomials in the right hand side are pairwise relatively prime. Since the coefficients of both sides are continuous functions of α which are equal outside a discrete subset of $\mathbb{R}$, they are equal for every $\alpha \in \mathbb{R}$].

c) Show that for $\alpha = 0$ both sides of the relation in (b) are squares. Extracting the square root of both sides, show that

$$1 - z^{2m} = (1-z)(1+z) \prod_{k=1}^{m-1} (1 - 2\cos(2k\pi/2m)z + z^2)$$

and $\quad 1 - z^{2m+1} = (1-z) \prod_{k=1}^{m} (1 - 2\cos(2k\pi/2m+1)z + z^2).$

For $\alpha = \pi/n$, derive similarly the formulas :

$$1 + z^{2m} = \prod_{k=0}^{m-1} (1 - 2\cos((2k+1)\pi/2m)z + z^2)$$

and $\quad 1 + z^{2m+1} = (1+z) \prod_{k=0}^{m-1} (1 - 2\cos((2k+1)\pi/2m+1)z + z^2).$

Cotes' formulas (3)-(6) follow by substituting x/a for z and multiplying both sides by an appropriate power of a to clear denominators.

The following exercise aims to show some of the remarkable properties of the polynomials f_n of exercise 1 :

2. a) Prove : $f_n(X) - 2\cos n\alpha = \prod_{k=0}^{n-1} (X - 2\cos(\alpha+(2k\pi/n)))$

$$\text{for } \alpha \in \mathbb{R}.$$

125

[Hint : argue as in exercise 1.(b)].
b) Prove : $f_m(f_n(X)) = f_{mn}(X)$ for every integers m,n.
[Hint : use (a)].
c) Prove : $f_m(X) \cdot f_n(X) = f_{m+n}(X) + f_{|m-n|}(X)$ for every integers m,n.

(To allow $m = n$, set $f_o(X) = 2$). [Hint : induction on n. For $n = 1$, use the recurrence relation (11)].

3. Let $\mu_r = \{\xi_1, \ldots, \xi_r\}$ and $\mu_s = \{\eta_1, \ldots, \eta_s\}$. Assuming that r and s are relatively prime, show :
$\mu_{rs} = \{\xi_i \eta_j \mid i=1,\ldots,r \text{ and } j=1,\ldots,s\}$.
[Hint : it suffices to prove that the products $\xi_i \eta_j$ are pairwise distinct. Use theorem 11].

4. Let ζ be a root of unity of exponent e and let k be an integer. Find the exponent of ζ^k. (Compare proposition 15).

5. Show that an n-th root of unity ζ is primitive if and only if $\zeta^d \neq 1$ for every proper factor d of n.

6. Prove : $\sum_{\omega \in \mu_p} \omega^i = 0$ if $i \in \mathbb{Z}$ is not divisible by the prime p

$= p$ if $i \in \mathbb{Z}$ is divisible by p.

[Hint : use Newton's formulas of chapter 4].

7. Let $P = a_n X^n + a_{n-1} X^{n-1} + \ldots + a_1 X + a_o$ and
$Q = b_m X^m + b_{m-1} X_{m-1} + \ldots + b_1 X + b_o$ (with $a_n, b_m \neq 0$) be polynomials over a field F. Let

$$y = \begin{pmatrix} X^{m-1}.P \\ X^{m-2}.P \\ \vdots \\ P \\ X^{n-1}.Q \\ X^{n-2}.Q \\ \vdots \\ Q \end{pmatrix} \quad \text{and} \quad X = \begin{pmatrix} X^{m+n-1} \\ X^{m+n-2} \\ \vdots \\ \vdots \\ X \\ 1 \end{pmatrix}$$

be column matrices with entries in $F[X]$. Show that

$$y = A.X$$

where A is the square matrix of order $m + n$ which appears in the definition of the resultant R of P and Q ($R = \det A$: see n° 5.29). Multiplying both sides of the preceding equation on the left by the transpose of the matrix of cofactors of A and then by $(0,\ldots,0,1)$, conclude that there exist polynomials $U,V \in F[X]$ such that $\deg U \leq m - 1$, $\deg V \leq n - 1$ and

$$PU + QV = R.$$

Using this result, give an alternative proof of theorem 5.30.

<u>This last exercise is an introduction to Euler's "totient function" φ :</u>

8. For any integer $n \geq 2$, let $\varphi(n)$ be the number of integers which are relatively prime to n between 0 and $n - 1$.
a) Show that $\varphi(n) = \deg \Phi_n$ and that $\varphi(n)$ is equal to the number of primitive n-th roots of unity.
b) Show that $\varphi(m.n) = \varphi(m).\varphi(n)$ if m and n are relatively

prime.

[Hint : compare exercise 3].

c) Show that $\varphi(p^k) = p^{k-1}(p-1)$ for any prime number p.

d) From (b) and (c), derive the following formula for $\varphi(n)$: if $n = p_1^{k_1}\ldots p_r^{k_r}$ ($p_1,\ldots,p_r$ distinct prime numbers), then $\varphi(n) = \Pi_{i=1}^{r} p_i^{k_i-1} (p_i-1) = n.\Pi_{i=1}^{r} (1-p_i^{-1})$.

e) Show that $n = \Sigma_d \varphi(d)$ where d runs over the factors of n (including n).

[Hint : compare definition 18].

8 Symmetric functions

§ 1. <u>Introduction</u>

During the first half of the eighteenth century, the structure of equations, as formerly investigated by Viète (§ 4.2), became clearer and clearer. Calculating formally with roots of equations, mathematicians became aware of the kind of information that can be gathered from the coefficients without solving equations. As Girard had shown (§ 4.2), for any polynomial

$$X^n - s_1 X^{n-1} + s_2 X^{n-2} - \ldots + (-1)^n s_n = (X-X_1)(X-X_2)\ldots(X-X_n) \tag{1}$$

we have :

$$\begin{aligned}
s_1 &= X_1 + \ldots + X_n \\
s_2 &= X_1 X_2 + \ldots + X_{n-1} X_n \\
s_3 &= X_1 X_2 X_3 + \ldots + X_{n-2} X_{n-1} X_n \\
&\ldots \\
s_n &= X_1 X_2 \ldots X_n
\end{aligned} \tag{2}$$

and the following question naturally arises : what kind of function of the roots $X_1, \ldots, X_n$ can be calculated from $s_1, \ldots, s_n$?

To translate properly the results of this period, laying on firm ground the formal calculations with roots of polynomials, the roots $X_1,\ldots,X_n$ should be considered as independent indeterminates over some base field F (usually, $F = \mathbb{Q}$, the field of rational numbers). Indeed, every calculation with indeterminates which does not involve divisions by non-constant polynomials can be done as well with arbitrary elements in a field K containing the base field F : this is a loose translation of the fact that any map from $\{X_1,\ldots,X_n\}$ to K can be (uniquely) extended to a ring homomorphism from the ring of polynomials $F[X_1,\ldots,X_n]$ to K, mapping a polynomial $P(X_1,\ldots,X_n)$ to the element of K obtained by replacing $X_1,\ldots,X_n$ by their assigned values in K. (If we wish to allow divisions by non-constant polynomials, caution is necessary since some denominators may vanish in K). Therefore, we introduce the following definition :

1. DEFINITION : If $X_1,\ldots,X_n$ are considered as independent indeterminates over some base field F, the polynomial (1) above (which lies in $F[X_1,\ldots,X_n,X]$) is called the **general** (or : **generic**) **monic polynomial of degree n** over F.

This polynomial is general (generic) in the following sense : if

$$P = X^n + a_{n-1}X^{n-1} + a_{n-2}X^{n-2} + \ldots + a_1X + a_0$$

is an arbitrary monic polynomial of degree n over some field K containing F, which splits* into a product of linear factors in some field L containing K :

$$P = (X-x_1)\ldots(X-x_n) \quad \text{with} \quad x_i \in L \text{ for } i=1,\ldots,n,$$

then there is a ring homomorphism from $F[X_1,\ldots,X_n]$ to L

(*) We shall see later that the condition that P splits into a product of linear factors in some field containing K is always fulfilled : see § 9.2, but at this point this provision cannot be disposed of. (Compare remark 13(a) below).

mapping X_i to x_i for $i=1,\ldots,n$, which translates any calculation with $X_1,\ldots,X_n$ into a calculation with $x_1,\ldots,x_n$. For instance, it is easily seen that this ring homomorphism maps $s_1, s_2, \ldots, s_n$ onto $-a_{n-1}, a_{n-2}, \ldots, (-1)^n a_o \in K$, which means that

$$s_1(x_1,\ldots,x_n) = -a_{n-1}, \text{ i.e. } x_1 + \ldots + x_n = -a_{n-1}$$

$$s_2(x_1,\ldots,x_n) = a_{n-2}, \text{ i.e. } x_1 x_2 + \ldots + x_{n-1} x_n = a_{n-2}$$

$$\ldots$$

$$s_n(x_1,\ldots,x_n) = (-1)^n a_o, \text{ i.e. } x_1 \ldots x_n = (-1)^n a_o.$$

After this slight change of viewpoint, from arbitrary elements to indeterminates, the question becomes : which are the rational fractions in n indeterminates $X_1,\ldots,X_n$ that can be expressed as rational fractions in $s_1,\ldots,s_n$ (where $s_1,\ldots,s_n$ are defined by the equalities (2) above) ?

The crucial condition turns out to be the following :

2. <u>DEFINITION</u> : A polynomial $P(X_1,\ldots,X_n)$ in n indeterminates is <u>symmetric</u> if it is not altered when the indeterminates are arbitrarily permuted among themselves; i.e. for every permutation σ of $\{1,\ldots,n\}$,

$$P(X_{\sigma(1)},\ldots,X_{\sigma(n)}) = P(X_1,\ldots,X_n).$$

Similarly, a rational fraction P/Q in n indeterminates is <u>symmetric</u> if it is not altered when the indeterminates are permuted; i.e. for every permutation σ of $\{1,\ldots,n\}$,

$$\frac{P(X_{\sigma(1)},\ldots,X_{\sigma(n)})}{Q(X_{\sigma(1)},\ldots,X_{\sigma(n)})} = \frac{P(X_1,\ldots,X_n)}{Q(X_1,\ldots,X_n)}.$$

Note that this does <u>not</u> imply that P and Q are both symmetric, since P and Q can be both multiplied by an arbitrary non-zero polynomial without changing the fraction P/Q, but we shall see in n° 10 that every symmetric rational fraction can be represented as a quotient of symmetric polynomials.

Since the polynomials $s_1,\ldots,s_n$ are symmetric, it is clear that every rational fraction in $s_1,\ldots,s_n$ is a symmetric rational fraction in $X_1,\ldots,X_n$. The converse turns out to be also true, so that the following result holds :

3. THEOREM : A rational fraction in n indeterminates $X_1,\ldots,X_n$ over a field F can be expressed as a rational fraction in $s_1,\ldots,s_n$ if and only if it is symmetric.

This theorem is in fact a consequence of the analogous result for polynomials :

4. THEOREM : A polynomial in n indeterminates $X_1,\ldots,X_n$ over a field F can be expressed as a polynomial in $s_1,\ldots,s_n$ if and only if it is symmetric.

These theorems are known as the <u>fundamental theorems</u> of symmetric fractions or polynomials, respectively, and $s_1,\ldots,s_n$ are sometimes called the <u>elementary</u> symmetric polynomials, since the others can be expressed in terms of these ones.

Because of their progressive emergence through the calculations of eighteenth century mathematicians, these theorems can hardly be credited to any specific author. (There is not much credit to give anyway, since the proofs are not difficult). It seems that they first appeared in print around 1770, in "Meditationes Algebraicae" of Edward Waring (1736-1798) and in "Mémoire sur la résolution des équations" of A.T. Vandermonde, and in presumably other works. It is noteworthy that Lagrange in 1770 qualifies the fundamental theorem of symmetric fractions as "évident par soi-même" [L, Art. 98, p. 372], "self evident".

Therefore, the most interesting feature that one may demand of a proof is that it be effective : it has to provide a method to express any symmetric polynomial as a polynomial in $s_1,\ldots,s_n$. In the next section, we explain the particularly simple method suggested by Waring, and thereafter we shall discuss some applications, but first we point out a convenient notation which allows to denote symmetric polynomials without writing out all the terms :

5. NOTATION : We write $\Sigma\, x_1^{i_1} x_2^{i_2} \ldots x_n^{i_n}$ for the symmetric polynomial whose terms are the various <u>distinct</u> monomials (without repetition) obtained from $x_1^{i_1} x_2^{i_2} \ldots x_n^{i_n}$ by permutation of the indeterminates. Observe that this notation is slightly ambiguous, since the total number of variables is not clear from the notation, as some of the exponents $i_1,\ldots,i_n$ may be zero. Therefore, the total number of variables should always be indicated, unless it is clear from the context. For example, as a symmetric polynomial in two variables :

$$\Sigma\, x_1^2 x_2 = x_1^2 x_2 + x_1 x_2^2$$

while, as a symmetric polynomial in three variables :

$$\Sigma\, x_1^2 x_2 = x_1^2 x_2 + x_1 x_2^2 + x_1^2 x_3 + x_1 x_3^2 + x_2^2 x_3 + x_2 x_3^2.$$

With this notation, the elementary symmetric polynomials can be written simply as :

$$s_1 = \Sigma\, x_1,\ s_2 = \Sigma\, x_1 x_2,\ldots, s_{n-1} = \Sigma\, x_1\ldots x_{n-1} \text{ and}$$

$$s_n = \Sigma\, x_1\ldots x_n\ (= x_1\ldots x_n).$$

§ 2. Waring's method

6. DEFINITIONS : In order to define the degree of a polynomial in n indeterminates, we endow the set $\mathbb{N}^n$ of n-tuples of integers with the lexicographic ordering; thus

$$(i_1,\ldots,i_n) \geq (j_1,\ldots,j_n)$$

if the first non-zero difference in the sequence $i_1-j_1,\ldots, i_n-j_n$ (if any) is positive. For any non-zero polynomial $P = P(X_1,\ldots,X_n)$ in n indeterminates $X_1,\ldots,X_n$ over a field, the <u>degree</u> of P is then defined as the greatest n-tuple $(i_1,\ldots,i_n) \in \mathbb{N}^n$ for which the coefficient of $X_1^{i_1}\ldots X_n^{i_n}$ in P is non-zero. The degree of P is denoted by deg P. For instance, we have :

$$\deg s_1 = (1,0,0,\ldots,0)$$
$$\deg s_2 = (1,1,0,\ldots,0)$$
$$\ldots \qquad (3)$$
$$\deg s_{n-1} = (1,1,\ldots,1,0)$$
$$\deg s_n = (1,1,\ldots,1,1).$$

By convention, we also set deg $0 = -\infty$, and the same relations as for polynomials in one indeterminate hold :

$$\deg(P+Q) \leq \max(\deg P, \deg Q) \qquad (4)$$

$$\deg(PQ) = \deg P + \deg Q. \qquad (5)$$

7. Proof of theorem 4 : Waring's method to express any symmetric polynomial $P(X_1,\ldots,X_n)$ as a polynomial in $s_1,\ldots,s_n$ is quite similar to Euclid's division algorithm (theorem 5.5) : the idea is to match P with a polynomial in $s_1,\ldots,s_n$ which has the same degree as P. Adjusting the leading coefficient, we can arrange that the degree of the difference be less than deg P, and we are finished by induction on the degree.

First, observe that if

$$\deg P = (i_1, i_2, \ldots, i_n)$$

for a non-zero symmetric polynomial P, then $i_1 \geq i_2 \geq \ldots \geq i_n$: indeed, if we can find among the terms of P a term like $aX_1^{i_1} \ldots X_n^{i_n}$ (with $a \in F^\times$), then we can also find all the terms obtained from this one by permutation of $X_1, \ldots, X_n$, since P is symmetric. The degrees of these terms are the various n-tuples obtained from $(i_1, \ldots, i_n)$ by permutation of the entries, and the greatest among these n-tuples is the one in which the entries are in not-increasing order.

We can thus consider

$$f = s_1^{i_1-i_2} s_2^{i_2-i_3} \ldots s_{n-1}^{i_{n-1}-i_n} s_n^{i_n}.$$

By relations (3) and (5), we have

$$\deg f = (i_1-i_2)\deg(s_1) + (i_2-i_3)\deg(s_2) + \ldots + i_n \deg(s_n)$$

$$= (i_1-i_2, 0, \ldots, 0) + (i_2-i_3, i_2-i_3, 0, \ldots, 0) + \ldots + (i_n, \ldots, i_n)$$

$$= (i_1, i_2, \ldots, i_n),$$

and it is readily verified that the leading coefficient of f, which is the coefficient of $X_1^{i_1} \ldots X_n^{i_n}$, is 1, so that

$$f = X_1^{i_1} \ldots X_n^{i_n} + \text{(terms of lower degree)}.$$

Therefore, if $a \in F^\times$ is the leading coefficient of P, so that

$$P = aX_1^{i_1} \ldots X_n^{i_n} + \text{(terms of lower degree)},$$

then, letting

$$P_1 = P - af,$$

we see that

$$\deg P_1 < \deg P.$$

Moreover, P_1 is symmetric (possibly zero), since P and f are symmetric. We can thus apply the same arguments to P_1, which has lower degree than P.

To complete the proof, it remains to prove that the process above, by which the degree of the initial symmetric polynomial has been reduced, terminates in a finite number of steps ; this readily follows from the following observation :

8. LEMMA : $\mathbb{N}^n$ satisfies the descending chain condition; i.e. it does not contain any infinite strictly decreasing sequence of elements.

As the lemma is obvious if n = 1, we argue by induction on n : if

$$(i_{11}, i_{12}, \ldots, i_{1n}) > (i_{21}, i_{22}, \ldots, i_{2n}) > \ldots > (i_{m1}, i_{m2}, \ldots, i_{mn}) > \ldots \quad (6)$$

is an infinite strictly decreasing sequence in $\mathbb{N}^n$, then the sequence of first entries is not increasing :

$$i_{11} \geqslant i_{21} \geqslant \ldots \geqslant i_{m1} \geqslant \ldots .$$

Therefore, this sequence is eventually constant : there is an index M such that

$$i_{m1} = i_{M1} \quad \text{for all} \quad m \geqslant M.$$

We then delete the first M-1 terms in sequence (6), and consider the last n-1 entries of the elements of the remaining (infinite) sequence :

$$(i_{M2}, i_{M3}, \ldots, i_{Mn}) > (i_{(M+1)2}, i_{(M+1)3}, \ldots, i_{(M+1)n}) > \ldots$$

We thus obtain a strictly decreasing sequence of elements in $\mathbb{N}^{n-1}$: the existence of such a sequence contradicts the induction hypothesis. □

9. <u>EXAMPLE</u> : Let us express the symmetric polynomial in three variables :

$$S = \Sigma\ x_1^4 x_2 x_3 + \Sigma\ x_1^3 x_2^3$$

(that is : $S = x_1^4 x_2 x_3 + x_1 x_2^4 x_3 + x_1 x_2 x_3^4 + x_1^3 x_2^3 + x_1^3 x_3^3 + x_2^3 x_3^3$: see notation 5) as a polynomial in s_1, s_2, s_3. Since $(4,1,1) > (3,3,0)$, the degree of S is $(4,1,1)$, so we first calculate $s_1^{4-1}\ s_2^{1-1}\ s_3^1$:

$$s_1^3 s_3 = (\Sigma\ X_1)^3 (X_1 X_2 X_3) = \Sigma\ x_1^4 x_2 x_3 + 3\ \Sigma\ x_1^3 x_2^2 x_3 + 6\ x_1^2 x_2^2 x_3^2,$$

whence

$$S = \Sigma\ x_1^3 x_2^3 - 3\ \Sigma\ x_1^3 x_2^2 x_3 - 6\ x_1^2 x_2^2 x_3^2 + s_1^3 s_3$$

and it remains to express in terms of s_1, s_2, s_3 a polynomial of degree $(3,3,0)$. Therefore, we calculate the cube of s_2:

$$s_2^3 = (\Sigma\ X_1 X_2)^3 = \Sigma\ x_1^3 x_2^3 + 3\ \Sigma\ x_1^3 x_2^2 x_3 + 6\ x_1^2 x_2^2 x_3^2.$$

Replacing $\Sigma\ x_1^3 x_2^3$ in the preceding expression of S, we obtain :

$$S = -6\ \Sigma\ x_1^3 x_2^2 x_3 - 12\ x_1^2 x_2^2 x_3^2 + s_1^3 s_3 + s_2^3.$$

Next, in order to eliminate $\Sigma\ x_1^3 x_2^2 x_3$, we calculate $s_1 s_2 s_3$:

$$s_1 s_2 s_3 = (\Sigma\ X_1)(\Sigma\ X_1 X_2)(X_1 X_2 X_3) = \Sigma\ X_1^3 X_2^2 X_3 + 3\ X_1^2 X_2^2 X_3^2$$

whence

$$S = 6\ X_1^2 X_2^2 X_3^2 + s_1^3 s_3 + s_2^3 - 6\ s_1 s_2 s_3.$$

Since $X_1^2 X_2^2 X_3^2 = s_3^2$, we finally obtain the required result :

$$S = s_1^3 s_3 + s_2^3 - 6\ s_1 s_2 s_3 + 6\ s_3^2.$$

From this brief example, it is already clear that the only difficulty in carrying out Waring's method is to write out the various monomials in products like $s_1^{i_1} \ldots s_n^{i_n}$ with their proper coefficient.

10. Rational fractions : proof of theorem 3 : Let P, Q be polynomials in n indeterminates $X_1, \ldots, X_n$ such that the rational fraction P/Q is symmetric. In order to prove that P/Q is a rational fraction in $s_1, \ldots, s_n$, we represent P/Q as a quotient of symmetric polynomials in $X_1, \ldots, X_n$, as follows : if Q is symmetric then P is symmetric too, since P/Q is symmetric, and there is nothing to do; otherwise, let $Q_1, \ldots, Q_r$ be the various distinct polynomials (other than Q) obtained from Q by permutation of the indeterminates. The product $QQ_1 \ldots Q_r$ is then symmetric since any permutation of the indeterminates merely permutes the factors. Since P/Q is symmetric and

$$P/Q = PQ_1 \ldots Q_r / QQ_1 \ldots Q_r,$$

it follows that the polynomial $PQ_1 \ldots Q_r$ is symmetric too. We have thus achieved the required representation of P/Q. Now, by the fundamental theorem of symmetric polynomials (n° 4), there are polynomials f,g such that

$$PQ_1 \ldots Q_r = f(s_1, \ldots, s_n) \text{ and } QQ_1 \ldots Q_r = g(s_1, \ldots, s_n),$$

whence

$$P/Q = f(s_1,\ldots,s_n)/g(s_1,\ldots,s_n).$$ □

11. Uniqueness : The fundamental theorem of symmetric polynomials asserts that every symmetric polynomial $P(X_1,\ldots,X_n)$ is of the form :

$$P(X_1,\ldots,X_n) = f(s_1,\ldots,s_n)$$

for some polynomial f in n indeterminates; in other words, there is a polynomial $f(Y_1,\ldots,Y_n)$ in n indeterminates which yields $P(X_1,\ldots,X_n)$ when the indeterminates $Y_1,\ldots,Y_n$ are replaced by $s_1,\ldots,s_n$. However, it is not clear <u>a priori</u> that the expression of P as a polynomial in $s_1, \ldots, s_n$ is unique, or in other words that there is only one polynomial f for which the above equality holds. Admittedly, the contrary would be surprising, but no one seems to have cared to prove the uniqueness of f before Gauss, who needed it for his second proof (1815) of the fundamental theorem of algebra [Gau3, § 5] (see also [Sm, vol. I, pp. 292-306]).

12. THEOREM : Let f and g be polynomials in n indeterminates $Y_1,\ldots,Y_n$ over a field F. If f and g yield the same polynomial in $X_1,\ldots,X_n$ when $Y_1,\ldots,Y_n$ are replaced by $s_1,\ldots,s_n$, i.e. if

$$f(s_1,\ldots,s_n) = g(s_1,\ldots,s_n) \text{ in } F[X_1,\ldots,X_n],$$

then $f = g$ in $F[Y_1,\ldots,Y_n]$.

<u>Proof</u> : We compare the degree $(i_1,\ldots,i_n)$ of a non-zero monomial

$$m(Y_1,\ldots,Y_n) = aY_1^{i_1}\ldots Y_n^{i_n}$$

to the degree of the monomial

$$m(s_1,\ldots,s_n) = a\, s_1^{i_1} \ldots s_n^{i_n} \in F[X_1,\ldots,X_n].$$

By relations (3) and (5), we have :

$$\deg m(s_1,\ldots,s_n) = (i_1+\ldots+i_n, i_2+\ldots+i_n,\ldots,i_{n-1}+i_n, i_n).$$

Since the map $(i_1,\ldots,i_n) \to (i_1+\ldots+i_n, i_2+\ldots+i_n,\ldots,i_n)$ from $\mathbb{N}^n$ to $\mathbb{N}^n$ is injective, it follows that monomials of different degrees in $F[Y_1,\ldots,Y_n]$ cannot cancel out in $F[X_1,\ldots,X_n]$ when $Y_1,\ldots,Y_n$ are replaced by $s_1,\ldots,s_n$. Therefore, every non-zero polynomial $h \in F[Y_1,\ldots,Y_n]$ yields a non-zero polynomial $h(s_1,\ldots,s_n)$ in $F[X_1,\ldots,X_n]$. Applying this result to $h = f - g$, the theorem follows. □

13. REMARKS : a) Let $\varphi : F[Y_1,\ldots,Y_n] \to F[X_1,\ldots,X_n]$ be the ring-homomorphism which maps every polynomial $h(Y_1,\ldots,Y_n)$ to $h(s_1,\ldots,s_n)$. The preceding theorem asserts that φ is injective; hence the image of φ, which is the subring $F[s_1,\ldots,s_n]$ of $F[X_1,\ldots,X_n]$ generated by $s_1,\ldots,s_n$, is isomorphic under φ to a ring of polynomials in n indeterminates. In other words, the polynomials $s_1,\ldots,s_n$ in $F[X_1,\ldots,X_n]$ can be considered as independent indeterminates : this fact is expressed by saying that $s_1,\ldots,s_n$ are <u>algebraically independent</u>.

The point of this remark is that the generic monic polynomial of degree n over F :

$$X^n - s_1 X^{n-1} + s_2 X^{n-2} - \ldots + (-1)^n s_n$$

is really generic for all monic polynomials of degree n over a field K containing F : indeed, if

$$X^n + a_{n-1} X^{n-1} + a_{n-2} X^{n-2} + \ldots + a_0$$

is such a polynomial, then, since $s_1,\ldots,s_n$ can be considered as independent indeterminates over F, there is a (unique) ring-homomorphism from $F[s_1,\ldots,s_n]$ to K which maps $s_1,s_2,\ldots,s_n$ to $-a_{n-1},a_{n-2},\ldots,(-1)^n a_0$. This homomorphism translates calculations with the coefficients of the generic polynomial into calculations with the coefficients of arbitrary polynomials. By contrast with the discussion following definition 1, we do not need to restrict here to polynomials which split into products of linear factors over some extension of the base field ; this is precisely why theorem 12 is significant in Gauss' paper [Gau3], since the purpose of this work was to prove the fundamental theorem of algebra that polynomials split into products of linear factors over the field of complex numbers.

b) Inspection shows that the hypothesis that the base ring F is a field has not been used in our exposition of Waring's method nor in the proof of theorem 12. Therefore, theorem 4 and theorem 12 are valid over any base ring. Theorem 3 also holds over any (commutative) domain, but to generalize it further, some caution is necessary in the very definition of a rational fraction.

§ 3. The discriminant

14. DEFINITION : Let

$$\Delta(X_1,\ldots,X_n) = \prod_{1 \leq i < j \leq n} (X_i - X_j) \in \mathbb{Z}[X_1,\ldots,X_n].$$

Every permutation of $X_1,\ldots,X_n$ permutes the factors $X_i - X_j$ among themselves, changing the sign of some of them. So, Δ is either left unchanged or changed into its opposite by a permutation, and Δ^2 is a symmetric polynomial. It follows from theorem 4 (and theorem 12) that

$$\Delta(X_1,\ldots,X_n)^2 = D(s_1,\ldots,s_n)$$

for some (well-defined) polynomial D with integral coefficients, called the <u>discriminant</u> of the generic polynomial of degree n.

The discriminant of an arbitrary polynomial over a field F

$$x^n + a_{n-1}x^{n-1} + a_{n-2}x^{n-2} + \ldots + a_0$$

is then defined as $D(-a_{n-1}, a_{n-2}, \ldots, (-1)^n a_0)$, the element in F obtained from $D(s_1, \ldots, s_n)$ by replacing $s_1, \ldots, s_n$ by the coefficients of the polynomial.

15. Calculation of the discriminant in degrees 2 and 3 :

In degree 2, we readily have :

$$\Delta(X_1, X_2)^2 = (X_1 - X_2)^2 = X_1^2 + X_2^2 - 2X_1X_2$$

and this symmetric polynomial can be expressed in terms of elementary symmetric polynomials as :

$$\Delta(X_1, X_2)^2 = (X_1 + X_2)^2 - 4X_1X_2 = s_1^2 - 4s_2.$$

Hence, the discriminant of the generic polynomial of degree 2 is :

$$D(s_1, s_2) = s_1^2 - 4s_2.$$

For degree 3, we shall use some artifices to simplify the calculations. Written out as a sum of monomials, $\Delta(X_1, X_2, X_3) = (X_1 - X_2)(X_1 - X_3)(X_2 - X_3)$ appears as :

$$\Delta(X_1, X_2, X_3) = A - B$$

where $A = X_1^2 X_2 + X_2^2 X_3 + X_3^2 X_1$

and $B = X_1 X_2^2 + X_2 X_3^2 + X_3 X_1^2.$

Therefore,

$$\Delta(X_1, X_2, X_3)^2 = A^2 + B^2 - 2AB = (A+B)^2 - 4AB. \tag{7}$$

Now, $A+B$ and AB are symmetric polynomials which are not difficult to express in terms of s_1, s_2, s_3. Straightforward calculations by Waring's method yield the following results (with notation 5; see example 9):

$$A + B = \Sigma\ X_1^2 X_2 = s_1 s_2 - 3 s_3$$

$$AB = \Sigma\ X_1^4 X_2 X_3 + \Sigma\ X_1^3 X_2^3 + 3 X_1^2 X_2^2 X_3^2 = s_1^3 s_3 + s_2^3 - 6 s_1 s_2 s_3 + 9 s_3^2,$$

and the discriminant $D(s_1, s_2, s_3)$, which is equal to $\Delta(X_1, X_2, X_3)^2$, is easily calculated from equation (7) above:

$$D(s_1, s_2, s_3) = s_1^2 s_2^2 + 18 s_1 s_2 s_3 - 27 s_3^2 - 4 s_1^3 s_3 - 4 s_2^3.$$

In particular, it follows that the discriminant of $X^3 + pX + q$ is $-27 q^2 - 4 p^3$. Denoting this discriminant by d, we thus have:

$$d = -2^2 \cdot 3^3 [(p/3)^3 + (q/2)^2].$$

We will now show what kind of information on the roots of polynomials with real coefficients can be obtained from the discriminant. We shall need the following easy result:

16. LEMMA: If $a \in \mathbb{C}$ is a root of a polynomial $P \in \mathbb{R}[X]$, then the conjugate complex number $\bar{a}$ also is a root of P.

Proof: Conjugating both sides of the equality $P(a) = 0$, we obtain: $P(\bar{a}) = 0$, since the coefficients of P are equal to their own conjugate. □

Let now $P \in \mathbb{R}[X]$ be a monic polynomial with real coefficients, which splits into a product of linear factors over $\mathbb{C}$:

$$P = (X-x_1)\ldots(X-x_n).$$

(The fundamental theorem of algebra 9.1 will show that this is no restriction on P). Let $d \in \mathbb{R}$ be the discriminant of P.

17. THEOREM : $d = 0$ if and only if P has a root of multiplicity at least 2 in $\mathbb{C}$. If all the roots of P are real, then $d \geq 0$. The converse is true if $n = 2,3$.

Proof : Since the calculations with the roots of the generic polynomial are also valid with the roots $x_1,\ldots,x_n$ of P (see the discussion following definition 1), we have :

$$d = \Delta(x_1,\ldots,x_n)^2 = \prod_{1\leq i<j\leq n} (x_i-x_j)^2.$$

This readily shows that $d \geq 0$ if all the roots $x_1,\ldots,x_n$ are real. If P has a root of multiplicity at least 2, then $x_i = x_j$ for some indices i,j with $i \neq j$, whence $d = 0$ since the product above has a zero factor. If the roots of P are all simple, then each factor is non-zero, whence $d \neq 0$.

For $n = 2$,

$$d = (x_1-x_2)^2.$$

If x_1 is not real, then the preceding lemma shows that $x_2 = \overline{x_1}$. It follows that

$$\overline{(x_1-x_2)} = -(x_1-x_2),$$

whence

$$(x_1-x_2)^2 = -(x_1-x_2)\overline{(x_1-x_2)} = -|x_1-x_2|^2,$$

and $d < 0$.

Similarly, for $n = 3$,

$$d = (x_1-x_2)^2(x_1-x_3)^2(x_2-x_3)^2.$$

If one of the roots, x_1 say, is not real, then its conjugate $\bar{x}_1$ is among x_2, x_3. Without loss of generality, we can assume $\bar{x}_1 = x_2$. Then $(X-x_1)(X-x_2) \in \mathbb{R}[X]$, whence

$$X - x_3 = \frac{P}{(X-x_1)(X-x_2)} \in \mathbb{R}[X].$$

This shows $x_3 \in \mathbb{R}$. Now

$$\overline{(x_1-x_2)} = -(x_1-x_2),$$

$$\overline{(x_1-x_3)} = (x_2-x_3)$$

and $\overline{(x_2-x_3)} = (x_1-x_3)$.

Therefore,

$$\overline{(x_1-x_2)(x_1-x_3)(x_2-x_3)} = -(x_1-x_2)(x_1-x_3)(x_2-x_3),$$

whence

$$d = -|(x_1-x_2)(x_1-x_3)(x_2-x_3)|^2 < 0. \qquad \square$$

18. REMARKS : a) For $n \geq 4$, the sign of d determines the number of real roots of P up to a multiple of 4; see exercise 4.

b) The first statement (about multiple roots) in the preceding theorem is valid over an arbitrary field. Thus, we now have two necessary and sufficient conditions for a (monic) polynomial to have at least one multiple root in

some extension of the base field : its discriminant has to
be zero or, equivalently, the polynomial has to have a
non-constant common divisor with its derivative (see proposition 5.24). The equivalence of these two conditions can
be verified directly with the aid of theorem 5.30; indeed,
the discriminant of a monic polynomial P is equal (up to
sign) to the resultant of P and its derivative ∂P; see
exercise 6.

19. COROLLARY : Let $p,q \in \mathbb{R}$. The equation

$$X^3 + pX + q = 0$$

has three distinct real solutions if and only if
$(p/3)^3 + (q/2)^2 < 0$.

Proof : This readily follows from the preceding theorem,
since the discriminant d of $X^3 + pX + q$ is :

$$d = -2^2 \cdot 3^3 \cdot [(p/3)^3 + (q/2)^2]$$

(see n° 15 above). □

This corollary shows that the "casus irreducibilis" of cubic equations (see § 2.3(c)) is precisely the case where
the equation has three distinct real roots.

Appendix to chapter 8 : Euler's summation of the series of reciprocals of perfect squares

Around 1735, Euler succeeded in finding the sum of the series $\sum_{k=1}^{\infty} 1/k^2$, thus achieving a result that had baffled
Leibniz and Jacques Bernoulli (see [Boy, ch. 21, n° 4] or
[Go, § 3.2]). His method was to apply to a certain power
series the relations between roots and coefficients of
polynomials.

For the generic polynomial

$$x^n - s_1 x^{n-1} + s_2 x^{n-2} - \ldots + (-1)^n s_n = (X-X_1)\ldots(X-X_n),$$

we have seen that

$$s_1 = \Sigma X_1, \quad s_2 = \Sigma X_1 X_2, \ldots, s_{n-1} = \Sigma X_1 \ldots X_{n-1} \text{ and}$$
$$s_n = X_1 \ldots X_n,$$

whence (using notation 5 for rational fractions) :

$$\Sigma X_1^{-1} = s_{n-1} s_n^{-1}, \; \Sigma (X_1 X_2)^{-1} = s_{n-2} s_n^{-1}, \ldots, \Sigma (X_1 \ldots X_{n-1})^{-1} = s_1 s_n^{-1}.$$

By the same calculations as for Newton's formulas in § 4.2, we then obtain :

$$\Sigma X_1^{-2} = [\Sigma X_1^{-1}]^2 - 2 \Sigma (X_1 X_2)^{-1} = (s_{n-1}^2 - 2 s_{n-2} s_n) s_n^{-2}$$

$$\Sigma X_1^{-3} = [\Sigma X_1^{-1}][\Sigma X_1^{-2}] - [\Sigma (X_1 X_2)^{-1}][\Sigma X_1^{-1}] + 3 \Sigma (X_1 X_2 X_3)^{-1}$$

$$= (s_{n-1}^3 - 3 s_{n-2} s_{n-1} s_n + 3 s_{n-3} s_n^2) s_n^{-3}$$

etc.

Now, if $x_1, \ldots, x_n$ are the roots of an arbitrary (not necessarily monic) polynomial

$$a_n X^n + a_{n-1} X^{n-1} + \ldots + a_1 X + a_0 \quad (a_n \neq 0)$$

then $x_1, \ldots, x_n$ are the roots of the monic polynomial

$$X^n + a_{n-1} a_n^{-1} X^{n-1} + \ldots + a_1 a_n^{-1} X + a_0 a_n^{-1}.$$

Replacing $s_1, s_2, \ldots, s_n$ by $-a_{n-1} a_n^{-1}, a_{n-2} a_n^{-1}, \ldots, (-1)^n a_0 a_n^{-1}$ in the calculations above, it follows (provided that $a_0 \neq 0$) that

$$x_1^{-1} + \ldots + x_n^{-1} = -a_1 a_o^{-1} \tag{8}$$

$$x_1^{-2} + \ldots + x_n^{-2} = (a_1^2 - 2a_2 a_o) a_o^{-2} \tag{9}$$

$$x_1^{-3} + \ldots + x_n^{-3} = (-a_1^3 + 3a_2 a_1 a_o - 3a_3 a_o^2) a_o^{-3}. \tag{10}$$

As Euler pointed out, these calculations yield interesting results when applied to the sine function, considered as an infinite polynomial :

$$\sin z = z - z^3/3! + z^5/5! - z^7/7! + \ldots$$

which has as roots : $0, \pm\pi, \pm 2\pi, \ldots, \pm k\pi, \ldots$. Dividing the series by z to get rid of the root 0, and changing the variable to $x = z^2$, we obtain the series :

$$1 - x/3! + x^2/5! - x^3/7! + \ldots$$

with roots : $\pi^2, (2\pi)^2, \ldots, (k\pi)^2, \ldots$. Then formulas (8), (9), (10), etc. with $n = \infty$, $a_o = 1$, $a_1 = -1/3!$, $a_2 = 1/5!$, $a_3 = -1/7!$, etc. yield :

$$\Sigma_{k=1}^{\infty} (k\pi)^{-2} = 1/6$$

$$\Sigma_{k=1}^{\infty} (k\pi)^{-4} = 1/90$$

$$\Sigma_{k=1}^{\infty} (k\pi)^{-6} = 1/945$$

etc.

whence, multiplying both sides by the appropriate power of π :

$$\Sigma_{k=1}^{\infty} 1/k^2 = \pi^2/6$$

$$\Sigma_{k=1}^{\infty} 1/k^4 = \pi^4/90$$

$$\Sigma_{k=1}^{\infty} 1/k^6 = \pi^6/945$$

etc.

That Euler's calculations are valid is of course not obvious, but they can be rigorously verified. The point is that the sine function can be expressed as an infinite product :

$$\sin z = z \ \pi_{k=1}^{\infty} (1 - z^2/k^2\pi^2) \quad \text{for} \quad z \in \mathbb{C}.$$

Note also that the above calculations do not yield any information on the values of Riemann's zeta function

$$\zeta(s) = \Sigma_{k=1}^{\infty} 1/k^s$$

at <u>odd</u> integers s. Indeed, very little is known about these values. Until very recently, it was not even known whether $\zeta(3)$ is rational or not : this question was answered in the negative by Roger Apéry in 1978 (see [VP]).

Exercises for chapter 8

The following exercise provides an alternative procedure for calculating the discriminant of a polynomial :
1) Let $P(X) = X^n - s_1 X^{n-1} + s_2 X^{n-2} - \ldots + (-1)^n s_n = (X-X_1)\ldots(X-X_n)$, let $\sigma_i = \Sigma_{j=1}^n X_j^i$ for $i=1,2,\ldots$ and let A and B be the $n \times n$ matrices :

$$A = \begin{pmatrix} 1 & 1 & \ldots & 1 \\ X_1 & X_2 & \ldots & X_n \\ X_1^2 & X_2^2 & \ldots & X_n^2 \\ \vdots & \vdots & & \vdots \\ X_1^{n-1} & X_2^{n-1} & \ldots & X_n^{n-1} \end{pmatrix} \quad B = \begin{pmatrix} n & \sigma_1 & \sigma_2 & \ldots & \sigma_{n-1} \\ \sigma_1 & \sigma_2 & \sigma_3 & \ldots & \sigma_n \\ \sigma_2 & \sigma_3 & \sigma_4 & \ldots & \sigma_{n+1} \\ \vdots & \vdots & \vdots & & \vdots \\ \sigma_{n-1} & \sigma_n & \sigma_{n+1} & \ldots & \sigma_{2n-2} \end{pmatrix}$$

a) Show that $\det A = \Pi_{i>j} (X_i - X_j)$. (The matrix A is called

a Vandermonde matrix).
b) Show that $B = AA^T$ (where A^T denotes the transpose of A).
c) Derive from (a) and (b) that the discriminant of $P(X)$ is :

$$D(s_1, \ldots, s_n) = \det B.$$

d) Use this result to prove that the discriminant of the polynomial $X^n + pX + q$ is :

$$(-1)^{n(n-1)/2} [(-1)^{n-1}(n-1)^{n-1} p_n + n^n q^{n-1}].$$

2) Let X_1, X_2, X_3 be the roots of the cubic equation : $X^3 + pX + q = 0$. Prove that the equation whose roots are $(X_1-X_2)^2$, $(X_1-X_3)^2$ and $(X_2-X_3)^2$ is :

$$Y^3 + 6pY^2 + 9p^2Y + (4p^3 + 27q^2) = 0.$$

(This equation is called the <u>equation of squared differences</u> of the given cubic equation).

3) Let $P(X) = X^4 - s_1 X^3 + s_2 X^2 - s_3 X + s_4 = (X-X_1)(X-X_2)(X-X_3)(X-X_4)$, and let

$$u_1 = (X_1+X_2)(X_3+X_4) \quad u_2 = (X_1+X_3)(X_2+X_4) \quad u_3 = (X_1+X_4)(X_2+X_3)$$

$$v_1 = X_1X_2 + X_3X_4 \quad v_2 = X_1X_3 + X_2X_4 \quad v_3 = X_1X_4 + X_2X_3.$$

a) Show that the equation which has as roots u_1, u_2 and u_3 is :

$$Q(Y) = Y^3 - 2s_2 Y^2 + (s_2^2 + s_1 s_3 - 4s_4)Y - (s_1 s_2 s_3 - s_1^2 s_4 - s_3^2) = 0$$

and that the equation which has as roots v_1, v_2 and v_3 is :

$$R(Z) = Z^3 - s_2 Z^2 + (s_1 s_3 - 4s_4)Z - (s_1^2 s_4 - 4s_2 s_4 + s_3^2) = 0.$$

(Compare equations (4) and (5) in chapter 3).
b) Show that the discriminants of P,Q and R are all equal.
[Hint : prove : $\Delta(X_1,X_2,X_3,X_4) = -\Delta(u_1,u_2,u_3) = -\Delta(v_1,v_2,v_3)$].

4) Let $P \in \mathbb{R}[X]$ be a monic polynomial with real coefficients, which splits into a product of linear factors over $\mathbb{C}$:

$$P = (X-x_1)\ldots(X-x_n) \qquad (x_i \in \mathbb{C}).$$

Assume that the roots $x_1,\ldots,x_n$ of P are pairwise distinct and denote by r the number of real roots among $x_1,\ldots,x_n$ and by d the discriminant of P. Show that n - r is an even integer, which is divisible by 4 if and only if d > 0.

5) Show that the resultant of

$$P = (X-x_1)\ldots(X-x_n)$$

and $\quad Q = (X-y_1)\ldots(X-y_n)$

is

$$\Pi_{i=1}^{n} \Pi_{j=1}^{m} (x_i-y_j) = \Pi_{i=1}^{n} Q(x_i) = (-1)^n \Pi_{j=1}^{m} P(y_j).$$

[Hint : consider $x_1,\ldots,x_n, y_1,\ldots,y_m$ as indeterminates and use theorem 5.30].

6) Let $P = (X-x_1)\ldots(X-x_n)$; let d denote the discriminant of P and R the resultant of P and its derivative ∂P. Show that

$$R = \Pi_{i \neq j} (x_i-x_j) = (-1)^{n(n-1)/2} d.$$

[Hint : calculate $\partial P(x_i)$ and use exercise 5].

9 The fundamental theorem of algebra

§ 1. Introduction

The title of this chapter refers to the following result:

1. THEOREM : The number of roots of a non-zero polynomial over the field $\mathbb{C}$ of complex numbers, each root being counted with its multiplicity, is equal to the degree of the polynomial.

Equivalently, by theorem 5.20, the fundamental theorem of algebra asserts that every polynomial splits into a product of linear factors in $\mathbb{C}[X]$. There is also an equivalent formulation in terms of real polynomials only:

2. THEOREM : Every (non-constant) real polynomial can be decomposed into a product of (real) polynomials of degree 1 or 2.

The equivalence of these statements will be proved in proposition 7 below.

Theorem 1 can be traced back to Girard, in a considerably looser form (see § 4.2). Indeed, Leibniz's proposed counter-example (in § 7.2) clearly shows how remote a proof still was at the beginning of the eighteenth century. Yet during the first half of this century, de Moivre's work had prompted a deeper understanding of the operations on complex numbers and opened the way to the first attempts of proof.

Since the ultimate structure of $\mathbb{R}$ (or $\mathbb{C}$) is analytical, it is not surprising to us that the first idea of a proof, published in 1746 by Jean le Rond d'Alembert (1717-1783), used analytic techniques. However, an analytic proof for what was perceived as an algebraic theorem was hardly satisfying, and Euler in 1749 tried a more algebraic method. Euler's idea was to prove the equivalent theorem 2 by an induction argument on the highest power of 2 which divides the degree. Omitting several key details, Euler failed to carry out his program completely, so that his proof is only a sketch. Some simplifications in Euler's proof were subsequently suggested by Daviet François de Foncenex (1734-1799), and Lagrange eventually gave in 1772 a complete proof, elaborating on Euler's and de Foncenex's ideas and correcting all the flaws in their proofs.

All, but one. As Gauss noticed in 1799 in his inaugural dissertation [Gau1, § 12], a critical flaw remained : lured by the custom inherited from seventeenth century mathematicians, Lagrange implicitly takes for granted the <u>existence</u> of n "imaginary" roots for any equation of degree n, and he thus only proves that the <u>form</u> of these imaginary roots is a + b $\sqrt{-1}$ with a,b $\in \mathbb{R}$. Gauss also shows that the same critic can be addressed to the earlier proofs of Euler, de Foncenex and d'Alembert [Gau1, §§ 6 ff] and he then proceeds to give the first essentially complete proof of the fundamental theorem of algebra, along the lines of d'Alembert's proof.

In 1815, Gauss also found a way to amend the Euler-de Foncenex-Lagrange proof [Gau3], and he subsequently gave two other proofs of the fundamental theorem 1.

The proof we give is based upon the Euler-de Foncenex-Lagrange ideas. However, instead of following Gauss' correction, we use some ideas from the late nineteenth century to prove first Girard's "theorem" on the existence of imaginary roots (of which nothing is known, except that operations can be performed on these roots as if they were numbers). Having thus justified the postulatum on which Euler's proof implicitly relied, we are then able to use

153

Euler's arguments in a more direct way, to prove that the imaginary roots are of the form $a + b\sqrt{-1}$ (with $a,b \in \mathbb{R}$). We thus obtain a streamlined and almost completely algebraic proof of the fundamental theorem of algebra, also found in [Sam].

§ 2. Girard's theorem

In this section, we let F be an arbitrary field. The modern translation of Girard's intuition is as follows :

3. THEOREM : For any non constant polynomial $P \in F[X]$, there is a field K containing F such that P splits over K into a product of linear factors :

$$P = a(X-x_1)\ldots(X-x_n) \quad \text{in} \quad K[X].$$

The field K is constructed abstractly by successive quotients of polynomial rings by ideals.

We first recall that an <u>ideal</u> in a commutative ring A is a subgroup I of the additive group of A which is stable under multiplication by elements in A, i.e. such that

$$ax \in I \quad \text{for} \quad a \in A \quad \text{and} \quad x \in I.$$

In order to define the quotient ring of A by I, we set, fo $a \in A$:

$$a + I = \{a + x \mid x \in I\}.$$

This is a subset of A, and from the hypothesis that I is a subgroup of the additive group of A, it easily follows that, for $a,b \in A$:

$$a + I = b + I \quad \text{if and only if} \quad a - b \in I. \tag{1}$$

We then set :

$$A/I = \{a + I \mid a \in A\};$$

the condition that I is an ideal ensures that the operations on A induce a ring structure on A/I, by :

$$(a+I) + (b+I) = (a+b) + I$$

$$(a+I)(b+I) = ab + I.$$

The ring A/I is called the <u>quotient ring</u> of A by the ideal.

The zero element in this quotient ring is $0 + I$ ($= I$), also denoted simply as 0; therefore, (1) shows that in A/I:

$$a + I = 0 \quad \text{if and only if} \quad a \in I. \tag{2}$$

We shall need the following instance of this construction: let $A = F[X]$ and let I be the set (P) of multiples of a polynomial P :

$$(P) = \{PQ \mid Q \in F[X]\}.$$

It is readily verified that (P) is an ideal of $F[X]$, so that we can construct the quotient ring $F[X]/(P)$. This construction is essentially due to Kronecker (1823-1897), although a special case had been considered earlier by Cauchy. (In 1847, Cauchy represented $\mathbb{C}$ as the quotient ring $\mathbb{R}[X]/(X^2+1)$).

The basic lemma required for the proof of Girard's theorem is the following :

. <u>LEMMA</u> : If $P \in F[X]$ is irreducible, then $F[X]/(P)$ is a field containing F and a root of P.

<u>Proof</u> : In order to show that $F[X]/(P)$ is a field, it suffices to prove that every non-zero element $Q + (P)$ in

155

$F[X]/(P)$ is invertible. Since $Q + (P) \neq 0$, it follows from (2) above that $Q \notin (P)$, i.e. that Q is not divisible by P. Since P is irreducible, P and Q are then relatively prime, whence, by corollary 5.9 :

$$PP_1 + QQ_1 = 1 \quad \text{for some} \quad P_1, Q_1 \in F[X].$$

This relation shows that $QQ_1 - 1$ is divisible by P, so that, by (1) above :

$$QQ_1 + (P) = 1 + (P).$$

Hence,

$$(Q+(P))(Q_1+(P)) = 1 + (P),$$

which means that $Q_1 + (P)$ is the inverse of $Q + (P)$ in $F[X]/(P)$.

The map $a \to a + (P)$ from F to $F[X]/(P)$ is injective since no non-zero element in F is divisible by P. Therefore, this map is an embedding of F in $F[X]/(P)$, and F is thus identified to a subfield of $F[X]/(P)$; we can thus consider P as a polynomial over $F[X]/(P)$, and it only remains to prove that $F[X]/(P)$ contains a root of P.

From the definition of the operations in $F[X]/(P)$ it follows that

$$P(X+(P)) = P(X) + (P).$$

Hence, by (2) above :

$$P(X+(P)) = 0,$$

which means that $X + (P)$ is a root of P in $F[X]/(P)$.

5. Proof of theorem 3 : We decompose P into a product of irreducible factors in F[X] :

$$P = P_1 \cdot \ldots \cdot P_r$$

and let s be the number of linear factors among $P_1, \ldots, P_r$. (Possibly s = 0 : this occurs when P has no root in F).

We then argue by induction on (deg P) - s, noting that this number is equal to the degree of the product of the non-linear factors among $P_1, \ldots, P_r$.

If (deg P) - s = 0, then each of the factors $P_1, \ldots, P_r$ is linear, and we can choose K = F.

If (deg P) - s > 0, then at least one of the factors $P_1, \ldots, P_r$ has degree greater than or equal to 2. Assume for instance that deg $P_1 \geq 2$, and let

$$F_1 = F[X]/(P_1).$$

Since P_1 has a root in F_1, the decomposition of P_1 over F_1 involves at least one linear factor, by theorem 5.17, whence the number s_1 of linear factors in the decomposition of P into irreducible factors over F_1 is at least s + 1. Therefore, (deg P) - s_1 < (deg P) - s, and the induction hypothesis implies that there is a field K containing F_1 and such that P splits into linear factors over K. Since F_1 contains F, the field K also contains F and satisfies all the requirements. □

§ 3. Proof of the fundamental theorem

Instead of proving theorem 1 directly, we shall prove an equivalent formulation in terms of real polynomials. We first note for later reference the following easy special case of the fundamental theorem 1 :

6. LEMMA : Every quadratic polynomial over ℂ splits into a product of linear factors in ℂ[X].

Proof : It suffices to show that the roots of every quadratic equation with complex coefficients are complex numbers This readily follows from the usual formula by radicals fo the roots (§ 1.1), since, by proposition 7.2, every comple number has square roots in $\mathbb{C}$. □

We now prove the equivalence of several formulations of th fundamental theorem :

7. PROPOSITION : The following statements are equivalent :
a) The number of roots of any non-zero polynomial over $\mathbb{C}$ is equal to its degree (each root being counted with its multiplicity).
b) Every non-constant polynomial over $\mathbb{R}$ has at least one root in $\mathbb{C}$.
c) Every non-constant real polynomial can be decomposed into a product of (real) polynomials of degree 1 or 2.

Proof : (a) ⇒ (b) is clear.
(b) ⇒ (c) : By theorem 5.13, it suffices to show, assuming (b), that every irreducible polynomial in $\mathbb{R}[X]$ has degree 1 or 2. Let P be an irreducible polynomial in $\mathbb{R}[X]$, and let $a \in \mathbb{C}$ be a root of P.

If $a \in \mathbb{R}$, then $X - a$ divides P in $\mathbb{R}[X]$, whence deg P = 1 since, by definition, an irreducible polynomial cannot be divided by a non-constant polynomial of strictly smaller degree.

If $a \notin \mathbb{R}$, then $\bar{a} \neq a$, and $\bar{a}$ is also a root of P, by lemma 8.16. Therefore, by proposition 5.15, P is divisible by $(X-a)(X-\bar{a})$ in $\mathbb{C}[X]$. But since

$$(X-a)(X-\bar{a}) = X^2 - (a+\bar{a})X + a\bar{a} \in \mathbb{R}[X],$$

P is also divisible by $(X-a)(X-\bar{a})$ in $\mathbb{R}[X]$ (see remark 5.10 (b)), whence the same argument as above implies : deg P = 2
(c) ⇒ (a) : Let $P \in \mathbb{C}[X]$ be a non-constant polynomial. We

extend to $\mathbb{C}[X]$ the complex conjugation map from $\mathbb{C}$ to $\mathbb{C}$ by setting $\overline{X} = X$; namely, we set :

$$\overline{a_0 + a_1 X + \ldots + a_n X^n} = \overline{a}_0 + \overline{a}_1 X + \ldots + \overline{a}_n X^n.$$

The invariant elements are readily seen to be the polynomials with real coefficients. Therefore, $P\overline{P} \in \mathbb{R}[X]$ and it follows from the hypothesis (c) that :

$$P\overline{P} = P_1 \cdot \ldots \cdot P_r$$

where $P_1, \ldots, P_r \in \mathbb{R}[X]$ are polynomials of degree 1 or 2. By lemma 6, the real polynomials of degree 2 split into products of linear factors in $\mathbb{C}[X]$, whence $P\overline{P}$ is a product of linear factors in $\mathbb{C}[X]$. Therefore, every irreducible factor of P in $\mathbb{C}[X]$ has degree 1, so that P splits into a product of linear factors in $\mathbb{C}[X]$, which proves (a). □

As we noted in the introduction, every proof of the fundamental theorem of algebra uses at some point an analytical (or topological) argument, since $\mathbb{R}$ (or $\mathbb{C}$) cannot be completely defined without reference to some of its topological properties. The only analytical result we shall need in our proof is the following :

8. LEMMA : Every real polynomial P of odd degree has at least one root in $\mathbb{R}$.

Proof : Since deg P is odd, the polynomial function $P(.) : \mathbb{R} \to \mathbb{R}$ changes sign when the variable runs from $-\infty$ to $+\infty$, so, by continuity, it must take the value 0 at least once. □

The continuity argument, according to which every continuous function which changes sign on an interval must take the value 0 at least once, may seem, and was for a long time considered as, evident by itself. It was first proved by Bolzano in 1817 (see [Di, p. 340] or [Kl, p. 952]), in an attempt to provide "arithmetical" proofs to the intuitive geometric arguments that Gauss used in his 1799 proof of the fundamental theorem.

9. Proof of the fundamental theorem

We shall prove the equivalent formulation (b) in proposition 7, that every real non-constant polynomial has at least one root in $\mathbb{C}$.

Let $P \in \mathbb{R}[X]$ be a non-constant polynomial. Dividing P by its leading coefficient if necessary, we may assume that P is monic. We write the degree of P in the form :

$$\deg P = n = 2^e m \quad \text{where} \quad e \geq 0 \text{ and } m \text{ is odd.}$$

If $e = 0$, then the degree of P is odd, and the preceding lemma shows that P has a root in $\mathbb{R}$. We then argue by induction on e, assuming that $e \geq 1$ and that the property holds when the exponent of the highest power of 2 which divides the degree of the polynomial is at most $e - 1$.

Let K be a field containing $\mathbb{C}$, over which P splits into a product of linear factors :

$$P = (X-x_1) \ldots (X-x_n).$$

(The existence of such a field K follows from theorem 3). For $c \in \mathbb{R}$ and for $i,j = 1,\ldots,n$ with $i < j$, let

$$y_{ij}(c) = (x_i + x_j) + c x_i x_j$$

and let

$$Q_c(Y) = \prod_{1 \leq i < j \leq n} (Y - y_{ij}(c)).$$

The coefficients of Q_c are the values of the elementary symmetric polynomials in the roots $y_{ij}(c)$; these coefficients are therefore the values of symmetric polynomials in $x_1,\ldots,x_n$ with real coefficients, hence they can be expressed in terms of the values of the elementary symmetric polynomials in $x_1,\ldots,x_n$, by the fundamental theorem of symmetric polynomials. Since the values of the elementary symmetric polynomials in $x_1,\ldots,x_n$ are the coefficients of P, which are real numbers, it follows that the coefficients of Q_c also are real numbers.

Moreover, the degree of Q_c is $n(n-1)/2$, whence

$$\deg Q_c = 2^{e-1}[m(2^e m-1)],$$

and the integer between brackets is odd. We may thus apply the induction hypothesis, to conclude that Q_c has at least one root in $\mathbb{C}$, i.e.

$$y_{r(c),s(c)}(c) \in \mathbb{C} \text{ for some indices } r(c),s(c).$$

If we let the real number c run over the set of real numbers, the indices $r(c),s(c)$ for which $y_{r(c),s(c)}(c) \in \mathbb{C}$ cannot be all distinct, since the set of indices is finite, while $\mathbb{R}$ is infinite. Therefore, we can find some distinct real numbers c_1,c_2 such that $r(c_1) = r(c_2)$ and $s(c_1) = s(c_2)$. Denoting by r and s these common indices, this means that :

$$(x_r+x_s) + c_1 x_r x_s \in \mathbb{C}$$

and $\quad (x_r+x_s) + c_2 x_r x_s \in \mathbb{C},$

with $c_1,c_2 \in \mathbb{R}$ and $c_1 \neq c_2$. By subtraction, these relations imply :

$$(c_1-c_2) x_r x_s \in \mathbb{C},$$

whence

$$x_r x_s \in \mathbb{C}.$$

Comparing this result with the relations from which it has been derived, we obtain moreover :

$$x_r + x_s \in \mathbb{C}.$$

This shows that the coefficients of the polynomial

$$X^2 - (x_r + x_s)X + x_r x_s$$

are complex numbers, and it then follows from lemma 6 that its roots x_r and x_s are complex numbers. We have thus shown that at least one of the roots $x_1, \ldots, x_n$ of P in K is a complex number, as was required. □

10. COROLLARY : Over $\mathbb{C}$, the irreducible polynomials are the polynomials of degree 1. Over $\mathbb{R}$, the irreducible polynomials are the polynomials of degree 1 and the polynomials of degree 2 which have no real root.

<u>Proof</u> : This readily follows from the fundamental theorem 1 or the equivalent theorem 2, since, by theorem 5.17, the irreducible polynomials which have a root in the base field have degree 1. □

10 Lagrange

§ 1. Introduction

The theory of equations comes of age.
In the second half of the eighteenth century, the algebraic theory of equations is ripe for new advances. All the more or less elementary facts on polynomials are well-known, and computational skills are very high, even by modern standards. Moreover, deeper insights on the ambiguity of roots of (complex) numbers become available through de Moivre's work. The relevance of these insights for the problem of solving equations by radicals is obvious (see the end of § 7.2), and one may venture the hypothesis that de Moivre's work provided an important stimulus to new research in the algebraic theory of equations.

Whatever its origin, it is clear that the spirit of the most significant research in this period is completely different from that of Cardano and his contemporaries : no direct application to the solution of numerical equations is expected, and no reference to any practical problem is made, even allusively. The subject has become pure mathematics, and is pursued for its own interest. Within less than a century, in the hands of several mathematicians of genius, it will undergo a rapid development which will dramatically change the whole subject of algebra.

The earliest works in this line appear in the sixties of the eighteenth century, when Euler and Bezout devise various new methods to solve equations of degree at most 4, which can seemingly be extended to equations of higher degree. One of these methods, proposed by Bezout in 1765, is particularly interesting by its explicit use of roots of unity; it is in fact very close to a method of Euler, and has deep resemblances to Tschirnhaus' method.

The idea[*] is to eliminate the indeterminate Y between the two equations :

$$X = a_0 + a_1 Y + a_2 Y^2 + \ldots + a_{n-1} Y^{n-1} \qquad (1)$$

$$Y^n = 1, \qquad (2)$$

producing an equation of degree n in X :

$$R_n(X) = 0,$$

as in Tschirnhaus' method (see § 6.4). Dividing R_n by its leading coefficient if necessary, we can assume that R_n is monic. The properties of $R_n(X)$ imply that if x,y are related by equations (1) and (2), i.e. if

$$x = a_0 + a_1 \omega + a_2 \omega^2 + \ldots + a_{n-1} \omega^{n-1}$$

for some n-th root of unity ω, then x is a root of $R_n(X)$, whence $R_n(X)$ is divisible by $X - (a_0 + a_1 \omega + \ldots + a_{n-1} \omega^{n-1})$. Regarding $a_0, a_1, \ldots, a_{n-1}$ as independent indeterminates, the values of x corresponding to the various n-th roots of unity ω are all different, so that, by proposition 5.15,

$$R_n(X) = \prod_\omega (X - (a_0 + a_1 \omega + \ldots + a_{n-1} \omega^{n-1})), \qquad (3)$$

where the product runs over the n different n-th roots of unity ω. The roots of $R_n(X) = 0$ are thus known.

Now, to solve an arbitrary monic equation of degree n :

---

[*] according to Bezout's presentation; in Euler's work, equation (2) is replaced by $Y^n = b$ and in (1), one of the coefficients a_i is chosen to be 1. Tschirnhaus' method can be presented in a similar way, replacing equation (2) by $Y^n = b$ and (1) by $Y = a_0 + a_1 X + \ldots + a_{n-2} X^{n-2} + X^{n-1}$.

P(X) = 0,

the method is to determine the parameters $a_0, a_1, \ldots, a_{n-1}$ in such a way that the polynomial $R_n(X)$ be identical to P(X). The solutions of P(X) = 0 are then readily obtained in the form: $a_0 + a_1\omega + \ldots + a_{n-1}\omega^{n-1}$.

Of course, whether it is possible to assign some value to $a_0, \ldots, a_{n-1}$ in such a way that R_n becomes identical to P is not clear at all, but this turns out to be the case for n = 2, 3 or 4, as we will now see.

The method for constructing $R_n(X)$ by elimination of Y between equations (1) and (2) has already been discussed in § 6.4. For the small values of n, the following results are found:

$$R_2(X) = (X-a_0)^2 - a_1^2$$

$$R_3(X) = (X-a_0)^3 - 3a_1a_2(X-a_0) - (a_1^3+a_2^3)$$

$$R_4(X) = (X-a_0)^4 - 2(a_2^2+2a_1a_3)(X-a_0)^2 - 4a_2(a_1^2+a_3^2)(X-a_0)$$
$$- (a_1^4-a_2^4+a_3^4+4a_1a_2^2a_3-2a_1^2a_3^2).$$

Alternatively, these results can be obtained from equation (3) by expanding the right-hand side.

To obtain the solutions of the cubic equation

$$X^3 + pX + q = 0 \tag{4}$$

(to which the general cubic equation can be reduced by a linear change of variable: see § 2.2), it now suffices to assign values to a_0, a_1 and a_2 in such a way that $R_3(X)$ takes the form $X^3 + pX + q$. We thus choose $a_0 = 0$ and determine a_1 and a_2 by:

$$-3a_1a_2 = p \tag{5}$$

$$-(a_1^3+a_2^3) = q \tag{6}$$

(compare § 2.2). The first equation gives the value of a_2 in function of a_1; substituting this value in the second equation yields a quadratic equation in a_1^3:

$$a_1^6 + qa_1^3 - (p/3)^3 = 0.$$

A root of this equation is easily found : one can choose for a_1 any cubic root of $-(q/2) + \sqrt{(p/3)^3 + (q/2)^2}$ or $-(q/2) - \sqrt{(p/3)^3 + (q/2)^2}$. Letting then $a_2 = -p/3a_1$, it follows that equations (5) and (6) both hold, whence $R_3(X) = X^3 + pX + q$. Then, equation (3) shows that the solutions of equation (4) are of the form $\omega a_1 + \omega^2 a_2$, where ω runs over the set of cubic roots of unity. If ζ denotes one of these cubic roots other than 1, then the cubic roots of unity are 1, ζ and ζ^2, and consequently the solutions of (4) are :

$$a_1 + a_2, \qquad \zeta a_1 + \zeta^2 a_2 \qquad \text{and} \qquad \zeta^2 a_1 + \zeta a_2.$$

(Note that $\zeta^4 = \zeta$).

Remark : The fact that one can choose $a_0 = 0$ obviously follows from the particular form of the proposed cubic equation (4), which lacks the term in X^2. The general case is in no way more difficult and could be treated in the same way, but the calculations are less transparent.

Similarly, for equations of degree 4 such as

$$X^4 + pX^2 + qX + r = 0, \tag{7}$$

we seek values of a_0, a_1, a_2 and a_3 for which $R_4(X) = X^4 + pX^2 + qX + r$. As above, we choose $a_0 = 0$ and we are left with the relations :

$$-2(a_2^2 + 2a_1a_3) = p \tag{8}$$

$$-4a_2(a_1^2 + a_3^2) = q \tag{9}$$

$$-(a_1^4 - a_2^4 + a_3^4 + 4a_1a_2^2a_3 - 2a_1^2a_3^2) = r. \tag{10}$$

Replacing in the third equation $a_1^4 + a_3^4$ by $(a_1^2 + a_3^2)^2 - 2a_1^2a_3^2$, we get:

$$-r = (a_1^2 + a_3^2)^2 - 4a_1^2a_3^2 + 4(a_1a_3)a_2^2 - a_2^4.$$

Equations (8) and (9) can then be used to eliminate a_1 and a_3 from this equation. The resulting equation is a cubic equation in a_2^2, from which a value of a_2 can be determined. Values for a_1 and a_3 are then easily found from equations (8) and (9), and the roots of the proposed quartic equation (7) are obtained in the form $a_1\omega + a_2\omega^2 + a_3\omega^3$, where ω runs over the set of 4-th roots of unity. Letting i denote (as usual) a square root of -1, the 4-th roots of unity are $1, i, -1, -i$ and the roots of the quartic equation (7) are:

$$a_1 + a_2 + a_3, \quad ia_1 - a_2 - ia_3, \quad -a_1 + a_2 - a_3 \text{ and}$$

$$-ia_1 - a_2 + ia_3.$$

As noted above, the principle of this method, whence also its difficulty, is not very different from that of Tschirnhaus' method. To its credit, one can nevertheless observe that the method of Euler and Bezout leads to easier calculations, that it is somewhat more direct and, what is more significant for later researches, that Bezout's method stresses the importance of the roots of unity. Altogether, it does not represent a very substantial progress.

The first really important burst of activity in the theory of equations takes place only a few years later, around 1770, with the almost simultaneous publication of

Lagrange's "Réflexions sur la résolution algébrique des équations" and Vandermonde's "Mémoire sur la résolution des équations", and of comparatively less important works such as Waring's "Meditationes Algebricae", which we already quoted in chapter 8. Among all the works of this period, Lagrange's massive paper clearly is the most lucid and the most comprehensive; therefore, it proved to be also the most influential. Moreover, Lagrange provides an almost unhoped-for link between the early stages of the theory of equations and the subsequent period, by first reviewing the various methods for equations of degree 3 and 4, and the attempts at equations of higher degree so far proposed, before making his own highly original observations.

We shall thus begin our study of the two critical works of this period with Lagrange's paper, and discuss Vandermonde's memoir in the next chapter.

§ 2. Lagrange's observations on previously known methods

Lagrange's discussion of the previously known methods is not a mere summary : it is a vast unification and reassessment of these methods. His very explicit aim is to determine not only how these methods work, but why.

> "I propose in this Memoir to examine the various methods found so far for the algebraic solution of equations, to reduce them to general principles, and to let see a priori why these methods succeed for the third and the fourth degree, and fail for higher degrees.
>
> This examination will have a double advantage : on one hand, it will shed a greater light on the known solutions of the third and the fourth degree; on the other hand, it will be useful to those who will want to deal with the solution of higher degrees, by providing them with various views to this end and above all by sparing them a large number of useless steps and attempts" [L, pp. 206-207].

The term "<u>a priori</u>" keeps recurring throughout Lagrange's work : it is the hallmark of his new fruitful methodology. Lagrange started from the rather obvious observation that the various methods for solving equations have a common feature : they all reduce the problem by some clever transformations to the solution of a certain auxiliary equation of smaller degree. <u>A posteriori</u>, when these clever transformations have been found, one can only ascertain that the method provides the required solutions from those of the auxiliary equation, but this does not give any valuable insight into the solution of equations of higher degree. Indeed, the only evidence that supports the belief that Tschirnhaus, Euler or Bezout's method could be applied to equations of higher degree is that the approach is the same in all cases, that the first calculations are parallel, and that it works for equations of degree 2, 3 or 4. This is rather scant evidence.

To find out <u>a priori</u> why a method works, Lagrange's highly original idea is to reverse the steps, and determine the roots of the auxiliary equations as functions of the roots of the proposed equation. The properties of the roots of the auxiliary equation then become apparent, and they clearly show why these roots provide the solution of the proposed equation.

We shall take for example Cardano's method for cubic equations, which is the first method scrutinized by Lagrange. Lagrange begins with a careful description of the method : the cubic equation

$$X^3 + aX^2 + bX + c = 0$$

is first reduced to the form :

$$X'^3 + pX' + q = 0$$

by the change of variable $X' = X + (a/3)$. Next, by setting

$X' = Y + Z$, the equation becomes :

$$(Y^3 + Z^3 + q) + (Y + Z)(3YZ + p) = 0.$$

The solutions of the cubic equation are then obtained from the solutions of the system :

$$\begin{cases} Y^3 + Z^3 + q = 0 \\ 3YZ + p = 0. \end{cases}$$

From the second equation comes :

$$Z = -p/3Y \qquad (11)$$

and, substituting in the first equation, one gets :

$$Y^3 - (p/3Y)^3 + q = 0,$$

whence

$$Y^6 + qY^3 - (p/3)^3 = 0. \qquad (12)$$

This is the auxiliary equation, which Lagrange terms the "reduced" equation, on which Cardano's method depends. From this equation, the values of Y are easily obtained, since it is really a quadratic equation in Y^3; the corresponding values of Z are derived from (11), and the solutions of the initial equation are then :

$$X = -(a/3) + Y + Z.$$

Since the reduced equation (12) has degree 6, it has six roots, so we end up with six roots for the initial equation of degree 3; in fact, it can be seen that these six values of X are pairwise equal, so that each root of the cubic equation is obtained twice. Indeed, let $Y_1, Y_2, \ldots, Y_6$ be the six roots of (12); since this equation is quadratic in Y^3, their cubes $y_1^3, \ldots, y_6^3$ take only two values v_1, v_2, whose product is

$$v_1 v_2 = -(p/3)^3,$$

since $-(p/3)^3$ is the independent term of (12). Changing the numbering if necessary, we may assume :

$$y_1^3 = y_2^3 = y_3^3 = v_1; \quad y_4^3 = y_5^3 = y_6^3 = v_2.$$

So, y_1, y_2, y_3 (resp. y_4, y_5, y_6) are the various cube roots of v_1 (resp. v_2), whence, denoting by ω a cube root of unity other than 1, we may assume :

$$y_2 = \omega y_1, \; y_3 = \omega^2 y_1 \text{ and } y_5 = \omega y_4, \; y_6 = \omega^2 y_4.$$

Since $v_1 v_2 = -(p/3)^3$, there are some determinations of the cube roots of v_1 and v_2 which multiply up to $-(p/3)$. Assume for instance (renumbering $y_1, \ldots, y_6$ if necessary) that

$$y_1 y_4 = -(p/3).$$

Then

$$y_2 y_6 = \omega^3 y_1 y_4 = -(p/3)$$

and similarly

$$y_3 y_5 = -(p/3).$$

Therefore, if we denote by z_i the value of Z which corresponds to y_i by (11), we have :

$$z_1 = y_4, \; z_2 = y_6, \; z_3 = y_5, \; z_4 = y_1, \; z_5 = y_3 \text{ and } z_6 = y_2,$$

and it follows that $y_i + z_i$ takes only three different values :

$$y_1 + y_4 \qquad y_2 + y_6 = \omega y_1 + \omega^2 y_4 \text{ and } y_3 + y_5 = \omega^2 y_1 + \omega y_4,$$

which yield the three roots of the initial cubic equation :

$$\begin{aligned} x_1 &= -(a/3) + y_1 + y_4 \\ x_2 &= -(a/3) + \omega y_1 + \omega^2 y_4 \\ x_3 &= -(a/3) + \omega^2 y_1 + \omega y_4. \end{aligned} \qquad (13)$$

So, we see <u>a posteriori</u> how the reduced equation provides the solution of the initial cubic equation.

To understand <u>a priori</u> why it does, Lagrange determines $y_1, \ldots, y_6$ as functions of x_1, x_2 and x_3. This is fairly easy : it suffices to solve the system (13) for y_1 and y_4, and the other y_i's are multiples of y_1 or y_4 by ω or ω^2. It is even easier if one notices that, ω being a cube root of unity other than 1, it is a root of

$$(X^3 - 1)/(X-1) = X^2 + X + 1,$$

so that $\omega^2 + \omega + 1 = 0$; therefore, multiplying the second equation of (13) by ω^2, the third by ω and adding to the first, one obtains :

$$x_1 + \omega^2 x_2 + \omega x_3 = -(a/3)(1+\omega+\omega^2) + 3y_1 + (1+\omega+\omega^2)y_4,$$

whence

$$y_1 = (1/3)(x_1 + \omega^2 x_2 + \omega x_3).$$

Likewise, one obtains :

$$y_4 = (1/3)(x_1 + \omega x_2 + \omega^2 x_3)$$

and the other roots of the reduced equation are easily obtained by multiplying y_1 or y_4 by ω or ω^2 :

$$y_2 = (1/3)(\omega x_1 + x_2 + \omega^2 x_3)$$
$$y_3 = (1/3)(\omega^2 x_1 + \omega x_2 + x_3)$$
$$y_5 = (1/3)(\omega x_1 + \omega^2 x_2 + x_3)$$
$$y_6 = (1/3)(\omega^2 x_1 + x_2 + \omega x_3).$$

Thus, the roots of the reduced equation are all the expressions obtained from

$$y_4 = (1/3)(x_1 + \omega x_2 + \omega^2 x_3)$$

by permutation of x_1, x_2 and x_3, and the purpose of solving the reduced equation is to determine some (whence all) of these expressions.

From this observation, Lagrange draws some clever conclusions.

First, it explains why the reduced equation has degree 6. Indeed, since the coefficients of the reduced equation are functions of the coefficients of the proposed equation, which are the elementary symmetric polynomials in x_1, x_2, x_3, it follows that these coefficients are symmetric. Therefore, if some expression of x_1, x_2, x_3 is a root of the reduced equation, every other expression obtained from this one by permutation of x_1, x_2, x_3 also is a root of this equation. Since y_4 takes six different values by permutation of x_1, x_2, x_3, these six values are the roots of the reduced equation, which has therefore degree 6.

Moreover, it explains why the reduced equation is a quadratic equation in y^3 : this is because y_4^3 takes only two values by permutation of x_1, x_2, x_3. Indeed, since $\omega^3 = 1$, one has for instance :

$$\omega x_1 + \omega^2 x_2 + x_3 = \omega(x_1 + \omega x_2 + \omega^2 x_3)$$

and it follows that

$$(\omega x_1 + \omega^2 x_2 + x_3)^3 = (x_1 + \omega x_2 + \omega^2 x_3)^3.$$

Likewise, we obtain:

$$(x_1 + \omega x_2 + \omega^2 x_3)^3 = (\omega x_1 + \omega^2 x_2 + x_3)^3 = (\omega^2 x_1 + x_2 + \omega x_3)^3$$

and

$$(x_1 + \omega^2 x_2 + \omega x_3)^3 = (\omega^2 x_1 + \omega x_2 + x_3)^3 = (\omega x_1 + x_2 + \omega^2 x_3)^3;$$

so the two values of y_4^3 are

$$(1/3)^3 (x_1 + \omega x_2 + \omega^2 x_3)^3 \quad \text{and} \quad (1/3)^3 (x_1 + \omega^2 x_2 + \omega x_3)^3,$$

which are the roots of a quadratic equation.

The general result behind these arguments is the following:

1. PROPOSITION: Let f be a rational fraction in n indeterminates $X_1, \ldots, X_n$. If f takes m different values [*] when the indeterminates $X_1, \ldots, X_n$ are permuted in all possible ways, then f is a root of a monic equation $\Theta = 0$ of degree m, whose coefficients are symmetric in $X_1, \ldots, X_n$, whence expressible as functions of the elementary symmetric polynomials (by the fundamental theorem of symmetric fractions 8.3).

Moreover, if f is a root of another equation $\Phi = 0$ with coefficients symmetric in $X_1, \ldots, X_n$, then

---

[*] Properly speaking, one should say "if the permutations of $X_1, \ldots, X_n$ in f give rise to m different rational fractions". However, Lagrange's use of the term "values of a rational fraction" will be retained in the sequel since it

deg $\Phi \geqslant m$.

Proof : Let $f_1, f_2, \ldots, f_m$ be the various values of f obtained by permutation of $X_1, \ldots, X_n$ (with $f = f_1$, say), and let

$$\Theta(Y) = (Y-f_1) \ldots (Y-f_m).$$

Since every permutation of $X_1, \ldots, X_n$ permutes the f_i's among themselves, the coefficients of Θ, which are the elementary symmetric polynomials in $f_1, \ldots, f_m$, are not altered when the indeterminates $X_1, \ldots, X_n$ are permuted. Hence, these coefficients are symmetric in $X_1, \ldots, X_n$ and the equation $\Theta = 0$ satisfies the required properties.
If $\Phi(Y)$ is another polynomial with symmetric coefficients such that $\Phi(f) = 0$, then, for any $i=1, \ldots, m$, the permutation of $X_1, \ldots, X_n$ which gives to f the value f_i transforms $\Phi(f)$ into $\Phi(f_i)$, since it does not change the coefficients of Φ. Thus,

$$\Phi(f_i) = 0 \qquad \text{for } i=1, \ldots, m$$

whence Φ has m different roots and it follows that deg $\Phi \geqslant m$ (and, in fact, Φ is divisible by Θ, by proposition 5.15). □

For instance, the polynomial $X_1 + \omega X_2 + \omega^2 X_3$ takes six different values by permutation of X_1, X_2, X_3 and is therefore a root of an equation of degree 6 with symmetric coefficients, and of no equation of smaller degree; but $(X_1 + \omega X_2 + \omega^2 X_3)^3$ takes only two different values, whence it is a root of a quadratic equation.

After Cardano's method, Lagrange investigates Tschirnhaus' method : if the change of variable

$$Y = b_0 + b_1 X + X^2$$

---

is more suggestive and should not cause any confusion.

transforms a given cubic equation in X with roots x_1, x_2, x_3 into an equation like

$$y^3 = c,$$

which has as roots $\sqrt[3]{c}$, $\omega\sqrt[3]{c}$ and $\omega^2\sqrt[3]{c}$ (where, as above, ω denotes a cube root of unity other than 1), then one can assume :

$$x_1^2 + b_1 x_1 + b_0 = \sqrt[3]{c}$$
$$x_2^2 + b_1 x_2 + b_0 = \omega\sqrt[3]{c} \qquad (14)$$
$$x_3^2 + b_1 x_3 + b_0 = \omega^2\sqrt[3]{c}.$$

Multiplying the second equation by ω, the third by ω^2 and adding to the first, one obtains :

$$(x_1^2 + \omega x_2^2 + \omega^2 x_3^2) + b_1(x_1 + \omega x_2 + \omega^2 x_3) + b_0(1+\omega+\omega^2) = \sqrt[3]{c}(1+\omega+\omega^2).$$

Since $1 + \omega + \omega^2 = 0$, it follows that

$$b_1 = -\frac{x_1^2 + \omega x_2^2 + \omega^2 x_3^2}{x_1 + \omega x_2 + \omega^2 x_3}. \qquad (15)$$

This rational fraction takes only two values under the permutations of x_1, x_2, x_3 :

$$b_1 = -\frac{x_1^2 + \omega x_2^2 + \omega^2 x_3^2}{x_1 + \omega x_2 + \omega^2 x_3} \quad \text{and} \quad b_1' = -\frac{x_1^2 + \omega^2 x_2^2 + \omega x_3^2}{x_1 + \omega^2 x_2 + \omega x_3}.$$

Therefore, b_1 can be determined by solving the quadratic equation :

$$Y^2 - (b_1 + b_1')Y + b_1 b_1' = 0$$

whose coefficients are symmetric in x_1, x_2, x_3, whence can be calculated from the coefficients of the proposed equation. On the other hand, adding the equations (14) and taking into account the fact that $1 + \omega + \omega^2 = 0$, we obtain :

$$(x_1^2 + x_2^2 + x_3^2) + b_1(x_1 + x_2 + x_3) + 3b_0 = 0.$$

Since $x_1^2 + x_2^2 + x_3^2$ and $x_1 + x_2 + x_3$ are symmetric in x_1, x_2, x_3, they can be calculated from the coefficients of the proposed equation; this last equation then shows that b_0 can be rationally calculated from b_1 and the coefficients of the proposed equation. Similarly, multiplying the equations (14), we obtain :

$$b = (x_1^2 + b_1 x_1 + b_0)(x_2^2 + b_1 x_2 + b_0)(x_3^2 + b_1 x_3 + b_0)$$

and since this expression is symmetric in x_1, x_2, x_3, it can be rationally calculated from b_1, b_0 and the coefficients of the proposed equation. Once b_0, b_1 and c have been calculated, the roots x_1, x_2 and x_3 can be rationally calculated as explained in § 6.4. Therefore, Tschirnhaus' method for the cubic equation requires only the solution of a quadratic equation, beyond rational calculations; ultimately, this is because the rational fraction b_1 in (15) takes only two values.

Thereafter, Lagrange successively scrutinizes Euler and Bezout's methods, and the various methods for equations of degree 4. Each time, he shows how the roots of the auxiliary equations can be expressed in terms of those of the proposed equation, and he observes that the number of values of these expressions is less than the degree of the proposed equation.

For degree 4, Ferrari's method reduces the quartic equation

$$x^4 + px^2 + qx + r = 0$$

to

$$[x^2 + (p/2) + u]^2 = [\sqrt{2u}\, x - (q/2\sqrt{2u})]^2$$

by the choice of a suitable u (see § 3.2). This last equation splits into two quadratic equations:

$$x^2 + (p/2) + u = [\sqrt{2u}\, x - (q/2\sqrt{2u})]$$

and

$$x^2 + (p/2) + u = -[\sqrt{2u}\, x - (q/2\sqrt{2u})],$$

which yield the roots x_1, x_2, x_3, x_4 of the proposed quartic equation. Renumbering the roots if necessary, we may assume that x_1 and x_2 are the roots of the first quadratic equation and that x_3 and x_4 are the roots of the second. Then, since the independent term is the product of the roots, it follows that:

$$x_1 x_2 = (p/2) + u + (q/2\sqrt{2u})$$

and

$$x_3 x_4 = (p/2) + u - (q/2\sqrt{2u})$$

whence

$$x_1 x_2 + x_3 x_4 = p + 2u.$$

Since p is the coefficient of x^2 in the quartic equation, it is the second symmetric polynomial in x_1, x_2, x_3, x_4:

$$p = x_1 x_2 + x_1 x_3 + x_1 x_4 + x_2 x_3 + x_2 x_4 + x_3 x_4;$$

replacing p by its expression in the preceding equation,

the following value of u is found :

$$u = -(x_1+x_2)(x_3+x_4)/2.$$

This expression takes only three values when x_1, x_2, x_3 and x_4 are permuted : this explains why u is a root of a cubic equation.

Lagrange concludes his review with the attempted applications of Tschirnhaus, Euler and Bezout's methods to equations of higher degree. He then observes that, according to Bezout's method (see § 1), the roots of the proposed equation of degree n are obtained in the form $a_0 + a_1\omega + a_2\omega^2 + \ldots + a_{n-1}\omega^{n-1}$, where ω runs over the set of n-th roots of unity. If ζ is a primitive n-th root of unity, then by proposition 7.14 the n-th roots of unity are $1, \zeta, \zeta^2, \ldots, \zeta^{n-1}$; replacing ω successively by $1, \zeta, \zeta^2, \ldots, \zeta^{n-1}$, we obtain the following expressions of the roots :

$$x_1 = a_0 + a_1 + a_2 + \ldots + a_{n-1}$$
$$x_2 = a_0 + a_1\zeta + a_2\zeta^2 + \ldots + a_{n-1}\zeta^{n-1}$$
$$x_3 = a_0 + a_1\zeta^2 + a_2\zeta^4 + \ldots + a_{n-1}\zeta^{2(n-1)}$$
$$\ldots\ldots$$
$$x_n = a_0 + a_1\zeta^{n-1} + a_2\zeta^{2(n-1)} + \ldots + a_{n-1}\zeta^{(n-1)^2}$$

whence, in general :

$$x_i = \sum_{j=0}^{n-1} a_j \zeta^{(i-1)j} \qquad \text{for } i=1,\ldots,n. \tag{16}$$

This system is easily solved for $a_0, a_1, \ldots, a_{n-1}$: to obtain the value of a_k, it suffices to multiply each of these equations by a suitable power of ζ so that the coefficient of a_k be 1, and to add the equations thus produced. We obtain :

$$\Sigma_{i=1}^{n} \zeta^{-(i-1)k} x_i = \Sigma_{j=0}^{n-1} a_j (\Sigma_{i=1}^{n} \zeta^{(j-k)(i-1)}). \qquad (17)$$

If $j \neq k$, then ζ^{j-k} is an n-th root of unity different from 1; therefore, ζ^{j-k} is a root of

$$(X^n-1)/(X-1) = X^{n-1} + X^{n-2} + \ldots + X + 1,$$

whence

$$\Sigma_{i=1}^{n} \zeta^{(j-k)(i-1)} = 0.$$

Therefore, in the right-hand side of (17), all the terms vanish except the term corresponding to the index $j = k$, which is $n.a_k$. Thus, equation (17) yields :

$$a_k = (1/n)(\Sigma_{i=1}^{n} \zeta^{-(i-1)k} x_i). \qquad (18)$$

It is easily seen that, $x_1,\ldots,x_n$ being considered as independent indeterminates, all the values of a_k obtained by the various permutations of $x_1,\ldots,x_n$ are distinct. Therefore, a_k is a root of an equation of degree n!. However, Lagrange shows[*] that a_k^n takes only (n-1)! values. Moreover, if n is prime, then a_k^n is a root of an equation of degree (n-1) whose coefficients can be determined from the solution of a single equation of degree (n-2)!. Thus, for n = 5, the determination of a_k^5 still requires the solution of an equation of degree 3! = 6.

If n is not prime, the result is more complicated. Using the same arguments as in the case where n is prime, Lagrange shows that if n = p.q with p prime, and if k is divisible by q, then a_k^p is a root of an equation of degree p - 1 whose

---

[*] Lagrange's arguments are elementary, but can be more easily explained when they are related to some subsequent results of Lagrange and with the help of an appropriate notation. Therefore, we postpone the proof of this result to the next section.

coefficients depend on a single equation of degree $n!/(p-1)p(q!)^p$. For $n = 4$, it follows that a_2^2 can be found by solving an equation of degree $4!/1.2.(2!)^2 = 3$, but for $n = 6$ the determination of a_3^2 requires the solution of an equation of degree $6!/1.2.(3!)^2 = 10$.

These results led Lagrange to doubt the possibility of solving algebraically the general equations of degree 5 or higher, although he cautiously avoided to reject this possibility too categorically. The conclusion of his investigations of previously known methods is given in the Article 86 [L, pp. 355-357], which is quoted here extensively to give an idea of Lagrange's leisurely style of writing :

"As should be clear from the analysis that we have just given of the main known methods for the solution of equations, all these methods reduce to the same general principle, namely to find functions of the roots of the proposed equation which are such : 1° that the equation or the equations by which they are given, i.e. of which they are the roots (equations that are usually called the <u>reduced</u> equations), happen to be of a degree smaller than that of the proposed equation, or at least decomposable in other equations of a degree smaller than this one; 2° that the values of the sought roots can be easily deduced from them.

The art of solving equations thus consists of discovering functions of the roots which have the above mentioned properties; but is it always possible to find such functions, for equations of any degree, i.e. for any number of roots ? That is a question which seems very difficult to decide in general.

As to equations which do not exceed the fourth degree, the simplest functions which yield their solution can be represented by the general formula :

$$x_1 + \omega x_2 + \omega^2 x_3 + \ldots + \omega^{n-1} x_n,$$

$X_1, X_2, X_3, \ldots, X_n$ being the roots of the proposed equation, which is assumed of degree n, and ω being an arbitrary root other than 1 of the equation

$$\omega^n - 1 = 0,$$

i.e. an arbitrary root of the equation

$$\omega^{n-1} + \omega^{n-2} + \omega^{n-3} + \ldots + 1 = 0,$$

as follows from what has been shown in the first two sections about the solution of equations of the third and the fourth degree. (...)

It thus seems that one could conclude from this by induction that every equation of any degree will also be solvable with the help of a reduced equation whose roots are represented by the same formula :

$$X_1 + \omega X_2 + \omega^2 X_3 + \omega^3 X_4 + \ldots .$$

However, after what has been proved in the preceding section about the methods of MM. Euler and Bezout, which readily lead to such reduced equations, one has, it seems, the occasion to convince oneself beforehand that this conclusion will be wrong from the fifth degree on; hence it follows that, if the algebraic solution of the equations of degree higher than four is not impossible, it must rely on some functions of the roots, other than the above".

The polynomials of the form :

$$X_1 + \omega X_2 + \omega^2 X_3 + \ldots + \omega^{n-1} X_n,$$

where ω is an n-th root of unity, were subsequently christened : "<u>Lagrange resolvents</u>". As we have seen, they originate from the works of Euler and Bezout, and it will be clear later that they play a prominent role in Galois' theory of equations. For the convenience of later references, we recapitulate the formula which yields the roots of equations of any degree n in terms of Lagrange resolvents :

2. FORMULA : If $t(\omega)$ denotes the Lagrange resolvent

$$t(\omega) = X_1 + \omega X_2 + \omega^2 X_3 + \ldots + \omega^{n-1} X_n$$

(ω an n-th root of unity), then for $i=1,\ldots,n$:

$$X_i = \frac{1}{n} [\Sigma_\omega \, \omega^{-(i-1)} t(\omega)],$$

where the sum runs over all the n-th roots of unity.

This was shown above : see equations (16) and (18).

§ 3. First results of group theory and Galois theory

In the final section of his paper, Lagrange draws from his investigations some general conclusions concerning the degree of the equations by which functions of the roots of a given equation can be determined. Proposition 1 above is a first instance of Lagrange's observations, but in his conclusions Lagrange goes much farther. In effect, he begins to calculate with permutations of the roots, obtaining first results in group theory and Galois theory.

Remarkably enough, these results were achieved without even devising a notation for permutations, which were indeed very new objects to calculate with. Unfortunately, this makes Lagrange's arguments sometimes hard to follow. To facilitate our exposition of Lagrange's results, we shall not refrain from using modern notations. Thus, for any integer $n \geqslant 1$, we denote by S_n the symmetric group on $\{1,\ldots,n\}$, i.e. the group of permutations of $1,\ldots,n$; for $\sigma \in S_n$ and for any rational fraction f in n indeterminates

$X_1, \ldots, X_n$, we set :

$$\sigma(f(X_1, \ldots, X_n)) = f(X_{\sigma(1)}, \ldots, X_{\sigma(n)}).$$

So, S_n can be considered as the group of permutations of $X_1, \ldots, X_n$, and S_n acts on the rational fractions in $X_1, \ldots, X_n$ by permuting the indeterminates. For any rational fraction f, we denote by I(f) the subgroup of permutations $\sigma \in S_n$ which leave f invariant, (sometimes called the isotropy group of f), i.e. :

$$I(f) = \{\sigma \in S_n \mid \sigma(f(X_1, \ldots, X_n)) = f(X_1, \ldots, X_n)\}$$

and we generally denote by |E| the number of elements of a finite set E, which is called the order of E, if E is a group. Thus, for instance,

$$|S_n| = n!.$$

In his article 97, Lagrange proves the following theorem :

3. THEOREM : Let $f = f(X_1, \ldots, X_n)$ be a rational fraction in n indeterminates. The number m of different values that f takes under the permutations of $X_1, \ldots, X_n$ is equal to the quotient of n! by the number of permutations which leave f invariant :

$$m = n!/|I(f)|.$$

Proof : Let $f_1, \ldots, f_m$ be the various values of f (with $f = f_1$, say). For $i=1, \ldots, m$, let $I(f \to f_i)$ be the set of permutations $\sigma \in S_n$ such that

$$\sigma(f) = f_i;$$

thus,

$$I(f \to f_1) = I(f).$$

Now, fix some $i=1,\ldots,m$ and some $\sigma \in I(f \to f_i)$. If $\tau \in I(f)$, then

$$\sigma\tau(f) = \sigma(f) = f_i,$$

whence $\sigma\tau \in I(f \to f_i)$. Conversely, if $\rho \in I(f \to f_i)$, then $\sigma^{-1}\rho \in I(f)$. Since

$$\rho = \sigma(\sigma^{-1}\rho),$$

it follows that every element in $I(f \to f_i)$ is of the form $\sigma\tau$ where $\tau \in I(f)$. Therefore, multiplication on the left by σ defines a bijection from $I(f)$ onto $I(f \to f_i)$; hence

$$|I(f)| = |I(f \to f_i)|.$$

Since every permutation in S_n maps f onto one of its values $f_1,\ldots,f_m$, we have a decomposition of S_n as a disjoint union :

$$S_n = \cup_{i=1}^m I(f \to f_i),$$

whence

$$|S_n| = \Sigma_{i=1}^m |I(f \to f_i)|.$$

Since $|S_n| = n!$ and since each term in the right-hand side is equal to $|I(f)|$, this last equation yields :

$$n! = m.|I(f)|. \qquad \square$$

Here now are Lagrange's own words; to understand this quotation, one needs to know that Θ is the polynomial :

$$\Theta(t) = \Pi_{\sigma \in S_n}(t - \sigma(f(X_1,\ldots,X_n))),$$

whose roots are the values of f under the permutations of
$X_1,\ldots,X_n$ (compare proposition 1). Lagrange denotes the
roots by x',x'',x''',... instead of $X_1,X_2,\ldots$; the number of
roots is denoted by μ instead of n.

97. Quoique l'équation $\Theta = 0$ doive être en général du degré $1.2.3 - - - \mu = \pi$, qui eſt égal au nombre des permutations dont les μ racines x', x'', x''' &c. font fufceptibles; cependant s'il arrive que la fonction propofée foit telle qu'elle ne reçoive aucun changement par quelqu'une ou quelques-unes de ces permutations, alors l'équation dont il s'agit s'abaiſſera néceſſairement à un degré moindre.

Car fuppofons, par exemple, que la fonction $f\colon (x')\ (x'')\ (x''')\ (x^{IV}) - - -$ foit telle qu'elle conferve la même valeur en échangeant x' en x'', x'' en x''', & x''' en x', en forte que l'on ait

$$f\colon (x')\ (x'')\ (x''')\ (x^{IV}) - - - = f\colon (x'')\ (x''')\ (x')\ (x^{IV}) - - -,$$

il eſt clair que l'équation $\Theta = 0$ aura déjà deux racines égales; mais je vais prouver que dans cette hypothefe toutes les autres racines feront auſſi égales deux à deux. En effet, confidérons une racine quelconque de la même équation, laquelle foit repréfentée par la fonction $f\colon (x^{IV})\ (x''')\ (x')\ (x'') - - - -$, comme celle-ci dérive de la fonction $f\colon (x')\ (x'')\ (x''')\ (x^{IV}) - - - -$, en échangeant x' en x^{IV}, x'' en x''', x''' en x', x^{IV} en x'', il s'enfuit qu'elle devra garder auſſi la même valeur en y changeant x^{IV} en x''', x''' en x' & x' en x^{IV}; de forte qu'on aura auſſi

$$f\colon (x^{IV})\ (x''')\ (x')\ (x'') - - - = f\colon (x''')\ (x')\ (x^{IV})\ (x'').$$

Donc, dans ce cas, la quantité Θ fera égale à un carré θ^2, & par conféquent l'équation $\Theta = 0$ fe réduira à celle-ci, $\theta = 0$, dont la dimenſion fera $\frac{\pi}{2}$.

On démontrera de la même maniere que, fi la fonction $f\colon (x')\ (x'')\ (x''')\ (x^{IV}) - - -$ eſt de fa propre nature telle qu'elle conferve la même valeur en faifant deux ou trois ou un plus grand nombre de permutations différentes entre les racines x', x'', x''', x^{IV} &c., les racines de l'équation $\Theta = 0$ feront égales trois à trois ou quatre à quatre ou &c.; en forte que la quantité Θ fera égale à un cube θ^3 ou à un carré-carré θ^4 ou &c., & que par conféquent l'équation $\Theta = 0$ fe réduira à celle-ci $\theta = 0$, dont le degré fera $= \frac{\pi}{3}$ ou $= \frac{\pi}{4}$ ou &c.

[L, Art. 97] (Univ. Cath. Louvain, Centre général de documentation

This quotation translates as follows, changing slightly Lagrange's notation to suit the notation of this section :

"Although the equation $\theta = 0$ must in general be of degree $1.2.3. \ldots n = n!$, which is equal to the number of permutations of $X_1, \ldots, X_n$, yet if it happens that the function be such that it does not receive any change by some or several permutations, then the equation in question will necessarily reduce to a smaller degree.

Assume, for instance, that the function $f(X_1, X_2, X_3, X_4, \ldots)$ be such that it keeps the same value when X_1 is changed into X_2, X_2 into X_3 and X_3 into X_1, so that :

$$f(X_1, X_2, X_3, X_4, \ldots) = f(X_2, X_3, X_1, X_4, \ldots),$$

it is clear that the equation $\theta = 0$ will already have two equal roots; but I am going to prove that with this hypothesis all the other roots will be pairwise equal too. Indeed, let us consider an arbitrary root of the same equation, such as :

$$f(X_4, X_3, X_1, X_2, \ldots).$$

As this one comes from the function

$$f(X_1, X_2, X_3, X_4, \ldots)$$

by changing X_1 into X_4, X_2 into X_3, X_3 into X_1, X_4 into X_2, it follows that it will have to keep the same value when we change in it X_4 into X_3, X_3 into X_1 and X_1 into X_4; so that we will also have :

$$f(X_4, X_3, X_1, X_2, \ldots) = f(X_3, X_1, X_4, X_2, \ldots).$$

Therefore, in this case, the quantity θ will be equal to a square θ^2 and consequently the equation $\theta = 0$ will be reduced to this one $\theta = 0$, which will have dimension $n!/2$.

Likewise, one proves that if the function

$$f(X_1, X_2, X_3, X_4, \ldots)$$

is by its own nature such that it keeps the same value when two or three or a greater number of different permutations are made among the roots $X_1, X_2, X_3, X_4, \ldots, X_n$, the roots of the equation $\Theta = 0$ will be equal three by three or four by four or, etc.; so that the quantity Θ will be equal to a cube θ^3 or to a square-square θ^4 or, etc., and therefore the equation $\Theta = 0$ will reduce to this one $\theta = 0$, whose degree will be equal to $n!/3$, or equal to $n!/4$, or, etc."

To see the link between Lagrange's argument in the quotation above and the preceding proof, denote $f = f(X_1, X_2, X_3, X_4, \ldots)$ and $f_i = f(X_4, X_3, X_1, X_2, \ldots)$, and let σ be the permutation : $X_1 \to X_4$, $X_2 \to X_3$, $X_3 \to X_1$, $X_4 \to X_2, \ldots$ so that

$$\sigma(f) = f_i \quad \text{or} \quad \sigma \in I(f \to f_i).$$

Lagrange's observation is that if $\tau : X_1 \to X_2 \to X_3 \to X_1$ is in $I(f)$, then

$$\sigma\tau : X_1 \to X_3, \; X_2 \to X_1, \; X_3 \to X_4, \; X_4 \to X_2, \ldots$$

is such that

$$\sigma\tau(f) = f_i, \text{ i.e. } \sigma\tau \in I(f \to f_i).$$

This is indeed the crucial step in the proof.

The theorem which is often referred to as "Lagrange's theorem" nowadays deals with the order (i.e. the number of elements) of subgroups of a group :

4. THEOREM : Let H be a subgroup of a finite group G. Then $|H|$ divides $|G|$.

Proof : For $g \in G$, define the (left) coset gH by :

$$gH = \{gh \mid h \in H\}.$$

We readily have

$$|gH| = |H|,$$

since the multiplication by g defines a bijection from H onto gH. Since $g = g.1 \in gH$, it is clear that every element of G is in some coset. Moreover, if two cosets have a common element, then they are equal : indeed, if there exists an element $x \in G$ such that

$$x \in g_1H \cap g_2H,$$

let $\quad x = g_1h_1 = g_2h_2 \qquad$ for some $h_1, h_2 \in H$.

Then every element $g_1h \in g_1H$ can be written as

$$g_1h = g_2(h_2h_1^{-1}h) \in g_2H,$$

so that $g_1H \subset g_2H$. Interchanging the indices 1 and 2, we obtain $g_2H \subset g_1H$, whence

$$g_1H = g_2H.$$

This shows that the group G decomposes into a disjoint union of cosets. Since the number of elements of each of these cosets is equal to $|H|$, it follows that $|H|$ divides $|G|$ (and the quotient of $|G|$ by $|H|$ is the number of different cosets in a decomposition of G, which is called the <u>index</u> of H in G). □

Although the pattern of this proof is quite similar to that of theorem 3 (observe that $I(f \to f_i)$ is a left coset of

$I(f)$), Lagrange did not reach this generality, nor did he need to. His primary concern was to obtain some information on the number of values of functions, whence, by proposition 1 above, on the degree of the equation by which a given function of the roots can be determined.

In this respect, his achievement is even more stunning : he proves a "relative version" of proposition 1 above, which can be seen as a part of the fundamental theorem of Galois theory for the splitting field of the general polynomial :

"Now, as soon as the value of a given function of the roots $X_1,\ldots,X_n$ has been found, either by the solution of the equation $\theta = 0$ or otherwise, I claim that the value of another arbitrary function of the same roots can be found, and that, generally speaking, simply by a linear equation, except some particular cases which demand an equation of the second degree, or of the third, etc. This Problem seems to me to be one of the most important of the theory of equations, and the general Solution that we are going to give will shed a new light on this part of Algebra" [L, Art. 100].

Lagrange's result can be stated more precisely as follows :

5. <u>THEOREM</u> : Let f and g be two rational fractions in n indeterminates $X_1,\ldots,X_n$. If f takes m different values by the permutations which leave g invariant, then f is a root of an equation of degree m whose coefficients are rational fractions in g and in the elementary symmetric polynomials $s_1,\ldots,s_n$.

In particular, if f is invariant by the permutations which leave g invariant, then f is a rational fraction in g and $s_1,\ldots,s_n$.

Proof: We begin with the special case above, i.e. we first assume $m = 1$. Then, let $g_1, \ldots, g_r$ be the different values of g under the permutations of $X_1, \ldots, X_n$ (with $g_1 = g$, say) and let $f_1 = f, f_2, \ldots, f_r$ be the corresponding values of f, in the sense that if a permutation of $X_1, \ldots, X_n$ gives to g the value g_i (for some i), then it gives to f the value f_i. The possibility of defining such a correspondence from the values of g to the values of f comes from the hypothesis that f is invariant under the permutations which leave g invariant. Indeed, this hypothesis means that $I(g) \subset I(f)$; thus, if σ and ρ both give to g the value g_i, then the proof of theorem 3 shows that $\rho = \sigma\tau$ for some $\tau \in I(g)$, and the hypothesis ensures that $\tau \in I(f)$, whence $\rho(f) = \sigma(f)$. We may therefore define $f_i = \sigma(f)$ for <u>any</u> $\sigma \in S_n$ such that $\sigma(g) = g_i$.

Consider then the following expressions, which are denoted by $a_0, \ldots, a_{r-1}$:

$$a_0 = f_1 + f_2 + \ldots + f_r$$
$$a_1 = f_1 g_1 + f_2 g_2 + \ldots + f_r g_r$$
$$a_2 = f_1 g_1^2 + f_2 g_2^2 + \ldots + f_r g_r^2 \qquad (19)$$
$$\ldots\ldots$$
$$a_{r-1} = f_1 g_1^{r-1} + f_2 g_2^{r-1} + \ldots + f_r g_r^{r-1}.$$

From the definition of f_i and g_i, it follows that every permutation of $X_1, \ldots, X_n$ merely permutes the terms in $a_0, \ldots, a_{r-1}$, so that each of these expressions is symmetric in $X_1, \ldots, X_n$ and can therefore be calculated as rational fractions in $s_1, \ldots, s_n$, by the fundamental theorem of symmetric fractions 8.3.

Now, the idea is to solve the system (19) for $f_1, \ldots, f_r$. However, the usual elimination method would yield f_1 in terms of $a_0, \ldots, a_{r-1}$ (thus, eventually, in terms of $s_1, \ldots,$

s_n) but also of $g_1,\ldots,g_r$, while we need an expression of f_1 in terms of $s_1,\ldots,s_n$ and g_1 only.

Lagrange then uses the following trick : let

$$\theta(Y) = (Y-g_1)(Y-g_2)\ldots(Y-g_r) = Y^r + b_{r-1}Y^{r-1} + \ldots + b_o.$$

Dividing this polynomial by $Y-g_1$, we obtain :

$$\psi(Y) = (Y-g_2)\ldots(Y-g_r) = Y^{r-1} + c_{r-2}Y^{r-2} + \ldots + c_o$$

and the coefficients $c_o, c_1, \ldots, c_{r-2}$ are rational fractions in $b_o, b_1, \ldots, b_{r-1}$ and g_1, as is easily seen by carrying out explicitly the division of $\theta(Y)$ by $Y-g_1$. Now, the coefficients $b_o, \ldots, b_{r-1}$ of θ are symmetric in $g_1, \ldots, g_r$ whence also in $X_1, \ldots, X_n$, and can therefore be calculated in terms of $s_1, \ldots, s_n$; thus, $c_o, c_1, \ldots, c_{r-2}$ are rational fractions in g_1 and $s_1, \ldots, s_n$.
Multiplying the first equation of (19) by c_o, the second by c_1, the third by c_2, etc. and the last by 1, and adding the equations thus obtained, we obtain an equation in which the coefficient of f_i (for $i=1,\ldots,r$) is the polynomial $\psi(Y)$ calculated at $Y = g_i$:

$$a_o c_o + a_1 c_1 + \ldots + a_{r-1} = f_1 \psi(g_1) + f_2 \psi(g_2) + \ldots + f_r \psi(g_r)$$

Since $\psi(g_2) = \ldots = \psi(g_r) = 0$, we thus end up with an expression of f_1 as a rational fraction in g and $s_1, \ldots, s_n$:

$$f_1 = (a_o c_o + a_1 c_1 + \ldots + a_{r-1}) \psi(g_1)^{-1},$$

as was required.
This proves theorem 5 in the special case where $m = 1$, but the general case follows easily : indeed, assume now that f takes m values $f_1, \ldots, f_m$ under the permutations which leave g invariant. Then f is a root of the equation :

$$(Y-f_1)\ldots(Y-f_m) = 0,$$

and this equation satisfies the required properties since its coefficients are symmetric in $f_1,\ldots,f_m$, whence invariant under the permutations which leave g invariant, whence, by the special case above, rational fractions in g and $s_1,\ldots,s_n$. □

Lagrange's result is even more general than the above, since he also considers the case where $X_1,\ldots,X_n$ are related by some algebraic relations (this occurs when $X_1,\ldots,X_n$ are the roots of particular equations instead of the general equation of degree n), but the theorem above gives the flavour of Lagrange's proof and covers the essential of the applications that Lagrange has in mind, since his purpose was to investigate the solution of general equations.

After Lagrange's preceding theorem 5, the solution of general equations is much enlightened indeed. The strategy appears as follows : to solve the general equation of degree n, one has to find a (finite) sequence of rational fractions $V_0,V_1,\ldots,V_r$ in n indeterminates $X_1,\ldots,X_n$ such that the first function V_0 is symmetric in $X_1,\ldots,X_n$, the last function V_r is one of the roots, say $V_r = X_1$, and for $i=1,\ldots,r$, the function V_i satisfies either :
  1) $V_i^n = V_{i-1}$, or
  2) the number of values of V_i under the permutations which leave V_{i-1} invariant is strictly less than n.
In case (1), the function V_i can be calculated from the preceding ones by extraction of an n-th root and in case (2) it can be found by solving an equation of degree less than n, by theorem 5. Since the last function is a root of the proposed equation, it means that a root can be found by successive extractions of roots and solutions of equations of lower degree. The sequence $V_0,V_1,\ldots,V_r$ indicates in which order the calculations can be arranged. The other roots can be found likewise, replacing $V_0,V_1,\ldots,V_r$ by similar functions. (More precisely,

for any $\sigma \in S_n$, the root $\sigma(V_r)$ can be found by the sequence $\sigma(V_0) = V_0$, $\sigma(V_1), \ldots, \sigma(V_r))$. For $n = 2$, one chooses :

$$V_0 = (X_1 - X_2)^2$$

$$V_1 = (X_1 - X_2)$$

$$V_2 = X_1.$$

Thus, V_1 can be found by extracting the square root of V_0, and V_2 can then be found rationally, since
$V_2 = \frac{1}{2}(V_1 + (X_1+X_2))$.
For $n = 3$, one can choose for V_0 any symmetric function, next

$$V_1 = (X_1+\omega X_2+\omega^2 X_3)^3 \quad \text{(where } \omega \text{ is a cube root of unity other than 1)}$$

$$V_2 = (X_1+\omega X_2+\omega^2 X_3)$$

$$V_3 = X_1.$$

Since V_1 takes only two values by all the permutations of X_1, X_2, X_3, it can be found by solving a quadratic equation; next, V_2 is found by extracting a cube root of V_1 and finally, since V_3 is invariant under the permutations which leave V_2 invariant (as only the identity leaves V_2 invariant), it can be determined rationally from V_2, i.e. by solving an equation of degree 1. Likewise, X_2 and X_3 can be determined rationally from V_2.
For $n = 4$, one can choose for V_0 any symmetric function, next

$$V_1 = (X_1+X_2)(X_3+X_4)$$

$$V_2 = (X_1+X_2)$$

$$V_3 = X_1.$$

Indeed, V_1 can be determined by solving a cubic equation since it takes only three values by the permutations of X_1, X_2, X_3, X_4; next, V_2 can be determined by a quadratic equation since it takes only two values under the permutations which leave V_1 invariant, and finally V_3 can be found by a quadratic equation since it takes only two values under the permutations which leave V_2 invariant. The root X_2 is then readily found, since it is the other root of the quadratic equation which yields the value of V_3 (= X_1), and the other roots X_3 and X_4 are found by similar calculations: $X_3 + X_4$ is the other root of the equation which yields the value of V_2 and X_3, X_4 are the roots of a quadratic equation. This is the pattern which is suggested by Ferrari's solution of quartic equations. Of course, other choices are possible; for instance, using Lagrange's resolvents, one could choose :

$$V_1 = (X_1 - X_2 + X_3 - X_4)^2$$

$$V_2 = (X_1 - X_2 + X_3 - X_4)$$

$$V_3 = X_1.$$

The first function V_1 is the root of a cubic equation, and V_2, V_3 are obtained by solving successively two quadratic equations.

"These are, if I am not mistaken, the genuine principles of the solution of equations and the analysis which is most suitable to lead to it; everything is reduced, as is seen, to a kind of calculus of combinations, by which the results to which one is led are found <u>a priori</u>. It would be opportune to apply it to the equations of the fifth degree and higher degrees, whose solution is so far unknown; but this application requires a too large amount of researches and combinations, whose success is, for that matter, still very dubious, for us to tackle this problem now; we hope however to come back to it at another time, and we will be content to have here set the foundations of a theory which seems to us new and general" [L, Art. 109].

To conclude this chapter, we now apply the results above to sketch a proof of the properties of "Lagrange resolvents", which we have pointed out in § 2, at least for the case where n is prime. We will need a result on the existence of rational fractions which are invariant under a prescribed group :

6. PROPOSITION : For any subgroup G in S_n, there exists a rational fraction f in n indeterminates such that $I(f) = G$.

Proof : Choose a monomial m which is not invariant under any (non-trivial) permutation of the indeterminates, for instance $m = X_1 X_2^2 X_3^3 \ldots X_n^n$, and let

$$f = \Sigma_{\sigma \in G}\, \sigma(m).$$

Since for any $\tau \in G$ the set of products $\{\tau\sigma \mid \sigma \in G\}$ is G, it follows that

$$\Sigma_{\sigma \in G}\, \tau\sigma(m) = \Sigma_{\sigma \in G}\, \sigma(m),$$

whence

$$\tau(f) = f \qquad \text{for any } \tau \in G.$$

Therefore,

$$G \subset I(f).$$

On the other hand, if $\rho \notin G$, then the monomial $\rho(m)$ appears in $\rho(f)$ but not in f, so $\rho(f) \neq f$. Thus,

$G = I(f)$. □

Henceforth, to simplify notations a little, we index the indeterminates from 0. We shall thus consider S_n as the group of permutations of $\{0,1,\ldots,n-1\}$, and we now define certain permutations which have interesting properties in relation with Lagrange's resolvents.

For any integer k relatively prime to n, we denote by σ_k the map from $\{0,1,\ldots,n-1\}$ into itself defined as follows: for any $i \in \{0,1,\ldots,n-1\}$, the image $\sigma_k(i)$ is the unique integer j between 0 and n-1 such that

ik - j is divisible by n

(i.e. : ik ≡ j (mod n), using subsequent notations); in other words, j is the remainder of the division of ik by n.

7. PROPOSITION : For any integer k relatively prime to n, the map σ_k is a permutation of $\{0,1,\ldots,n-1\}$.

<u>Proof</u> : By theorem 7.11, it is possible to find integers ℓ,m such that

$$k\ell + mn = 1. \qquad (20)$$

For any $i \in \{0,1,\ldots,n-1\}$, the definition of $\sigma_k(i)$ shows that

$ik - \sigma_k(i)$ is divisible by n,

hence

$ik\ell - \sigma_k(i)\ell$ is divisible by n,

and, adding imn, which is clearly divisible by n, we see that :

$i(k\ell+mn) - \sigma_k(i)\ell$ is divisible by n.

By (20), it follows that

$i - \sigma_k(i)\ell$ is divisible by n. (21)

This last relation means that $\sigma_\ell(\sigma_k(i)) = i$, so that $\sigma_\ell \sigma_k$ is the identity on $\{0,1,\ldots,n-1\}$. Interchanging k and ℓ in the above discussion, it follows that $\sigma_k \sigma_\ell$ also is the identity on $\{0,1,\ldots,n-1\}$. Therefore, σ_ℓ and σ_k are reciprocal bijections of $\{0,1,\ldots,n-1\}$ onto itself. □

From now on, we assume that <u>n is a prime number</u>, so that σ_i is defined for any $i=1,\ldots,n-1$. We denote by τ the cyclic permutation : $0 \to 1 \to 2 \to \ldots \to n-1 \to 0$ and by GA(n) the subgroup of S_n generated by $\sigma_1,\ldots,\sigma_{n-1}$ and τ. It can be shown that

$|GA(n)| = n(n-1)$

(see exercise 5); in fact, from a less elementary point of view, GA(n) can be identified to the group of affinities of the affine line over the field with n elements : τ generates the group of translations while $\sigma_1,\ldots,\sigma_{n-1}$ are homotheties.

Let V be a rational fraction in $X_0, X_1, \ldots, X_{n-1}$ such that I(V) = GA(n) (the existence of such a function is ensured by proposition 6) and, for any n-th root of unity ω, let $t(\omega)$ denote the following Lagrange resolvent :

$t(\omega) = X_0 + \omega X_1 + \omega^2 X_2 + \ldots + \omega^{n-1} X_{n-1}.$

<u>8. THEOREM</u> : Assume n is prime. If $\omega \neq 1$, then $t(\omega)^n$ is a root of an equation of degree n-1 whose coefficients are rational fractions in V and in the elementary symmetric polynomials in $X_0,\ldots,X_{n-1}$, and V is a root of an equation

of degree $(n-2)!$ whose coefficients are rational fractions in the elementary symmetric polynomials.

Proof : The fact that V is a root of an equation of degree $(n-2)!$ readily follows from proposition 1 and theorem 3, since $I(V)$ has $n(n-1)$ elements. To prove the rest, it suffices, by theorem 5, to show that $t(\omega)^n$ takes $n-1$ values by the permutations which leave f invariant, i.e. by the permutations in $GA(n)$.

First, we consider the action of σ_k :

$$\sigma_k(t(\omega)) = X_o + \omega X_{\sigma_k(1)} + \omega^2 X_{\sigma_k(2)} + \ldots + \omega^{n-1} X_{\sigma_k(n-1)}.$$

Since $\omega^n = 1$, relation (21) yields :

$$\omega^i = (\omega^\ell)^{\sigma_k(i)} \quad \text{for all } i=0,1,\ldots,n-1;$$

therefore,

$$\sigma_k(t(\omega)) = X_o + (\omega^\ell)^{\sigma_k(1)} X_{\sigma_k(1)} + (\omega^\ell)^{\sigma_k(2)} X_{\sigma_k(2)} + \ldots + (\omega^\ell)^{\sigma_k(n-1)} X_{\sigma_k(n-1)},$$

which shows :

$$\sigma_k(t(\omega)) = t(\omega^\ell). \tag{22}$$

Next, we consider the action of τ. Since

$$\tau(t(\omega)) = X_1 + \omega X_2 + \omega^2 X_3 + \ldots + \omega^{n-1} X_o,$$

we have

$$\tau(t(\omega)) = \omega^{-1} t(\omega),$$

for any n-th root of unity ω. Since $\omega^{-n}=1$, this last equation yields :

$$\tau(t(\omega)^n) = t(\omega)^n.$$

This result, together with (22), shows that under any product of the permutations $\sigma_1,\ldots,\sigma_{n-1},\tau$ the function $t(\omega)^n$ takes one of the values $t(\omega)^n, t(\omega^2)^n,\ldots,t(\omega^{n-1})^n$, which are pairwise different if $\omega \neq 1$. Since GA(n) is generated by $\sigma_1,\ldots,\sigma_{n-1},\tau$, this means that $t(\omega)^n$ takes n-1 values under the permutations in GA(n), and the proof is complete. □

Simpler arguments yield the number of values of $t(\omega)^n$:

9. PROPOSITION : The function $t(\omega)^n$ takes (n-1)! values under the permutations of $X_0,\ldots,X_{n-1}$.

Proof : Let k be the number of values of $t(\omega)^n$. At the end of the preceding proof, it was shown that $t(\omega)^n$ is invariant under τ, whence also under all the powers $\tau^2,\tau^3,\ldots,\tau^{n-1}$. Thus, $|I(t(\omega)^n)| \geq n$ and theorem 3 shows that

$$k \leq (n-1)!.$$

On the other hand, it follows from proposition 1 that $t(\omega)^n$ is a root of an equation $\theta(Y) = 0$ of degree k. Then, $t(\omega)$ is a root of $\theta(Y^n) = 0$, which has degree kn; but since $t(\omega)$ takes n! different values under the permutations of the variables, it cannot be a root of an equation of degree less than n!, by proposition 1. Therefore, kn $\geq$ n!, hence

$$k \geq (n-1)!.$$ □

This proposition remains valid with the same proof when n is not prime, since the permutations σ_k were not used.

Exercises for chapter 10

1) Let $u_1 = (X_1+X_2)(X_3+X_4)$ $u_2 = (X_1+X_3)(X_2+X_4)$

$$u_3 = (X_1+X_4)(X_2+X_3)$$

$v_1 = X_1X_2 + X_3X_4$ $v_2 = X_1X_3 + X_2X_4$

$$v_3 = X_1X_4 + X_2X_3.$$

Show that v_1, v_2 and v_3 are rational fractions in u_1, u_2, u_3 with symmetric coefficients. Use this result to show how the cubic equation which has as roots v_1, v_2, v_3 is related to the equation with roots u_1, u_2, u_3. (Compare exercise 3 of chapter 8).
Same questions with $w_1 = (X_1-X_2+X_3-X_4)^2$, $w_2 = (X_1+X_2-X_3-X_4)^2$, $w_3 = (X_1-X_2-X_3+X_4)^2$ instead of v_1, v_2, v_3.

2) Use the arguments in the proof of Lagrange's theorem 5 to express X_1X_2 as a rational fraction of X_1+X_2 with coefficients symmetric in the three indeterminates X_1, X_2, X_3. Is this expression unique?

3) Find all the polynomials $f = AX_1 + BX_2 + CX_3$ (with $A,B,C \in \mathbb{C}$) which have the property that X_1, X_2, X_3 can be rationally expressed from f with symmetric coefficients and such that f^3 takes only two values by the permutations of X_1, X_2, X_3.

4) Let n be a prime number. For any n-th root of unity ω, let

$$t(\omega) = X_0 + \omega X_1 + \ldots + \omega^{n-1} X_{n-1}.$$

Show that $t(\omega^k) \cdot t(\omega)^{-k}$ is a rational fraction in $t(\omega)^n$ with symmetric coefficients, for any integer k.

5) Let n be a prime number and let the notations be as in proposition 7 and after. Prove that $\tau\sigma_i = \sigma_i \tau^k$ for some k. Deduce that

$$GA(n) = \{\sigma_i \tau^j \mid i=1,\ldots,n-1 \text{ and } j=0,\ldots,n-1\}$$

and that $|GA(n)| = n(n-1)$.

6) Show that for any group G and any subgroup H, the map $g \to g^{-1}$ (which is an anti-automorphism of G) induces a bijection between the set of left cosets of H in G and the set of right cosets of H.

11 Vandermonde

§ 1. Introduction

Alexandre-Théophile Vandermonde (1735-1796) is not a mathematician of the same class as Lagrange or Euler. His contributions to mathematics were scarce and hardly influential. Ironically enough, he is most often remembered nowadays for a determinant which bears his name but is not to be found in his papers : Vandermonde determinants may have been so christened because someone misread indices for exponents [Le, pp. 206-207].

Nevertheless, his work, remarkably described by Lebesgue [Le], shows that brilliant ideas and deep insights come not only from first-class mathematicians. Several of Lagrange's ideas were indeed discovered simultaneously or perhaps even a little earlier by Vandermonde. Most notably, Vandermonde performed calculations with permutations and singled out the functions known as Lagrange resolvents, but his exposition is less clear, less authoritative than Lagrange's. Moreover, the delay in publication was such that Vandermonde's "Mémoire sur la résolution des équations" [VM] appeared two years after the first part of Lagrange's "Réflexions sur la résolution algébrique des équations". Lagrange was already famous at that time, and Vandermonde's self-effacing comment (in a footnote added in proof) :

> "One will notice some conformities between this [Lagrange's] work and mine, of which I cannot but feel flattered" [VM, p. 365]

did not help to secure notoriety for his paper. However,

Vandermonde can be credited with a real breakthrough in the theory of equations : the solution of cyclotomic equations. This was definitely not obtained previously by Lagrange.

We will thus divide up our discussion of Vandermonde's memoir into two parts : the discussion of general equations, which is somewhat analogous to Lagrange's, and the solution of cyclotomic equations.

§ 2. The solution of general equations

Vandermonde's starting point is that the formula which yields the solutions of an equation in function of the coefficients is necessarily ambiguous, since it must take as values the various roots. He then separates the solution into three "heads" :

"1° To find a function of the roots, of which it can be said, in some sense, that it equals such of the roots that one wants.
2° To put this function in such a form that it be indifferent to interchange the roots in it.
3° To replace in it the values of the sum of the roots, the sum of pairwise products, etc." [VM, p. 370].

Consider for instance the solution of quadratic equations :

$$x^2 - s_1 x + s_2 = 0, \text{ with roots } x_1, x_2.$$

The function

$$F_2(x_1, x_2) = \tfrac{1}{2} [(x_1+x_2) + \sqrt{(x_1-x_2)^2}]$$

satisfies the condition in 1°, since its value is x_1 or x_2, depending on the choice of the square root of $(x_1-x_2)^2$:

$$\sqrt{(x_1-x_2)^2} = \pm (x_1-x_2).$$

Since moreover $F_2(x_1,x_2)$ is not altered when the roots x_1 and x_2 are interchanged, it is already in the form which is called for by (2°). Finally, (3°) requires the evaluation of $F_2(x_1,x_2)$ in terms of s_1 and s_2 : this is quite easy :

$$x_1 + x_2 = s_1 \quad \text{and} \quad (x_1-x_2)^2 = s_1^2 - 4s_2,$$

whence

$$F_2(x_1,x_2) = \tfrac{1}{2} [s_1 + \sqrt{s_1^2 - 4s_2}].$$

Vandermonde first solves in full generality problem (3°) : he thus proves the fundamental theorem of symmetric functions, which says that every symmetric function can be evaluated from the elementary symmetric polynomials (theorem 8.3).

He then solves problem (1°), displaying the following formula :

$$F_n(X_1,\ldots,X_n) = \tfrac{1}{n} [(X_1+\ldots+X_n) + \sum_{i=1}^{n-1} \sqrt[n]{V_i^n}] \tag{1}$$

where

$$V_i = \rho_1^i X_1 + \ldots + \rho_n^i X_n$$

and $\rho_1,\ldots,\rho_n$ denote the n-th roots of unity (including 1).

To see that this function indeed answers to head (1°), we have to prove that for any $k=1,\ldots,n$, some determination of the n-th roots $\sqrt[n]{V_i^n}$ can be chosen in such a way that $F_n(X_1,\ldots,X_n) = X_k$. This can be done as follows : choose :

$$\sqrt[n]{V_i^n} = \rho_k^{-i} V_i,$$

i.e.

205

$$\sqrt[n]{V_i^n} = X_k + \Sigma_{j \neq k} (\rho_k^{-1} \rho_j)^i X_j;$$

then

$$F_n(X_1, \ldots, X_n) = \frac{1}{n} [n X_k + \Sigma_{j \neq k} [\Sigma_{i=1}^{n-1} (\rho_k^{-1} \rho_j)^i] X_j]. \quad (2)$$

Now, $\rho_k^{-1} \rho_j$ is an n-th root of unity, different from 1 if $k \neq j$, hence it is a root of

$$(X^n - 1)/(X - 1) = 1 + X + \ldots + X^{n-1};$$

therefore,

$$\Sigma_{i=1}^{n-1} (\rho_k^{-1} \rho_j)^i = 0$$

and equation (2) simplifies to :

$$F_n(X_1, \ldots, X_n) = X_k.$$

Of course, if $n \geq 3$ the function $F_n(X_1, \ldots, X_n)$ also has other determinations besides $X_1, \ldots, X_n$, but this does not seem to matter to Vandermonde.

It is instructive to compare the formula (1) above to Lagrange's formula 10.2 : it turns out that the functions V_i are none others than Lagrange resolvents. To enlighten this point, choose a primitive n-th root of unity ω; the various n-th roots of unity are then powers of ω, and we can set $\rho_k = \omega^{k-1}$ for $k = 1, \ldots, n$. Then

$$V_i = (\omega^0)^i X_1 + (\omega^1)^i X_2 + (\omega^2)^i X_3 + \ldots + (\omega^{n-1})^i X_n$$

hence

$$V_i = X_1 + \omega^i X_2 + (\omega^i)^2 X_3 + \ldots + (\omega^i)^{n-1} X_n$$

and it follows that V_i is the Lagrange resolvent which was denoted by $t(\omega^i)$ in formula 10.2.

The problems (1°) and (3°) are thus completely solved by Vandermonde; the real stumbling-block is of course problem (2°). For $n = 3$, Vandermonde observes that, choosing $\rho_1 = 1$, $\rho_2 = \omega$ and $\rho_3 = \omega^2$, where ω is a cube root of unity other than 1, the functions involved in $F_3(X_1, X_2, X_3)$, which are

$$V_1^3 = (X_1 + \omega X_2 + \omega^2 X_3)^3 \text{ and } V_2^3 = (X_1 + \omega^2 X_2 + \omega X_3)^3,$$

are not invariant under all the permutations of X_1, X_2, X_3, but every permutation either leaves V_1^3 and V_2^3 invariant or interchanges V_1^3 and V_2^3. Therefore, in order to make the function $F_3(X_1, X_2, X_3)$ invariant under all the permutations, it suffices to replace V_1^3 and V_2^3 by an ambiguous function which takes the values V_1^3 and V_2^3 : such a function has been found previously in the solution of quadratic equations : it is

$$F_2(V_1^3, V_2^3) = \tfrac{1}{2} [(V_1^3 + V_2^3) + \sqrt{(V_1^3 - V_2^3)^2}].$$

So, problem (2°) is solved for $n = 3$.
Vandermonde argues similarly for $n = 4$, using $F_4(X_1, X_2, X_3, X_4)$. He also points out that in this case, since n is not a prime number, other functions can be chosen instead of $F_4(X_1, X_2, X_3, X_4)$, for instance :

$$G_4(X_1, X_2, X_3, X_4) = \tfrac{1}{4}[(X_1 + X_2 + X_3 + X_4) + \sqrt{W_1^2} + \sqrt{W_2^2} + \sqrt{W_3^2}]$$

with $W_1 = X_1 + X_2 - X_3 - X_4$

$W_2 = X_1 - X_2 + X_3 - X_4$

and $W_3 = X_1 - X_2 - X_3 + X_4$.

It is easy to put G_4 in such a form that it is not altered when X_1, X_2, X_3, X_4 are permuted, since every permutation interchanges W_1^2, W_2^2 and W_3^2 : it therefore suffices to replace them by $F_3(W_1^2,W_2^2,W_3^2)$, which takes the values W_1^2, W_2^2 and W_3^2 and can be put in symmetric form, as previously observed.

For $n \geq 5$, the problem is that the functions V_i^n for $i=1,\ldots,n-1$ are not interchanged among themselves when the indeterminates are permuted : indeed, the function V_1^n takes $(n-1)!$ values under the permutations of the indeterminates (see proposition 10.9). Nevertheless, for $n=5$ Vandermonde succeeds in reducing the determination of V_1^5 to the solution of an equation of degree 6 (compare theorem 10.8). For $n=6$, he shows that his method requires the solution of an equation of degree 10 or 15.

Inconclusive, as it is, this section is not devoid of interest, since it prompts Vandermonde to initiate rather explicit calculations with permutations. He decomposes the symmetric polynomials (which he calls "types") into sums of "partial types" which are, in fact, sums of the values that a monomial takes under a subgroup of the symmetric group (especially, but not exclusively, cyclic subgroups). For instance, for three variables a,b,c, he denotes :

$$[\begin{smallmatrix} \alpha & \beta & \gamma \\ ii & iii & i \end{smallmatrix}] = a^\alpha b^\beta c^\gamma + a^\gamma b^\alpha c^\beta + a^\beta b^\gamma c^\alpha$$

(where α,β,γ are pairwise distinct integers). The Latin subscripts indicate that in the second term the exponents α,β,γ must be changed in such a way that γ takes the first place, α the second and β the third; the third term is obtained from the second as the second was obtained from the first, and so on for the next terms, as long as this process yields new monomials. The function thus produced is obviously invariant under the (cyclic) subgroup of S_3 generated by the permutation $a \to b \to c \to a$. (Compare the proof of proposition 10.6).

Sometimes, Vandermonde also uses a more general notation,

which comprehends all the partial types which are invariant under the same group of permutations, but he stops short of devising a notation for permutations. For instance,

$$[\begin{array}{ccccc} a & b & c & d & e \\ v & i & iv & ii & iii \end{array}]$$

(where a,b,c,d,e are the indeterminates) is a generic notation for the various partial types which are invariant under the permutation which sets the letters a,b,c,d,e in the order b,d,e,c,a indicated by the Latin numerals, i.e. under the permutation $a \to b \to d \to c \to e \to a$.

These notations enable Vandermonde to perform coherently some very complicated explicit calculations, but he cannot escape the conclusion that his method for equations of degree at least 5 leads to equations of ever higher degree, and that it may therefore not work eventually.

"That is all that the calculations taught me on this object, and I do not have enough faith in conjectures in such a thorny matter to dare try one here. I will only add that I have not found any partial type involving five letters which depends on an equation of the fourth or the third degree, and I am convinced that such a type does not exist" [VM, p. 414].

However, that is not the end of the story. In the final two articles of his paper, Vandermonde briefly considers cyclotomic equations.

§ 3. Cyclotomic equations

Recall from § 7.3 that the problem of determining radical expressions for the roots of unity had been reduced to the solution by radicals of the cyclotomic equations

$$\Phi_p(X) = X^{p-1} + X^{p-2} + \ldots + X + 1 = 0$$

for p prime. Moreover, for p odd, de Moivre had shown that the change of variable $Y = X + X^{-1}$ converts $\phi_p(X) = 0$ into an equation of degree $(p-1)/2$. Thus, for $p = 11$, the solution of the cyclotomic equation requires the solution of

$$Y^5 + Y^4 - 4Y^3 - 3Y^2 + 3Y + 1 = 0,$$

which de Moivre had been unable to solve by radicals. We also recall from remark 7.9 that, since the roots of ϕ_p are the complex numbers $e^{2k\pi i/11}$ for $k=1,\ldots,10$, it follows that the roots of the equation in Y are the values $2\cos(2k\pi/11)$ for $k=1,\ldots,5$.

Vandermonde in fact uses the (obviously equivalent) change of variable $Z = -(X+X^{-1})$, which yields the equation:

$$Z^5 - Z^4 - 4Z^3 + 3Z^2 + 3Z - 1 = 0. \qquad (3)$$

The roots of this equation, which are denoted by a,b,c,d,e, are chosen as :

$$a = -2\cos(2\pi/11) \qquad b = -2\cos(4\pi/11)$$

$$c = -2\cos(6\pi/11) \qquad d = -2\cos(8\pi/11) \qquad e = -2\cos(10\pi/11).$$

It is useful to note, with Vandermonde, that the trigonometric formula

$$2\cos\alpha\cos\beta = \cos(\alpha+\beta) + \cos(\alpha-\beta) \qquad (4)$$

yields relations between a,b,c,d and e. For instance, replacing α and β by $2\pi/11$, we obtain :

$$(\cos(2\pi/11))^2 = \cos(4\pi/11) + \cos 0,$$

whence

$$a^2 = -b + 2.$$

Likewise, replacing α by $2\pi/11$ and β by $4\pi/11$, we find :

$$ab = -c - a,$$

and so on. Thus, replacing α and β successively by the various angles $2k\pi/11$ for $k=1,\ldots,5$, the trigonometric formula (4) yields linear expressions for the products of roots. Observing these expressions, Vandermonde draws an amazing conclusion, which enables him to find expressions by radicals for the eleventh roots of unity. Here are Vandermonde's own words, in the penultimate article of his paper :

"In the particular cases where there are equations between the roots, the method just explained may be used to solve, without resorting to the general solution formulas. The equation $r^{11} - 1 = 0$ will provide us with an example : it leads (article VI) to this one :

$$x^5 - x^4 - 4x^3 + 3x^2 + 3x - 1 = 0,$$

and denoting its roots by a,b,c,d,e, it will readily follow from article XI :

$a^2 = -b+2 \quad b^2 = -d+2 \quad c^2 = -e+2 \quad d^2 = -c+2 \quad e^2 = -a+2$

$ab = -a-c \quad bc = -a-e \quad cd = -a-d \quad de = -a-b$

$ac = -b-d \quad bd = -b-e \quad ce = -b-c$

$ad = -c-e \quad be = -c-d$

$ae = -d-e$

and all the partial types of the form [a b c d e]
                                      v i iv ii iii
will have a purely rational value; thus, taking everywhere

in article XXVIII [α β ε δ γ] instead of
                   v i iv ii iii

[α β γ δ ε], we will find : (...)
 v iii iv i ii

$$X = \frac{1}{5} [1 + \Delta' + \Delta'' + \Delta''' + \Delta^{IV}]$$

with

$$\Delta' = \sqrt[5]{\frac{11}{4}(89+25\sqrt{5} - 5\sqrt{-5+2\sqrt{5}} + 45\sqrt{-5-2\sqrt{5}})}$$

$$\Delta'' = \sqrt[5]{\frac{11}{4}(89+25\sqrt{5} + 5\sqrt{-5+2\sqrt{5}} - 45\sqrt{-5-2\sqrt{5}})}$$

$$\Delta''' = \sqrt[5]{\frac{11}{4}(89-25\sqrt{5} - 5\sqrt{-5+2\sqrt{5}} - 45\sqrt{-5-2\sqrt{5}})}$$

$$\Delta^{IV} = \sqrt[5]{\frac{11}{4}(89-25\sqrt{5} + 5\sqrt{-5+2\sqrt{5}} + 45\sqrt{-5-2\sqrt{5}})}" \text{ [VM, pp.}$$
415-416].

Vandermonde's brilliant (but not quite explicit) observation is that the permutation a → b → d → c → e → a preserves the relations between the roots. For instance, applying this permutation to the relation

$$a^2 = -b + 2,$$

i.e. changing a in b and b in d, we obtain :

$$b^2 = -d + 2,$$

and this relation actually holds! (Compare exercise 2).

This is very significant since the relations between the roots can be used to lower the degree of any polynomial in a,b,c,d,e, eventually providing a linear expression. Thus, suppose f is a polynomial in five variables (of any degree). Using the relations between the roots, we can eventually find :

$$f(a,b,c,d,e) = Aa + Bb + Cc + Dd + Ee + F \qquad (5)$$

for some numbers $A, B, \ldots, F$ which can be explicitly determined from the coefficients of f. Now, since the permutation $a \to b \to d \to c \to e \to a$ preserves the relations which have been used in simplifying the expression of $f(a,b,c,d,e)$, we can perform the <u>same</u> simplification procedure changing a in b, b in d, d in c, c in e and e in a at each step, and we end up with :

$$f(b,d,e,c,a) = Ab + Bd + Ce + Dc + Ea + F.$$

This point is somewhat delicate, since the expression (5) of $f(a,b,c,d,e)$ is <u>not</u> unique : indeed, since the coefficient of z^4 in (3) is the opposite of the sum of the roots, it follows that

$$a + b + c + d + e = 1; \qquad (6)$$

therefore, we have for instance

$$Aa + Bb + Cc + Dd + Ee + F = (A+F)a + (B+F)b + (C+F)c + (D+F)d + (E+F)e.$$

This does not matter, however, as long as we use the same procedure to simplify $f(a,b,c,d,e)$ and (after the permutation $a \to b \to d \to c \to e \to a$) $f(b,d,e,c,a)$.

If we apply this observation to the Vandermonde (- Lagrange) resolvents :

$$V_i(a,b,c,d,e)^5 = (\rho_1^i a + \rho_2^i b + \rho_3^i c + \rho_4^i d + \rho_5^i e)^5$$

where $\rho_1, \ldots, \rho_5$ are the 5-th roots of unity, we find :

$$V_i(a,b,c,d,e)^5 = Aa + Bb + Cc + Dd + Ee + F \qquad (7)$$

where $A, B, \ldots, F$ are rational expressions in $\rho_1, \ldots, \rho_5$.

Applying the permutation four times, we obtain :

$$V_i(b,d,e,c,a)^5 = Ab + Bd + Ce + Dc + Ea + F$$

$$V_i(d,c,a,e,b)^5 = Ad + Bc + Ca + De + Eb + F$$

$$V_i(c,e,b,a,d)^5 = Ac + Be + Cb + Da + Ed + F$$

$$V_i(e,a,d,b,c)^5 = Ae + Ba + Cd + Db + Ec + F.$$

(8)

Now, if we choose $\rho_1 = 1$, $\rho_2 = \omega$, $\rho_3 = \omega^3$, $\rho_4 = \omega^2$ and $\rho_5 = \omega^4$, where ω is some primitive 5-th root of unity, then

$$V_i(a,b,c,d,e) = a + \omega^i b + \omega^{2i} d + \omega^{3i} c + \omega^{4i} e,$$

and the permutation $a \to b \to d \to c \to e \to a$ then leaves $V_i(a,b,c,d,e)^5$ invariant; indeed, this permutation changes $V_i(a,b,c,d,e)$ into

$$V_i(b,d,e,c,a) = b + \omega^i d + \omega^{2i} c + \omega^{3i} e + \omega^{4i} a$$

and since $\omega^5 = 1$, we have

$$V_i(b,d,e,c,a) = \omega^{-i} V_i(a,b,c,d,e)$$

(compare the proof of theorem 10.8) hence

$$V_i(b,d,e,c,a)^5 = V_i(a,b,c,d,e)^5.$$

Likewise, the left-hand sides of the equalities (7) and (8) are all equal to $V_i(a,b,c,d,e)^5$. Therefore, summing up all these equalities, we obtain :

$$5V_i(a,b,c,d,e)^5 = (A+B+C+D+E)(a+b+c+d+e) + 5F$$

and using (6) we conclude :

$$V_i(a,b,c,d,e)^5 = \frac{1}{5}(A+B+C+D+E) + F.$$

Since $A, B, \ldots, F$ can be rationally calculated from ω, which is already expressed by radicals (see § 7.3), it follows that the functions V_i^5 can be expressed by radicals. Therefore, a,b,c,d and e can also be expressed by radicals, using the formula $F_5(a,b,c,d,e)$ of § 2 (and (6)) :

$$a,b,c,d,e = \frac{1}{5}[1 + \sum_{i=1}^{4} \sqrt[5]{V_i^5}].$$

With hindsight, and with a view towards a possible generalization to cyclotomic equations of higher degree, the crucial steps in the calculations above appear to be :
a) the existence of relations among the roots, which can be used to reduce to degree 1 each polynomial expression in the roots.
b) the existence of a cyclic permutation of the roots which preserves the relations above.

Given (a) and (b), we number the roots in such a way that the cyclic permutation be

$$x_1 \to x_2 \to x_3 \to \ldots \to x_n \to x_1,$$

and the same arguments as above then show that the n-th power of each Lagrange resolvent of the form

$$t(\omega) = x_1 + \omega x_2 + \omega^2 x_3 + \ldots + \omega^{n-1} x_n,$$

for any n-th root of unity ω, is a rational expression in ω. Arguing inductively, we may assume that an expression by radicals has been found for ω; whence $t(\omega)^n$ can be expressed by radicals and the roots $x_1, \ldots, x_n$ can also be found by radicals, by Lagrange's formula 10.2 :

$$x_i = \frac{1}{n}[\Sigma_\omega\, \omega^{-(i-1)} t(\omega)].$$

Now, (a) is clear for the cyclotomic equations Φ_p for any prime p or, rather, for the equations obtained by de Moivre's change of variable $Y = X + X^{-1}$, whose roots are $2\cos(2k\pi/p)$ for $k=1,\ldots,(p-1)/2$; indeed, the same trigonometric formula (4) yields the required relations. But (b) is very far from clear! Yet Vandermonde simply states :

"Since to solve the equation

$$X^m - X^{m-1} - (m-1)X^{m-2} + \text{etc.} = 0,$$

the question is at most to determine (article VI) the quantity which is indifferently one of its roots, and by no means to arrange it in such a way that it be indifferent to interchange the roots among themselves, this solution will always be very easy" [VM, p. 416].

And he leaves it at that.

Yet, he had certainly noticed that the relations were not preserved by *any* permutation of the roots. The existence of a cyclic permutation which does preserve the relations is a very remarkable, and quite mysterious, property of cyclotomic equations, which should have awaken Vandermonde's curiosity. If he had investigated this property, he could have preceded Gauss by about thirty years.

Moreover, Vandermonde had pinpointed the very basic idea of Galois theory : in order to determine the "structure" of an equation, deciding eventually whether it is solvable by radicals, and more generally to evaluate its difficulty, one has to look at the permutations of the roots; but one needs only to consider those permutations which preserve the relations between the roots[*].

This is a conspicuous example of a deep insight which was completely wasted. To conclude this chapter, I could do no better than to quote Lebesgue :

> "Surely, any man who discovers something truly important is left behind by his own discovery; he himself hardly understands it, and only by pondering over it for a long time. But Vandermonde never came back to his algebraic investigations because he did not realize their importance in the first place, and if he did not understand them afterwards, it is precisely because he did not reflect deeply on them; he was interested in everything, he was occupied by everything; he was not able to go slowly to the bottom of anything (...).
>
> To assess exactly what Vandermonde saw, understood and what he did not catch, one would have to reconstruct not only the mind of a man from the eighteenth century, but Vandermonde's mind, and at the moment when he had a glimpse of genius and went ahead of his age. When trying to do so, one will always give too much or too little credit to Vandermonde" [Le, p. 222-223].

Exercises for chapter 11

1) List all the terms in the partial types :
$[\begin{smallmatrix}\alpha & \beta & \gamma & \delta & \epsilon \\ v & iii & iv & i & ii\end{smallmatrix}]$, $[\begin{smallmatrix}\alpha & \epsilon & \delta & \beta & \gamma \\ v & iii & iv & i & ii\end{smallmatrix}]$, $[\begin{smallmatrix}\alpha & \gamma & \beta & \epsilon & \delta \\ v & iii & iv & i & ii\end{smallmatrix}]$
and $[\begin{smallmatrix}\alpha & \delta & \epsilon & \gamma & \beta \\ v & iii & iv & i & ii\end{smallmatrix}]$. Show that the sum of all these partial types is also a partial type. What is the subgroup of S_5 which leaves this new partial type invariant ? [VM, Art. 24, p. 391].

---

[*] This restriction does not appear for general equations since their roots are independent indeterminates : in this case, there is no relation to preserve, so every permutation is admissible.

2) Show that the permutation $a \to b \to c \to d \to e \to a$ does not preserve the relations among $a = 2 \cos 2\pi/11$, $b = 2 \cos 4\pi/11$, $c = 2 \cos 6\pi/11$, $d = 2 \cos 8\pi/11$ and $e = 2 \cos 10\pi/11$.

3) Show that the permutation $a \to b \to c \to a$ preserves the relations among $a = 2 \cos 2\pi/7$, $b = 2 \cos 4\pi/7$ and $c = 2 \cos 6\pi/7$.

4) Find a permutation of the numbers $2 \cos(2k\pi/13)$ for $k=1,\ldots,6$, which preserves the relations among them.

12 Gauss on cyclotomic equations

§ 1. Introduction

The contributions of Carl Friedrich Gauss (1777-1855) to the theory of equations measure up to the outstanding advances he made in many other research areas. They occupy a special place in his work however, since they were among his earliest achievements. They can be divided up into two main topics : the fundamental theorem of algebra (1799) which we have already discussed in chapter 9, and the solution of cyclotomic equations.

Gauss' results on this last topic show how to complete Vandermonde's arguments to provide inductively expressions by radicals for the roots of unity (see corollary 39), but they far exceed this goal. In effect, they yield a thorough description of the possible reductions of cyclotomic equations of prime index to equations of smaller degree. Thus, in a brilliant way, they carry out for cyclotomic equations the program envisioned by Lagrange : to solve an equation by determining successively certain functions of the roots. As Gauss shows, the solution of $\Phi_p(X) = 0$ can be reduced to the solution of equations of degree equal to the prime factors of $p - 1$. In particular, the 17-th roots of unity can be determined by solving successively four quadratic equations, since $17 - 1 = 2^4$. As an application of this result, it follows that the regular polygon with 17 sides can be constructed by ruler and compass; this result was obtained by Gauss as early as 1796, and it is said to have been decisive in his vocation [Bu, p. 10]. This application will be discussed in an appendix to this chapter.

A definitive account of his results on cyclotomic equations was published by Gauss as the seventh and final section of his epoch-making treatise on number theory : "Disquisitiones Arithmeticae" (1801). The inclusion of such algebraic results in a book on number theory was commented by Gauss himself in the preface :

> "The theory of the division of the circle, or of regular polygons, which is treated in section VII, does not belong <u>by itself</u> to Arithmetic, but its <u>principles</u> cannot be found but in Higher Arithmetic : this may appear to geometers as unexpected as the new truths that follow from it, and which they will see, I hope, with pleasure" [Gau2, p. 8].

Accordingly, our review of Gauss' results will be preceded by some number-theoretic preliminaries. Thereafter, we divide the contents of the seventh section of the "Disquisitiones Arithmeticae" into three sections : first, we prove the irreducibility of cyclotomic equations of prime index, which is a key result in Gauss' investigations; next, we discuss the possible reductions of cyclotomic equations and we finish with the solvability by radicals of the cyclotomic equations and auxiliary equations. Some extra results which are needed to justify some of the steps in Gauss' proofs will be found in the final section.

It should be noted that we only review those results of section VII of the Disquisitiones Arithmeticae which directly concern the theory of equations. Several details which are meaningful in view of applications to number theory will be omitted.

§ 2. <u>Number-theoretic preliminaries</u>

<u>1</u>. At the very beginning of the Disquisitiones Arithmeticae [Gau2, Art. 2], Gauss introduces the following notation, which has gained wide acceptance and will be used repeated-

ly in the sequel : if a, b and n are integers and n ≠ 0, one denotes

$$a \equiv b \pmod{n}$$

whenever a - b is divisible by n. The integers a and b are then said to be <u>congruent modulo n</u>. The explicit reference to the modulus n is sometimes omitted when no confusion is likely to arise. It is readily verified that this relation is an equivalence relation which is compatible with the sum and the product of integers, i.e. :

if $a_1 \equiv b_1$ and $a_2 \equiv b_2 \pmod{n}$, then $a_1 + a_2 \equiv b_1 + b_2$

and $a_1 a_2 \equiv b_1 b_2 \pmod{n}$.

In the sequel, we focus on the case where the modulus n is prime. This case has very distinctive features. We will prove in particular the following result, which plays a key role in Gauss' investigations on cyclotomic equations:

2. THEOREM : For any prime number p, there exists an integer g whose various powers $g^0, g^1, g^2, \ldots, g^{p-2}$ are congruent to $1, 2, \ldots, p-1$ modulo p (not necessarily in that order).

In the course of proving this theorem, we shall see that an integer g satisfies the condition of the theorem if and only if

$$g^{p-1} \equiv 1 \pmod{p} \text{ and } g^i \not\equiv 1 \pmod{p} \text{ for } i=1,\ldots,p-2.$$

Therefore, any such integer g is called a <u>primitive root of p</u>. (It would be more accurate, if not shorter, to call it a primitive (p-1)th root of 1 modulo p). For instance, 2 is a primitive root of 11, since modulo 11 :
$2^0 \equiv 1$, $2^1 \equiv 2$, $2^2 \equiv 4$, $2^3 \equiv 8$, $2^4 \equiv 5$, $2^5 \equiv 10$, $2^6 \equiv 9$, $2^7 \equiv 7$, $2^8 \equiv 3$, $2^9 \equiv 6$.

On the contrary, 3 is not a primitive root of 11, as $3^5 \equiv 1 \bmod 11$, whence $3^6 \equiv 3$, $3^7 \equiv 3^2$, $3^8 \equiv 3^3$,... and the powers of 3 take modulo 11 only the values $3^0 \equiv 1$, $3^1 \equiv 3$, $3^2 \equiv 9$, $3^3 \equiv 5$ and $3^4 \equiv 4$.

The proof of theorem 2 occupies the rest of this section. We shall closely follow one of the two proofs which Gauss included in the Disquisitiones Arithmeticae [Gau2, Art. 55].

3. LEMMA : Let a,b be integers and let p be a prime number. If

$$ab \equiv 0 \pmod{p},$$

then $a \equiv 0 \pmod{p}$ or $b \equiv 0 \pmod{p}$.
[Gau2, Art. 14].

This amounts to the well-known fact that if a prime number divides a product of integers, then it divides one of the factors. To prove it, it suffices to mimick the proof of lemma 5.14, replacing polynomials by integers.

4. PROPOSITION : Let p be a prime number and let $a_0, a_1, \ldots, a_d$ be integers. If $a_d \not\equiv 0 \pmod{p}$, then the congruence equation

$$a_d X^d + a_{d-1} X^{d-1} + \ldots + a_1 X + a_0 \equiv 0 \pmod{p} \qquad (1)$$

has at most d incongruent solutions modulo p. [Gau2, Art. 43].

Proof : We argue by induction on d. The case $d = 0$ being trivial, we may assume inductively that every congruence equation of degree $(d-1)$ has at most $(d-1)$ solutions modulo p.
If equation (1) has $(d+1)$ solutions $x_1, \ldots, x_{d+1}$ pairwise distinct modulo p, then the change of variable $Y = X - x_1$

transforms equation (1) into another equation of degree d :

$$a_d Y^d + a'_{d-1} Y^{d-1} + \ldots + a'_1 Y + a'_0 \equiv 0 \pmod{p}$$

which has the same leading coefficient a_d and has the d+1 solutions :

$$0, x_2 - x_1, \ldots, x_{d+1} - x_1.$$

Since 0 is a root, it follows that $a'_0 \equiv 0 \pmod{p}$ and the equation in Y can be written :

$$Y(a_d Y^{d-1} + a'_{d-1} Y^{d-2} + \ldots + a'_1) \equiv 0 \pmod{p}.$$

Now, $x_2 - x_1, x_3 - x_1, \ldots, x_{d+1} - x_1$ are non-zero roots of this equation; hence, by lemma 3, they are roots of the second factor : $a_d Y^{d-1} + \ldots + a'_1$, which is a polynomial of degree d - 1. This contradicts the induction hypothesis. □

5. REMARK : As with any other equivalence relation, the congruence relation defines a partition of the set on which it is defined into equivalence classes. The congruence class modulo n of an integer m consists of all the integers which are congruent to m modulo n or, in other words, of all the integers of the form kn + m, for $k \in \mathbb{Z}$.

Since the congruence relation is compatible with the sum and the product of integers, these operations induce well-defined operations on the set of congruence classes of integers, and it is readily verified that this set inherits the commutative ring structure of $\mathbb{Z}$. The ring of congruence classes modulo n is denoted by $\mathbb{Z}/n\mathbb{Z}$. (Compare § 9.2). This ring has only finitely many elements, namely the congruence classes modulo n of $0, 1, \ldots, n-1$.

Lemma 3 asserts that $\mathbb{Z}/p\mathbb{Z}$ is a domain, for p prime, and the arguments in proposition 4 show, more generally, that an equation of degree d with coefficients in a domain has at most d solutions in the domain.

In fact, since $\mathbb{Z}/n\mathbb{Z}$ is finite, it is easily seen that this ring is a field whenever it is a domain : indeed, if $aX \equiv 0 \pmod{n}$ implies $X \equiv 0 \pmod{n}$, then the products ax are pairwise distinct modulo n when x runs over $0,1,\ldots,n-1$. Therefore, one of these products is congruent to 1 modulo n : this proves that a is invertible modulo n. Consequently, for p prime, the ring $\mathbb{Z}/p\mathbb{Z}$ is a field, which is denoted by $\mathbb{F}_p$.

For the proof of theorem 2, we also need the following result, due to Pierre de Fermat (1601-1665) :

6. THEOREM (Fermat) : Let p be any prime number and let a be an integer. If $a \not\equiv 0 \pmod{p}$, then $a^{p-1} \equiv 1 \pmod{p}$. [Gau2, Art. 50].

Proof (Euler) : We shall prove :

$$a^p \equiv a \pmod{p} \qquad \text{for every integer a.} \qquad (2)$$

It will then follow that

$$a(a^{p-1}-1) \equiv 0 \pmod{p} \qquad \text{for every integer a,}$$

hence, by lemma 3 :

$$a^{p-1}-1 \equiv 0 \pmod{p} \qquad \text{whenever } a \not\equiv 0 \pmod{p}.$$

The basic observation is that the binomial coefficients $\binom{p}{i} = \frac{p!}{i!(p-i)!}$ are all divisible by p for $i=1,\ldots,p-1$; therefore,

$$(a+1)^p \equiv a^p + 1 \pmod{p} \text{ for every integer a,}$$

and it easily follows by induction on a that (2) holds for every positive integer a. The property for negative a is then readily proved, since

$$(-a)^p \equiv -a^p \pmod{p}$$

for every integer a and every prime p. (For p = 2, observe that $-1 \equiv 1 \mod 2$). □

7. COROLLARY : Let p be a prime number. The following conditions on an integer g are equivalent :

<u>a</u>) $g^{p-1} \equiv 1 \pmod{p}$ and $g^i \not\equiv 1 \pmod{p}$ for $i=1,\ldots,p-2$

<u>b</u>) the powers $g^0, g^1, \ldots, g^{p-2}$ take the values $1, 2, \ldots, p-1$ modulo p.

<u>Proof</u> : (<u>a</u>) ⇒ (<u>b</u>) : If $g^{p-1} \equiv 1 \pmod{p}$, then $g \not\equiv 0 \pmod{p}$ and it follows from lemma 3 that the values modulo p of the powers $g^0, g^1, \ldots, g^{p-2}$ range in $\{1, 2, \ldots, p-1\}$. Therefore, to prove (<u>b</u>), it suffices to show that the powers g^i are pairwise distinct modulo p for $i=0, 1, \ldots, p-2$.
Assume on the contrary :

$$g^i \equiv g^j \pmod{p}$$

for some integers i,j between 0 and p-2, and i < j. Then,

$$g^i(1 - g^{j-i}) \equiv 0 \pmod{p},$$

hence, by lemma 3 :

$$g^{j-i} \equiv 1 \pmod{p}.$$

Since j - i is an integer between 1 and p-2, this relation contradicts (<u>a</u>).
(<u>b</u>) ⇒ (<u>a</u>) : From (<u>b</u>), it clearly follows that $g \not\equiv 0 \pmod{p}$, whence

$$g^{p-1} \equiv 1 \pmod{p},$$

by theorem 6. Moreover, condition (b) ensures that $g^i \not\equiv g^0$ for $i=1,\ldots,p-2$, hence

$$g^i \not\equiv 1 \pmod{p} \qquad \text{for } i=1,\ldots,p-2. \qquad \square$$

(Compare proposition 7.14).

As previously noted, every integer g satisfying the equivalent conditions (a) and (b) in corollary 7 is called a primitive root of p.

We now introduce the following technical definition : for any prime number p and any integer a relatively prime to p, the exponent (modulo p) of a is the smallest positive integer e such that

$$a^e \equiv 1 \pmod{p}.$$

Thus, the integers of exponent p-1, which will be shown to exist for every prime p, are the primitive roots of p. (Compare definitions 7.10).

8. LEMMA : Let e be the exponent (modulo a prime number p) of an integer a relatively prime to p, and let m be an integer. Then

$$a^m \equiv 1 \pmod{p}$$

if and only if e divides m. In particular, e divides (p-1).

Proof : The same arguments as in the proof of lemma 7.12 apply. $\qquad \square$

Of course, it is not clear a priori that there exist integers of exponent e for every divisor e of p-1. As a first step in the proof of theorem 2, we now show the existence of such integers in the case where e is the power of a prime number :

9. LEMMA : Let p and q be prime numbers. If some power q^m of q divides p-1, then there exists an integer of exponent q^m (modulo p). [Gau2, Art. 55].

Proof : By proposition 4, the congruence equation

$$x^{(p-1)/q} \equiv 1 \pmod{p}$$

has at most (p-1)/q solutions (modulo p). Therefore, one can find an integer x which is <u>not</u> a root of this equation and is relatively prime to p. Let then $a = x^{(p-1)/q^m}$. By Fermat's theorem 6, we have $x^{p-1} \equiv 1 \pmod{p}$, hence

$$a^{q^m} \equiv 1 \pmod{p}.$$

Lemma 9 then shows that the exponent of a divides q^m. On the other hand,

$$a^{q^{m-1}} = x^{(p-1)/q} \not\equiv 1 \pmod{p},$$

so the exponent of a does not divide q^{m-1}. Since q is prime, the only (positive) integer which divides q^m but not q^{m-1} is q^m, and the exponent of a is therefore equal to q^m. □

Proof of theorem 2 : Let

$$p - 1 = q_1^{m_1} \ldots q_r^{m_r}$$

be the decomposition of p-1 into a product of prime factors, where $q_1, \ldots, q_r$ are pairwise distinct prime numbers. By lemma 9, one can find integers $a_1, \ldots, a_r$ of respective exponents $q_1^{m_1}, \ldots, q_r^{m_r}$ modulo p. To prove the theorem, we show that the product $a_1 \ldots a_r$ has exponent p-1, and is thus a primitive root of p.

Let e be the exponent of $a_1 \ldots a_r$. Lemma 8 shows that e divides p-1. If $e \neq p-1$, then e lacks at least one of the prime factors of p-1 and divides therefore $(p-1)/q_i$ for some $i=1,\ldots,r$. Assume for instance e divides $(p-1)/q_1$. Then, by lemma 8,

$$(a_1 \ldots a_r)^{(p-1)/q_1} \equiv 1 \pmod{p}.$$

Now, since the exponent $q_i^{m_i}$ of a_i divides $(p-1)/q_1$ for $i=2,\ldots,r$, we have

$$a_i^{(p-1)/q_1} \equiv 1 \pmod{p} \qquad \text{for } i=2,\ldots,r,$$

hence the previous equation yields :

$$a_1^{(p-1)/q_1} \equiv 1 \pmod{p}.$$

This last relation shows, by lemma 8, that the exponent $q_1^{m_1}$ of a_1 divides $(p-1)/q_1$. Therefore, p-1 is divisible by $q_1^{m_1+1}$: this is a contradiction, which shows that is was absurd to assume $e \neq p-1$. (Alternatively, the claim that $e = p-1$ can also be proved by the same arguments as in proposition 7.13). □

10. REMARKS :
a) If g is a primitive root of an odd prime p, then $g^{(p-1)/2} \equiv -1 \pmod{p}$. Indeed, since

$$[g^{(p-1)/2}]^2 \equiv g^{(p-1)} \equiv 1 \pmod{p},$$

it follows that $g^{(p-1)/2}$ is a root of the congruence equation :

$$x^2 \equiv 1 \pmod{p}.$$

Now, this equation has two roots modulo p (see proposition 4), which are 1 and -1, so

$$g^{(p-1)/2} \equiv 1 \quad \text{or} \quad g^{(p-1)/2} \equiv -1 \pmod{p}.$$

Since g is primitive, no power of exponent smaller than p-1 is congruent to 1; therefore, the first relation is impossible and thus $g^{(p-1)/2} \equiv -1 \pmod{p}$.

b) Fermat's theorem 6 can also be proved by the following elaboration on Lagrange's theorem 10.4 : for any element a of a (multiplicative) group G, define the <u>order</u> (or the <u>exponent</u>) of a as the smallest positive integer e such that $a^e = 1$ in G. If the order of a is finite, and denoted by e, then it is easily seen that the set

$$S = \{1, a, a^2, \ldots, a^{e-1}\}$$

is a subgroup of G, and the arguments in corollary 7 or proposition 7.14 show that the elements $1, a, \ldots, a^{e-1}$ are pairwise distinct, whence $|S| = e$. By Lagrange's theorem 10.4, it follows that e divides $|G|$ (if G is finite). In particular, $a^{|G|} = 1$ for all $a \in G$, if G is finite. Fermat's theorem follows by applying this result to the multiplicative group $\mathbb{F}_p^\times = \mathbb{F}_p - \{0\}$ of the field with p elements. (Compare remark 5).

c) Tracing back through the proof of theorem 2, it appears that the only information on $\mathbb{F}_p$ which was needed (besides the fact that $\mathbb{F}_p^\times$ is an abelian group) was that the equation $x^{(p-1)/q} = 1$ does not have more than $(p-1)/q$ solutions. Therefore, the arguments in this proof provide the following result : if a finite abelian group G is such that for every integer n the equation $x^n = 1$ has at most n solutions in G, then G is cyclic, i.e. G is generated by a single element :

$$G = \{1, a, a^2, \ldots, a^{|G|-1}\} \quad \text{for some } a \in G.$$

In particular, every finite subgroup of the multiplicative group of a field is cyclic.

§ 3. Irreducibility of the cyclotomic polynomials of prime index

The aim of this section is to provide a proof of and develop some of the consequences of the following theorem :

11. THEOREM : For every prime p, the cyclotomic polynomial

$$\Phi_p(X) = X^{p-1} + X^{p-2} + \ldots + X + 1$$

is irreducible over the field of rational numbers.

This theorem was first proved by Gauss in Article 341 of the Disquisitiones Arithmeticae. Since then, it has been generalized, and the proofs have been simplified by several mathematicians, including Eisenstein, Dedekind, Kronecker, Mertens, Landau and I. Schur. Instead of following Gauss' own proof, which requires a careful analysis of several cases, we shall follow Eisenstein's ideas, which are simpler and in some sense more general, in that they provide a useful sufficient condition for the irreducibility of polynomials over $\mathbb{Q}$. In § 6, we prove some generalizations of this theorem, after Dedekind and Kronecker. The proofs given there yield an alternative proof of theorem 11. The starting point of all the proofs is a result known as "Gauss' lemma" :

12. LEMMA (Gauss) : [Gau2, Art. 42] : If a monic polynomial in $\mathbb{Q}[X]$ divides a monic polynomial with integral coef-

ficients, then its coefficients are all integral.

<u>Proof</u> : Let $f = X^n + a_{n-1}X^{n-1} + \ldots + a_1X + a_0 \in \mathbb{Q}[X]$ be a monic polynomial which divides a monic polynomial $P \in \mathbb{Z}[X]$, and let $g \in \mathbb{Q}[X]$ be the quotient :

$$fg = P \in \mathbb{Z}[X].$$

Since P and f are monic, g is monic too : let

$$g = X^m + b_{m-1}X^{m-1} + \ldots + b_1X + b_0 \in \mathbb{Q}[X].$$

We have to prove that the coefficients $a_0, a_1, \ldots, a_{n-1}$ are all integers. Suppose the contrary and let d be the least common multiple of the denominators of $a_0, \ldots, a_{n-1}$; then

$$f = \tfrac{1}{d}(dX^n + a'_{n-1}X^{n-1} + \ldots + a'_1X + a'_0)$$

where $d, a'_{n-1}, \ldots, a'_0$ are relatively prime integers. Similarly, let

$$g = \tfrac{1}{e}(eX^m + b'_{m-1}X^{m-1} + \ldots + b'_1X + b'_0)$$

where $e, b'_{m-1}, \ldots, b'_0$ are relatively prime integers. Let p be a prime number which divides d. Since $d, a'_{n-1}, \ldots, a'_0$ are relatively prime, there is a greatest index k such that p does not divide a'_k. Let also ℓ be the greatest index such that p does not divide b'_ℓ (if p does not divide e, let $\ell = m$ and $b'_m = e$). Since $fg = P \in \mathbb{Z}[X]$, it follows that

$$(dX^n + a'_{n-1}X^{n-1} + \ldots + a'_0)(eX^m + b'_{m-1}X^{m-1} + \ldots + b'_0) \in d\,e\,\mathbb{Z}[X].$$

In particular, the coefficient of $X^{k+\ell}$ in the product is divisible by p since p divides d; i.e.

$$\Sigma_{i+j=k+\ell}\, a'_i b'_j \equiv 0 \mod p.$$

Since $a_i' \equiv 0$ mod p for $i > k$ and $b_j' \equiv 0$ mod p for $j > \ell$, we have :

$$\Sigma_{i+j=k+\ell}\, a_i'b_j' \equiv a_k'b_\ell' \mod p;$$

therefore, the previous equation yields :

$$a_k'b_\ell' \equiv 0 \mod p.$$

This is a contradiction, since p does not divide a_k' nor b_ℓ'. □

13. PROPOSITION (Eisenstein's irreducibility criterion) :
Let P be a monic polynomial with integral coefficients :

$$P = X^t + c_{t-1}X^{t-1} + \ldots + c_1X + c_0 \in \mathbb{Z}[X].$$

If there is a prime number p which divides c_i for $i=0,\ldots,t-1$ but such that p^2 does not divide c_0, then P is irreducible over $\mathbb{Q}$.

Proof : Assume on the contrary that

$$P = fg$$

for some non-constant polynomials $f, g \in \mathbb{Q}[X]$, of degree n and m respectively. Since P is monic, we may assume that f and g are both monic, whence, by Gauss' lemma 12, that f and g have integral coefficients. Let

$$f = X^n + a_{n-1}X^{n-1} + \ldots + a_1X + a_0 \in \mathbb{Z}[X]$$

and $\quad g = X^m + b_{m-1}X^{m-1} + \ldots + b_1X + b_0 \in \mathbb{Z}[X].$

Since

$$a_0 b_0 = c_0,$$

it follows that p divides a_0 or b_0, but not both since p^2 does not divide c_0. Since f and g are interchangeable, we may assume without loss of generality that p divides a_0 but not b_0.

Let then i be the largest index such that p divides a_i; thus, p divides a_k for $k \leq i$ but not for $k = i+1$ (possibly, $i = n-1$; we then let $a_{i+1} = 1$). Now, the idea, as in the proof of Gauss' lemma, is to look at a well-chosen coefficient in the product fg; namely, we consider the coefficient of X^{i+1} :

$$c_{i+1} = a_{i+1}b_0 + a_i b_1 + a_{i-2}b_2 + \ldots + a_0 b_{i+1}. \qquad (3)$$

(We let $b_m = 1$ and $b_j = 0$ if $j > m$). Since $i+1 \leq n < t$, the hypothesis ensures that c_{i+1} is divisible by p; but since $a_i, a_{i-1}, \ldots, a_0$ are all divisible by p, it follows from (3) above that p divides $a_{i+1}b_0$; this is a contradiction since it was assumed that p does not divide a_{i+1} nor b_0. □

Proof of theorem 11 : If $\Phi_p(X)$ is not irreducible, then there is a factorization :

$$\Phi_p(X) = f(X)g(X)$$

for some non-constant polynomials f,g in $\mathbb{Q}[X]$. The change of variable $X = Y+1$ converts this equation into :

$$\Phi_p(Y+1) = f(Y+1)g(Y+1),$$

and since $f(Y+1)$ and $g(Y+1)$ are non-constant polynomials in $\mathbb{Q}[Y]$, it follows that $\Phi_p(Y+1)$ is reducible in $\mathbb{Q}[Y]$. But since

$$\Phi_p(Y+1) = \frac{(Y+1)^p - 1}{(Y+1) - 1},$$

it follows by expanding the numerator that

$$\Phi_p(Y+1) = Y^{p-1} + pY^{p-2} + \binom{p}{2}Y^{p-3} + \ldots + \binom{p}{2}Y + p$$

(where $\binom{p}{i} = \frac{p!}{i!(p-i)!}$ is the binomial coefficient), and this polynomial is easily seen to be irreducible by Eisenstein's criterion. Therefore, $\Phi_p(X)$ is irreducible. □

14. The importance of this theorem lies in the fact that it enables us to reduce to a standard form every rational expression in the p-th roots of unity. We denote by μ_p the set of p-th roots of unity, as in 7.3, and by $\mathbb{Q}(\mu_p)$ the set of complex numbers which are rational expressions in these p-th roots of unity. Thus, letting

$$\mu_p = \{\rho_1, \ldots, \rho_p\},$$

we have :

$$\mathbb{Q}(\mu_p) = \left\{ \frac{f(\rho_1, \ldots, \rho_p)}{g(\rho_1, \ldots, \rho_p)} \mid f, g \in \mathbb{Q}[X_1, \ldots, X_p] \text{ and } g(\rho_1, \ldots, \rho_p) \neq 0 \right\}.$$

This set is clearly a subfield of $\mathbb{C}$, since it is closed under sums, differences, products and division by non-zero elements.

Recall from proposition 7.14 (and corollary 7.17) that if ζ is any p-th root of unity other than 1, then every p-th root of unity is a power of ζ with an exponent between 0 and $p-1$: thus

$$\mu_p = \{1, \zeta, \zeta^2, \ldots, \zeta^{p-1}\}$$

and the complex numbers which are denoted by $\rho_1, \ldots, \rho_p$ above are powers of ζ. Therefore, every rational expression in $\rho_1, \ldots, \rho_p$ is a rational expression in ζ, and conversely, so that

$$\mathbb{Q}(\mu_p) = \left\{\frac{f(\zeta)}{g(\zeta)} \mid f,g \in \mathbb{Q}[X] \text{ and } g(\zeta) \neq 0\right\}.$$

15. THEOREM : Every element in $\mathbb{Q}(\mu_p)$ can be expressed in one and only one way as a linear combination with rational coefficients of the p-th roots of unity other than 1 :

$$a_1\zeta + a_2\zeta^2 + \ldots + a_{p-1}\zeta^{p-1} \qquad (a_i \in \mathbb{Q}).$$

Since some of the arguments used in the proof will be useful in various contexts, we shall quote them in full generality :

16. LEMMA : Let P and Q be polynomials with coefficients in some field F, and assume P is irreducible in F[X]. If P and Q have a common root in some field K containing F, then P divides Q.

(Compare [Gau2, Art. 346]).

<u>Proof</u> : If P does not divide Q, then P and Q are relatively prime, since P is irreducible. Corollary 5.9 then shows that there exist polynomials U,V in F[X] such that

$$P(X)U(X) + Q(X)V(X) = 1.$$

Replacing in this equality the indeterminate X by the common root u of P and Q, we obtain :

$$P(u)U(u) + Q(u)V(u) = 1 \quad \text{in K}.$$

Since $P(u) = Q(u) = 0$, this equality yields :

$$0 = 1 \quad \text{in K}.$$

This contradiction shows that P divides Q. □

__17.__ Now, consider a field F and an element u in a field K containing F. We generalize the above definition of $\mathbb{Q}(\zeta)$ (in n° 14), denoting by F(u) the set of elements in K which are rational expressions of u with coefficients in F :

$$F(u) = \left\{\frac{f(u)}{g(u)} \in K \mid f, g \in F[X] \text{ and } g(u) \neq 0\right\}.$$

The set F(u) is obviously closed under sums, differences, products and divisions by non-zero elements, and is therefore a subfield of K.

If u is a root of some non-zero polynomial $P \in F[X]$, then it is a root of some monic irreducible factor of P, since if $P = cP_1 \ldots P_r$ is the decomposition of P into prime factors according to theorem 5.13, then the relation P(u) = 0 implies that $P_i(u) = 0$ for at least one index i. Therefore, replacing P by a suitable monic irreducible factor if necessary, we may assume P itself is irreducible and monic.

__18. PROPOSITION__ : If $u \in K$ is a root of an irreducible polynomial $P \in F[X]$ of degree d, then every element in F(u) can be uniquely written in the form :

$$a_0 + a_1 u + a_2 u^2 + \ldots + a_{d-1} u^{d-1} \qquad \text{(with } a_i \in F\text{)}.$$

__Proof__ : Let f(u)/g(u) be an arbitrary element in F(u). Since $g(u) \neq 0$, the polynomial g is not divisible by P, and is therefore relatively prime to P, since P is irreducible. By corollary 5.9, there exist polynomials h and U such that

$$g(X)h(X) + P(X)U(X) = 1 \qquad \text{in } F[X].$$

Substituting u for the indeterminate X in this equation, and taking into account the fact that P(u) = 0, we get :

$$g(u)h(u) = 1 \qquad \text{in } K.$$

This shows that $f(u)/g(u)$ can be written as a __polynomial__ expression in u :

$$f(u)/g(u) = f(u)h(u) \quad \text{in } K.$$

Now, let R be the remainder of the division of fh by P :

$$fh = PQ + R \quad \text{in } F[X], \text{ with } \deg R \leq d - 1.$$

Since $P(u) = 0$, it follows that

$$f(u)h(u) = R(u) \quad \text{in } K,$$

and since $R \in F[X]$ is a polynomial of degree at most $(d-1)$, we have converted an arbitrary rational expression $f(u)/g(u) \in F(u)$ into a polynomial expression of the type :

$$a_0 + a_1 u + \ldots + a_{d-1} u^{d-1} \quad \text{(with } a_i \in F\text{)}.$$

To prove the uniqueness of this expression, assume

$$a_0 + a_1 u + \ldots + a_{d-1} u^{d-1} = b_0 + b_1 u + \ldots + b_{d-1} u^{d-1}$$

for some $a_0, \ldots, a_{d-1}, b_0, \ldots, b_{d-1} \in F$. Collecting all the terms on one side, we see that u is then a root of the polynomial

$$V(X) = (a_0 - b_0) + (a_1 - b_1)X + \ldots + (a_{d-1} - b_{d-1})X^{d-1} \in F[X].$$

From lemma 16, it follows that P divides V, but since $\deg V \leq d - 1$, this is impossible unless $V = 0$. Therefore,

$$a_0 - b_0 = a_1 - b_1 = \ldots = a_{d-1} - b_{d-1} = 0. \quad \square$$

19. REMARK : There is only one monic irreducible polynomial $P \in F[X]$ which has u as a root : indeed, if $Q \in F[X]$ is another polynomial with the same properties, then lemma 16 shows that P divides Q and, reversing the roles of P and Q, that Q divides P also. Since P and Q are both monic, it follows that P = Q. Moreover, among the non-zero polynomials in F[X] which have u as a root, P is the polynomial of least degree since it divides all the others. Therefore, this (unique) monic irreducible polynomial P ∈ F[X] which has u as a root is called the <u>minimum polynomial of u over F</u>.

<u>Proof of theorem 15</u> : We already observed in n° 14 that $\mathbb{Q}(\mu_p) = \mathbb{Q}(\zeta)$. Since ζ is a root of Φ_p, which is irreducible by theorem 11 and has degree $p-1$, it follows from proposition 18 that every element $a \in \mathbb{Q}(\mu_p)$ can be uniquely expressed in the form :

$$a = a_0 + a_1 \zeta + a_2 \zeta^2 + \ldots + a_{p-2} \zeta^{p-2} \quad (a_i \in \mathbb{Q}). \qquad (4)$$

To obtain the required form, it now suffices to use the fact that

$$\Phi_p(\zeta) = 1 + \zeta + \zeta^2 + \ldots + \zeta^{p-1} = 0; \qquad (5)$$

this shows that

$$a_0 = -a_0(\zeta + \zeta^2 + \ldots + \zeta^{p-1}),$$

whence, substituting in (4) above :

$$a = (a_1 - a_0)\zeta + (a_2 - a_0)\zeta^2 + \ldots + (a_{p-2} - a_0)\zeta^{p-2} + (-a_0)\zeta^{p-1}.$$

The uniqueness of this expression follows from the uniqueness of expression (4); indeed, if

$$a_1 \zeta + \ldots + a_{p-1} \zeta^{p-1} = b_1 \zeta + \ldots + b_{p-1} \zeta^{p-1},$$

then, using (5) to eliminate ζ^{p-1}, we get :

$$-a_{p-1} + (a_1-a_{p-1})\zeta + \ldots + (a_{p-2}-a_{p-1})\zeta^{p-2} = -b_{p-1} +$$
$$(b_1-b_{p-1})\zeta + \ldots + (b_{p-2}-b_{p-1})\zeta^{p-2}.$$

Hence, by the uniqueness of expression (4), it follows that the coefficients of $1,\zeta,\zeta^2,\ldots,\zeta^{p-2}$ in both sides are equal, and consequently :

$$a_{p-1} = b_{p-1}, \; a_1 = b_1, \ldots, a_{p-2} = b_{p-2}. \qquad \square$$

20. REMARK : For later use, we note the expression of rational numbers in the form indicated by theorem 15 : any element $a \in \mathbb{Q}$ is expressed as :

$$a = (-a)\zeta + (-a)\zeta^2 + \ldots + (-a)\zeta^{p-1},$$

as it easily follows from (5).

§ 4. The periods of cyclotomic equations

21. Let p be a prime number and let $\zeta \in \mathbb{C}$ be a primitive p-th root of unity (i.e. a p-th root of unity other than 1 : see corollary 7.17). If m and n are integers such that

$$m \equiv n \pmod{p},$$

then $\zeta^m = \zeta^n$.

Now, let g be a primitive root of p. Since the integers $g^0, g^1, g^2, \ldots, g^{p-2}$ are congruent modulo p to $1,2,3,\ldots,p-1$ (in some order), it follows that the complex numbers

$$\zeta^{g^0}, \zeta^{g^1}, \zeta^{g^2}, \ldots, \zeta^{g^{p-2}}$$

are the same as

$$\zeta, \zeta^2, \zeta^3, \ldots, \zeta^{p-1},$$

in some order. Therefore, letting $\zeta_i = \zeta^{g^i}$ for $i=0,\ldots,p-2$ to facilitate notations, the set of p-th roots of unity is:

$$\mu_p = \{1, \zeta_0, \zeta_1, \ldots, \zeta_{p-2}\}.$$

It turns out that this new ordering $\zeta_0, \zeta_1, \ldots, \zeta_{p-2}$ on the primitive p-th roots of unity is exactly the ordering for which Vandermonde's arguments in § 11.3 can be carried out. More precisely, the point is that the cyclic permutation

$$\sigma : \zeta_0 \to \zeta_1 \to \cdots \to \zeta_{p-2} \to \zeta_0,$$

extended to μ_p by setting $\sigma(1) = 1$, preserves the relations among the roots $\zeta_0, \ldots, \zeta_{p-2}$. This is the essence of the following proposition :

<u>22. PROPOSITION</u> : $\sigma(\rho\omega) = \sigma(\rho)\sigma(\omega)$ for $\rho, \omega \in \mu_p$.

<u>Proof</u> : For $i=0,\ldots,p-3$, we have $\sigma(\zeta_i) = \zeta_{i+1}$, hence, by definition of ζ_i and ζ_{i+1} :

$$\sigma(\zeta_i) = \zeta_i^g.$$

This relation also holds for $i=p-2$, since $\zeta_{p-2}^g = \zeta^{g^{p-1}}$ and $g^{p-1} \equiv g^0 \pmod{p}$ by Fermat's theorem 6. It also holds trivially with 1 instead of ζ_i; thus,

$$\sigma(\rho) = \rho^g \qquad \text{for all } \rho \in \mu_p.$$

Since $\rho\omega \in \mu_p$ for $\rho, \omega \in \mu_p$, we have :

$$\sigma(\rho\omega) = (\rho\omega)^g = \sigma(\rho)\sigma(\omega) \text{ for } \rho, \omega \in \mu_p. \qquad \square$$

<u>23</u>. For example, if $p = 11$, then we can choose $g = 2$, as we

observed in the example after the statement of theorem 2. The corresponding ordering on primitive 11-th roots of unity is the following :

$\zeta_0 = \zeta, \zeta_1 = \zeta^2, \zeta_2 = \zeta^4, \zeta_3 = \zeta^8, \zeta_4 = \zeta^5, \zeta_5 = \zeta^{10}, \zeta_6 = \zeta^9, \zeta_7 = \zeta^7,$

$\zeta_8 = \zeta^3, \zeta_9 = \zeta^6$

where we can choose for instance $\zeta = \cos(2\pi/11) + i \sin(2\pi/11)$:

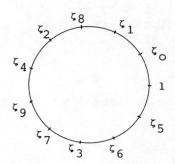

The values of $2\cos(2k\pi/11)$, which were denoted by a,b,c, d,e in § 11.3, are thus given by :

$a = 2\cos(2\pi/11) \quad = \zeta_0 + \zeta_5$

$b = 2\cos(4\pi/11) \quad = \zeta_1 + \zeta_6$

$c = 2\cos(6\pi/11) \quad = \zeta_3 + \zeta_8$

$d = 2\cos(8\pi/11) \quad = \zeta_2 + \zeta_7$

$e = 2\cos(10\pi/11) = \zeta_4 + \zeta_9$

and the permutation $\sigma : \zeta_0 \to \zeta_1 \to \ldots \to \zeta_9 \to \zeta_0$ induces the following permutation of $a,\ldots,e$:

$a \to b \to d \to c \to e \to a.$

This is the permutation which played a crucial role in § 11.3.

Thus, mimicking the arguments in our comments to Vandermonde's solution of $\Phi_{11}(X) = 0$, it is not hard to see that the cyclotomic equation $\Phi_p(X) = 0$ is solvable by radicals for every prime p. However, this result appears as secondary in Gauss' investigations and we will leave it for the next section. Gauss' primary concern is to decompose the solution of cyclotomic equations into the simplest steps as possible.

24. This decomposition is achieved as follows : for any two positive integers e,f such that

$$ef = p - 1,$$

Gauss defines e complex numbers which he calls the <u>periods of f terms</u> :

$$\eta_0 = \zeta_0 + \zeta_e + \zeta_{2e} + \ldots + \zeta_{e(f-1)}$$
$$\eta_1 = \zeta_1 + \zeta_{e+1} + \zeta_{2e+1} + \ldots + \zeta_{e(f-1)+1}$$
$$\eta_2 = \zeta_2 + \zeta_{e+2} + \zeta_{2e+2} + \ldots + \zeta_{e(f-1)+2}$$
$$\ldots\ldots$$
$$\eta_{e-1} = \zeta_{e-1} + \zeta_{2e-1} + \zeta_{3e-1} + \ldots + \zeta_{p-2}.$$

In particular, the periods of 1 term are the roots $\zeta_0, \zeta_1, \ldots, \zeta_{p-2}$, and the (unique) period of p - 1 terms is the sum of all the p-th roots of unity other than 1 or, in other words, the sum of all the roots of Φ_p. This period is therefore rational : it is the opposite of the first coefficient of Φ_p :

$$\zeta_0 + \zeta_1 + \ldots + \zeta_{p-2} = -1.$$

As a further example, for $p \geq 3$, the periods of two terms can be seen to be the values of $2\cos(2k\pi/p)$ for k=1,...,

$(p-1)/2$. This was already shown for $p = 11$ in n° 23 above, but can be proved in general by considering the form of these periods:

$$\eta_j = \zeta_j + \zeta_{j+(p-1)/2};$$

by definition of the indexing (see n° 21), we have

$$\zeta_{j+(p-1)/2} = (\zeta_j)^{g^{(p-1)/2}}$$

and since $g^{(p-1)/2} \equiv -1 \pmod{p}$ by remark 10(a), it follows that

$$\eta_j = \zeta_j + \zeta_j^{-1}.$$

Therefore, the $(p-1)/2$ periods of two terms are the roots of the equation of degree $(p-1)/2$ obtained from $\Phi_p(X) = 0$ by setting $Y = X + X^{-1}$: this proves the claim, in view of remark 7.9.

As Gauss shows, the periods of f terms thus defined have the following remarkable properties:

25. PROPERTY: Any period of f terms can be determined rationally from any other period of f terms.

26. PROPERTY: If f and g are two divisors of $p-1$ and if f divides g, then any period of f terms is a root of an equation of degree g/f whose coefficients are rational expressions of a period of g terms.

These properties will be proved in n° 34 and 36 below.

27. Thus, the periods can be used to provide remarkable examples of the step-by-step solution of equations as envisioned by Lagrange: fix a sequence of integers:

$$f_o = p-1, f_1, \ldots, f_{r-1}, f_r = 1$$

such that f_i divides f_{i-1} for $i=1,\ldots,r$, and define V_i to be a period of f_i terms (arbitrarily chosen) for $i=0,\ldots,r$. Then V_o is rational and for $i=1,\ldots,r$, the complex number V_i can be determined by solving an equation of degree f_{i-1}/f_i whose coefficients are rational expressions in V_{i-1}. Since V_r is a period of 1 term, this process eventually yields a primitive p-th root of unity. The other p-th roots of unity are then readily obtained as powers of this one.

The choice of V_i among the periods of f_i terms does not affect essentially the solution, since property 25 shows that the periods of f_i terms are rational expressions of each other.

Of course, it is not clear <u>a priori</u> that the equation which is used to determine V_i from V_{i-1} is solvable by radicals, since f_{i-1}/f_i might exceed 5, but Gauss further proves that these equations are indeed solvable by radicals for any value of f_{i-1}/f_i, including p-1. (This case occurs if r = 1). If one wants to deal with equations of the smallest degrees as possible, one can choose the sequence $f_o, f_1, \ldots, f_r$ in such a way that the successive quotients f_{i-1}/f_i are the prime factors which divide p-1, but this is in no way compulsory.

Take for instance p = 37 and look at the lattice of divisors of $p-1 = 2^2 \cdot 3^2$:

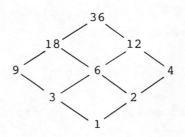

(In this diagram, a straight line indicates a relation of divisibility). To every path going down from 36 to 1 (without going up at any step) corresponds a pattern of solution of $\Phi_{37}(X) = 0$ by successive equations, whose degrees are shown by the successive quotients. For instance, if we choose the path : 36,12,6,1, then we first determine a period of 12 terms by an equation of degree $36/12 = 3$, next a period of 6 terms by an equation of degree $12/6 = 2$ and finally a period of 1 term, i.e. a primitive 37-th root of unity, by an equation of degree 6.

Instead of solving directly this last equation, one could determine a period of 3 terms by an equation of degree $6/3 = 2$ and a period of 1 term by an equation of degree 3 : this amounts to refine the proposed path into : 36,12,6, 3,1.

For $p = 17$, the lattice of divisors of $p - 1 = 2^4$ is much simpler :

```
16
 |
 8
 |
 4
 |
 2
 |
 1.
```

Thus, a primitive 17-th root of unity can be determined by solving successively 4 quadratic equations. This is the key fact which leads to the construction of the regular polygon with 17 sides by ruler and compass (see the appendix).

28. We now turn to the proof of properties 25 and 26, which we adapt from Gauss' own arguments with the added thrust

of some elementary linear algebra[*]. First, we define a map from the field $\mathbb{Q}(\mu_p)$ (defined in n° 14 above) onto itself, extending by linearity the map σ defined on μ_p in n° 21; we thus set :

$$\sigma(a_0 \zeta_0 + \ldots + a_{p-2} \zeta_{p-2}) = a_0 \sigma(\zeta_0) + \ldots + a_{p-2} \sigma(\zeta_{p-2}),$$

i.e.

$$\sigma(a_0 \zeta_0 + a_1 \zeta_1 + \ldots + a_{p-3} \zeta_{p-3} + a_{p-2} \zeta_{p-2}) = a_0 \zeta_1 + a_1 \zeta_2 + \ldots + a_{p-3} \zeta_{p-2} + a_{p-2} \zeta_0$$

and theorem 15 shows that this is sufficient to define σ on the whole of $\mathbb{Q}(\mu_p)$.

29. PROPOSITION : The map σ is a field automorphism of $\mathbb{Q}(\mu_p)$ which leaves every element of $\mathbb{Q}$ invariant.

<u>Proof</u> : That σ is bijective and that

$$\sigma(ua + vb) = u\sigma(a) + v\sigma(b) \qquad \text{for } a,b \in \mathbb{Q}(\mu_p) \text{ and } u,v \in \mathbb{Q}$$

(i.e. that σ is $\mathbb{Q}$-linear) readily follow from the definition of σ. Moreover, since by remark 20 the rational numbers $a \in \mathbb{Q}$ are written as :

$$a = (-a)\zeta_0 + (-a)\zeta_1 + \ldots + (-a)\zeta_{p-2},$$

the definition also shows that every rational number is invariant under σ. Thus, it only remains to prove :

$$\sigma(ab) = \sigma(a)\sigma(b) \qquad \text{for all } a,b \in \mathbb{Q}(\mu_p).$$

---

[*] Gauss' original arguments also use linear algebra, but expressed in an elementary way via systems of linear equations : see [Gau2, Art. 346].

This was already proved in proposition 22 in the particular case where $a, b \in \mu_p$. From this case, the general case can be derived as follows : let

$$a = \Sigma_{i=0}^{p-2} a_i \zeta_i \text{ and } b = \Sigma_{j=0}^{p-2} b_j \zeta_j \text{ (with } a_i, b_j \in \mathbb{Q} \text{ for all } i,j).$$

Then

$$ab = \Sigma_{i,j=0}^{p-2} a_i b_j \zeta_i \zeta_j$$

whence, since σ is $\mathbb{Q}$-linear :

$$\sigma(ab) = \Sigma_{i,j=0}^{p-2} a_i b_j \, \sigma(\zeta_i \zeta_j).$$

On the other hand, we have

$$\sigma(a)\sigma(b) = \Sigma_{i,j=0}^{p-2} a_i b_j \, \sigma(\zeta_i) \sigma(\zeta_j);$$

therefore, proposition 22 shows that $\sigma(ab) = \sigma(a)\sigma(b)$. □

30. REMARK : The irreducibility of Φ_p was used above in an essential, but rather implicit, way : indeed, that the map σ is well-defined on $\mathbb{Q}(\mu_p)$ results from the fact that the expression $a_0 \zeta_0 + \ldots + a_{p-2} \zeta_{p-2}$ for the elements in $\mathbb{Q}(\mu_p)$ is unique; the proof of this fact, in theorem 15, ultimately relies on the irreducibility of Φ_p.

31. Let now e and f be (positive) integers such that

$$ef = p - 1.$$

Denote by K_f the set of elements in $\mathbb{Q}(\mu_p)$ which are invariant under σ^e. Since σ, whence also σ^e, is a field automorphism of $\mathbb{Q}(\mu_p)$ which leaves $\mathbb{Q}$ elementwise invariant, the set K_f is clearly closed under sums, differences, products and divisions by non-zero elements, and contains $\mathbb{Q}$. In other words, K_f is a subfield of $\mathbb{Q}(\mu_p)$ containing $\mathbb{Q}$. Using the standard form of the elements in $\mathbb{Q}(\mu_p)$, a standard form for the elements of K_f is easily found :

32. PROPOSITION : Every element in K_f can be written in a unique way as a linear combination with rational coefficients of the e periods of f terms.

<u>Proof</u> : Let a be an arbitrary element in $\mathbb{Q}(\mu_p)$, which we write as follows :

$$a = a_0\zeta_0 \quad + a_1\zeta_1 \quad + \ldots + a_{e-1}\zeta_{e-1}$$
$$+ a_e\zeta_e \quad + a_{e+1}\zeta_{e+1} \quad + \ldots + a_{2e-1}\zeta_{2e-1}$$
$$+ \quad \ldots\ldots$$
$$+ a_{e(f-1)}\zeta_{e(f-1)} + a_{e(f-1)+1}\zeta_{e(f-1)+1} + \ldots + a_{p-2}\zeta_{p-2}.$$

Then, by definition of σ :

$$\sigma^e(a) = a_0\zeta_e \quad + a_1\zeta_{e+1} \quad + \ldots + a_{e-1}\zeta_{2e-1}$$
$$+ a_e\zeta_{2e} \quad + a_{e+1}\zeta_{2e+1} \quad + \ldots + a_{2e-1}\zeta_{3e-1}$$
$$+ \quad \ldots\ldots$$
$$+ a_{e(f-1)}\zeta_0 + a_{e(f-1)+1}\zeta_1 + \ldots + a_{p-2}\zeta_{e-1}.$$

If $\sigma^e(a) = a$, then, by theorem 15, the coefficients of ζ_i in the two expressions above are the same, for $i=0,\ldots,p-2$; hence :

$$a_0 = a_e = a_{2e} = \ldots = a_{e(f-1)}$$
$$a_1 = a_{e+1} = a_{2e+1} = \ldots = a_{e(f-1)+1}$$
$$\ldots\ldots$$
$$a_{e-1} = a_{2e-1} = a_{3e-1} = \ldots = a_{p-2}.$$

Therefore, every element $a \in K_f$ can be written as:

$$a = a_0(\zeta_0 \quad + \zeta_e \quad + \ldots + \zeta_{e(f-1)})$$
$$+ a_1(\zeta_1 \quad + \zeta_{e+1} \quad + \ldots + \zeta_{e(f-1)+1})$$
$$+ \quad \ldots\ldots$$
$$+ a_{e-1}(\zeta_{e-1} + \zeta_{2e-1} + \ldots + \zeta_{p-2}).$$

This proves that a is a linear combination of the periods, since the expressions between brackets are the periods of f terms.

The uniqueness of this expression of a readily follows from theorem 15, which asserts that every element in $\mathbb{Q}(\mu_p)$ can be written in only one way as a linear combination of $\zeta_0, \ldots, \zeta_{p-2}$. □

33. PROPOSITION: Let η be a period of f terms. Then every element in K_f can be written as:

$$a_0 + a_1\eta + a_2\eta^2 + \ldots + a_{e-1}\eta^{e-1} \quad \text{for some}$$
$$a_0, \ldots, a_{e-1} \in \mathbb{Q}.$$

Proof: Since K_f is a field containing $\mathbb{Q}$, it can be considered as a vector space over $\mathbb{Q}$ in a natural way: the vector space operations are induced by the operations in the field. To prove the proposition, it obviously suffices to show that $1, \eta, \ldots, \eta^{e-1}$ is a basis of K_f over $\mathbb{Q}$.
In fact, it even suffices to prove that $1, \eta, \ldots, \eta^{e-1}$ are linearly independent over $\mathbb{Q}$, since proposition 32 show that the e periods of f terms form a basis of K_f over $\mathbb{Q}$, hence that $\dim_{\mathbb{Q}} K_f = e$.
In order to prove this linear independence, suppose

$$a_0 \cdot 1 + a_1\eta + \ldots + a_{e-1}\eta^{e-1} = 0 \tag{6}$$

for some rational numbers $a_0,\ldots,a_{e-1}$. Then η is a root of the polynomial

$$P(X) = a_0 + a_1 X + \ldots + a_{e-1} X^{e-1}.$$

Applying σ, next σ^2, σ^3 and so on till σ^{e-1} to both sides of (6), and taking into account the fact that the coefficients a_i are invariant under σ, we observe that $\sigma(\eta), \sigma^2(\eta), \ldots, \sigma^{e-1}(\eta)$ are roots of $P(X)$ too.

Now, $\eta, \sigma(\eta), \ldots, \sigma^{e-1}(\eta)$ are the e periods of f terms, which are pairwise distinct by proposition 32. Since the polynomial $P(X)$ has degree at most $e-1$, it cannot have as roots the e periods of f terms, unless it is the zero polynomial. Therefore,

$$a_0 = \ldots = a_{e-1} = 0,$$

and this proves the linear independence of $1, \eta, \ldots, \eta^{e-1}$. □

34. COROLLARY : If η and η' are periods of f terms, then

$$\eta' = a_0 + a_1 \eta + \ldots + a_{e-1} \eta^{e-1}$$

for some rational numbers $a_0, \ldots, a_{e-1}$.

<u>Proof</u> : This readily follows from the proposition, since $\eta' \in K_f$. □

This corollary proves property 25 of the periods. In order to prove property 26, we now introduce another pair of integers g,h such that

$$gh = p - 1,$$

and assume that f divides g. Then, denoting $k = g/f = e/h$, we have :

$$\sigma^e = (\sigma^h)^k ;$$

therefore, every element invariant under σ^h is also invariant under σ^e, which means :

$$K_g \subset K_f.$$

35. PROPOSITION : Let f and g be divisors of p - 1. If f divides g, then every element in K_f is a root of a polynomial of degree g/f with coefficients in K_g.

<u>Proof</u> : For a $\in K_f$, we consider the polynomial :

$$P(X) = (X-a)(X-\sigma^h(a))(X-\sigma^{2h}(a))\ldots(X-\sigma^{h(k-1)}(a))$$

(with the same notations as above). This polynomial has degree k = g/f, and its coefficients are the elementary symmetric polynomials in $a, \sigma^h(a), \sigma^{2h}(a), \ldots, \sigma^{h(k-1)}(a)$. Since

$$\sigma^h(\sigma^{h(k-1)}(a)) = \sigma^e(a) = a,$$

the map σ^h permutes $a, \sigma^h(a), \ldots, \sigma^{h(k-1)}(a)$ among themselves and leaves therefore the coefficients of P invariant. This shows that the coefficients of P are in K_g. The polynomial P thus satisfies the required properties. □

The proof of property 26 can now be completed :

36. COROLLARY : Let f and g be divisors of p - 1 and let η and ξ be periods of f and g terms respectively. If f divides g, then η is a root of a polynomial of degree g/f whose coefficients are rational expressions of ξ.

<u>Proof</u> : Since $\xi \in K_g$ and $\eta \in K_f$, this corollary readily follows from propositions 35 and 33. □

It is instructive to note, with a view towards the modern framework of Galois theory, that the subfields K_f form a lattice of subfields of $\mathbb{Q}(\mu_p)$, which is anti-isomorphic to the lattice of divisors of $p-1$, since $K_g \subset K_f$ if and only if f divides g. Thus, if for instance $p = 37$, the periods define the following lattice of subfields of $\mathbb{Q}(\mu_{37})$:

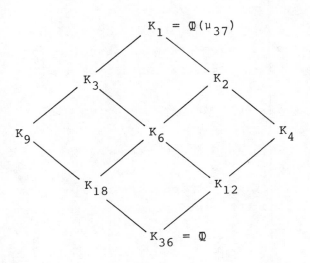

(a straight line indicates a relation of inclusion).

§ 5. Solvability by radicals

After his careful analysis of the periods of cyclotomic equations and their properties, Gauss shows in Art. 359-360 of "Disquisitiones Arithmeticae" that the equations by which the periods are determined can be solved by radicals. His exposition in this part is more sketchy and slurs at some points over a non-trivial difficulty which will be pinpointed below.

We use the notations of the preceding section. In particular, we let e,f and g,h be two pairs of integers such that

$$ef = gh = p - 1.$$

We assume that f divides g and set

$$k = g/f = e/h.$$

We denote by $\eta_0, \ldots, \eta_{e-1}$ (resp. $\xi_0, \ldots, \xi_{h-1}$) the periods of f (resp. g) terms :

$$\eta_i = \zeta_i + \zeta_{e+i} + \zeta_{2e+i} + \ldots + \zeta_{e(f-1)+i}$$

$$\xi_j = \zeta_j + \zeta_{h+j} + \zeta_{2h+j} + \ldots + \zeta_{h(g-1)+j}.$$

In corollary 36, we have seen that, when the periods $\xi_0, \ldots, \xi_{h-1}$ are considered as known, then any period η_i can be determined by an equation of degree g/f. Our aim in this section is to show that this equation is solvable by radicals.

Consider for instance the equation which yields η_0. (The arguments for the other periods is exactly the same, but the notations are more complicated). We denote this equation of degree k by $P(X) = 0$. Since the coefficients of P are in K_g, they are invariant under σ^h; hence, by repeatedly applying σ^h to both sides of the equation

$$P(\eta_0) = 0$$

we find :

$$P(\sigma^h(\eta_0)) = 0, \ P(\sigma^{2h}(\eta_0)) = 0, \ \ldots, \ P(\sigma^{h(k-1)}(\eta_0)) = 0.$$

Therefore, the roots of P are η_0 and its images under $\sigma^h, \sigma^{2h}, \ldots, \sigma^{h(k-1)}$, which are $\eta_h, \eta_{2h}, \ldots, \eta_{h(k-1)}$.

In order to prove that $P(X) = 0$ is solvable by radicals, it suffices, after Lagrange's formula 10.2, to show that the k-th power of the Lagrange resolvent :

$$t(\omega) = \eta_0 + \omega\eta_h + \omega^2\eta_{2h} + \ldots + \omega^{k-1}\eta_{h(k-1)}$$

253

(ω a k-th root of unity) can be calculated from the periods of g terms.

37. PROPOSITION : For every k-th root of unity ω, the complex number $t(\omega)^k$ has a rational expression in terms of ω and of the periods of g terms.

Proof : First, we observe that the product of any two periods of f terms can be expressed as a linear combination of the periods of f terms : this is shown by proposition 32. We thus have relations among the periods, which can be used to reduce to 1 the degree of any polynomial expression in the periods. In particular :

$$\begin{aligned}
t(\omega)^k &= (\eta_o + \omega\eta_h + \ldots + \omega^{k-1}\eta_{h(k-1)})^k \\
&= a_o\eta_o + \ldots + a_{h-1}\eta_{h-1} \\
&\quad + a_h\eta_h + \ldots + a_{2h-1}\eta_{2h-1} \\
&\quad + \ldots\ldots \\
&\quad + a_{h(k-1)}\eta_{h(k-1)} + \ldots + a_{e-1}\eta_{e-1}
\end{aligned} \quad (7)$$

where the coefficients $a_o, \ldots, a_{e-1}$ are rational (in fact : polynomial) expressions in ω over $\mathbb{Q}$.

Since the relations among the periods $\eta_o, \ldots, \eta_{e-1}$ are preserved under σ^h, by proposition 29, we can replace η_o by $\sigma^h(\eta_o) = \eta_h$, η_1 by $\sigma^h(\eta_1) = \eta_{h+1}$, etc. in the above calculation of $t(\omega)^k$. We thus find :

$$\begin{aligned}
(\eta_h + \omega\eta_{2h} + \ldots + \omega^{k-1}\eta_o)^k &= a_o\eta_h + \ldots + a_{h-1}\eta_{2h-1} \\
&\quad + a_h\eta_{2h} + \ldots + a_{2h-1}\eta_{3h-1} \\
&\quad + \ldots\ldots \\
&\quad + a_{h(k-1)}\eta_o + \ldots + a_{e-1}\eta_{h-1}.
\end{aligned} \quad (8)$$

This yields an expression of $(\sigma^h(t(\omega)))^k$. However, since

$$\sigma^h(t(\omega)) = \omega^{-1} t(\omega),$$

we have

$$(\sigma^h(t(\omega)))^k = t(\omega)^k,$$

so that (7) and (8) are two expressions of $t(\omega)^k$. Replacing in the initial calculation of $t(\omega)^k$ the period η_i by $\sigma^{2h}(\eta_i)$, next by $\sigma^{3h}(\eta_i),\ldots,\sigma^{h(k-1)}(\eta_i)$ (for $i=0,\ldots,e-1$), we still find $k-2$ other expressions of $t(\omega)^k$. Inspection shows that the coefficients of a given period η_i in these various expressions are $a_i, a_{i+h}, a_{i+2h}, \ldots, a_{i+h(k-1)}$. Therefore, if we sum up all these expressions, we get:

$$\begin{aligned}
k \cdot t(\omega)^k = &(a_0 + \ldots + a_{h(k-1)})(\eta_0 + \ldots + \eta_{h(k-1)}) \\
&+ (a_1 + \ldots + a_{h(k-1)+1})(\eta_1 + \ldots + \eta_{h(k-1)+1}) \\
&+ \quad\quad\quad \ldots\ldots \\
&+ (a_{h-1} + \ldots + a_{e-1})(\eta_{h-1} + \ldots + \eta_{e-1}).
\end{aligned}$$

Since $\eta_i + \eta_{h+i} + \ldots + \eta_{h(k-1)+i} = \xi_i$ for $i=0,\ldots,h-1$, it follows that $t(\omega)^k$ is rationally expressed in terms of ω and $\xi_0, \ldots, \xi_{h-1}$:

$$t(\omega)^k = \frac{1}{k}[(a_0 + \ldots + a_{h(k-1)})\xi_0 + \ldots + (a_{h-1} + \ldots + a_{e-1})\xi_{h-1}].$$

□

<u>38. REMARK</u> : The above proof is quite similar to that of Gauss, but the final arguments are different. Gauss argues as follows : after observing that the right-hand sides of (7) and (8) are equal, since both are expressions of $t(\omega)^k$, he draws the conclusion that the coefficients of any given period are the same in both expressions, hence :

$$a_o = a_h = a_{2h} = \cdots = a_{h(k-1)}$$
$$a_1 = a_{h+1} = a_{2h+1} = \cdots = a_{h(k-1)+1}$$
$$\cdots\cdots$$
$$a_{h-1} = a_{2h-1} = a_{3h-1} = \cdots = a_{e-1}.$$

These equalities can be used to simplify (7) to :

$$t(\omega)^k = a_o(\eta_o + \eta_h + \cdots + \eta_{h(k-1)}) + a_1(\eta_1 + \eta_{h+1} + \cdots + \eta_{h(k-1)+1}) + \cdots + a_{h-1}(\eta_{h-1} + \eta_{2h-1} + \cdots + \eta_{e-1}),$$

and this completes the proof of the proposition, since the expressions between brackets in the right-hand side are the periods of g terms.

However, the comparison of coefficients, which was also used in the proof of proposition 32 above, is justified only insofar as the expression of an element as a linear combination of $\eta_o, \ldots, \eta_{e-1}$ (or, more generally, of $\zeta_o, \ldots, \zeta_{p-2}$) is known to be unique. This was shown in theorem 15 for linear combinations with rational coefficients, which was sufficient to prove proposition 32, but here the scalars are rational expressions of a k-th root of unity ω, so new arguments are needed.

From the proof of theorem 15, it is clear that the crucial fact on which this uniqueness property ultimately relies is the irreducibility of Φ_p. Therefore, in order to justify Gauss' argument, we need to prove the irreducibility of Φ_p not only over the field $\mathbb{Q}$ of rational numbers, but over $\mathbb{Q}(\omega)$, where ω is a k-th root of unity for some integer k dividing p - 1. This will be done in the next section : see corollary 43.

To complete this section, we observe with Gauss [Gau2, Art. 360] that the full generality of periods is not needed if we only aim to show that the roots of unity can be expressed by radicals.

39. COROLLARY : For every integer n, the n-th roots of unity have expressions by radicals.

Proof : We argue by induction on n. The corollary is trivial if n = 1 or 2, so we may assume that for every integer k < n the k-th roots of unity are expressible by radicals. If n is not prime, then theorem 7.4 and the induction hypothesis readily show that the n-th roots of unity can be expressed by radicals. We may thus assume that n is prime. We then order the n-th roots of unity other than 1 as in n° 21 with the aid of a primitive root of n and we consider the Lagrange resolvent :

$$t(\omega) = \zeta_0 + \omega\zeta_1 + \ldots + \omega^{n-2}\zeta_{n-2}$$

(ω an (n-1)th root of unity). By the induction hypothesis, ω can be expressed by radicals. The preceding proposition (with k = g = n - 1) then shows that $t(\omega)^{n-1}$ has a rational expression in terms of ω, whence an expression by radicals. Lagrange's formula 10.2 now yields expressions by radicals for the n-th roots of unity :

$$\zeta_i = \frac{1}{n-1}[\Sigma_\omega \, \omega^{-i} \sqrt[n-1]{t(\omega)^{n-1}}]. \qquad \square$$

§ 6. Irreducibility of the cyclotomic polynomials

The aim of this section is to justify Gauss' argument (see remark 38), by proving the irreducibility of the cyclotomic polynomial Φ_p over $\mathbb{Q}(\mu_k)$, when p is a prime number and k is an integer which is relatively prime to p.

A proof of this result was first published by Kronecker in 1854. The proof we give is inspired by some ideas of Dedekind (see [VW1, § 60], [Web, § 174]). It is in fact valid for any integer n instead of p. Its essential step is to prove the irreducibility of Φ_n over $\mathbb{Q}$, which was first established for nonprime n by Gauss in 1808 (see [Bu, p. 74]).

40. **LEMMA** : Let f be a monic irreducible factor of Φ_n in $\mathbb{Q}[X]$ and let p be a prime number which does not divide n. If $\omega \in \mathbb{C}$ is a root of f, then ω^p also is a root of f :

$$f(\omega) = 0 \Rightarrow f(\omega^p) = 0.$$

Proof : Assume on the contrary that $f(\omega) = 0$ but $f(\omega^p) \neq 0$. Since Φ_n divides $X^n - 1$, we have :

$$X^n - 1 = f \cdot g \qquad (9)$$

for some monic polynomial $g \in \mathbb{Q}[X]$. Since $f(\omega) = 0$, it follows that

$$\omega^n = 1$$

whence also, raising both sides to the p-th power :

$$(\omega^p)^n = 1.$$

In other words, ω^p is a root of $X^n - 1$. Since on the other hand it was assumed that $f(\omega^p) \neq 0$, equation (9) implies :

$$g(\omega^p) = 0.$$

This last equality shows that ω is a root of $g(X^p)$, and it follows from lemma 16 that $f(X)$ divides $g(X^p)$:

$$g(X^p) = f(X) \cdot h(X) \qquad (10)$$

for some monic polynomial $h(X) \in \mathbb{Q}[X]$.
Gauss' lemma 12 and equations (9) and (10) show that f, g and h have integral coefficients. Therefore, we may consider the polynomials $\bar{f}$, $\bar{g}$ and $\bar{h}$ whose coefficients are the

congruence classes modulo p of the coefficients of f, g and h respectively, i.e. the images of these coefficients in $\mathbb{F}_p$ ($= \mathbb{Z}/p\mathbb{Z}$: see remark 5). By reduction modulo p, equations (9) and (10) yield :

$$X^n - 1 = \overline{f}(X).\overline{g}(X) \qquad \text{in } \mathbb{F}_p[X] \qquad (11)$$

and $\quad \overline{g}(X^p) = \overline{f}(X).\overline{h}(X) \qquad \text{in } \mathbb{F}_p[X]. \qquad (12)$

Now, Fermat's theorem 6 says that $a^p = a$ for all $a \in \mathbb{F}_p$. Therefore, if

$$\overline{g}(X) = a_0 + a_1 X + \ldots + a_{r-1} X^{r-1} + X^r,$$

then $\overline{g}(X) = a_0^p + a_1^p X + \ldots + a_{r-1}^p X^{r-1} + X^r,$

whence

$$\overline{g}(X^p) = a_0^p + a_1^p X^p + \ldots + a_{r-1}^p X^{p(r-1)} + X^{pr}$$

and since $(u+v)^p = u^p + v^p$ in $\mathbb{F}_p$ (because the binomial coefficients $\binom{p}{i}$ are divisible by p for $i=1,\ldots,p-1$), it follows that

$$\overline{g}(X^p) = (a_0 + a_1 X + \ldots + a_{r-1} X^{r-1} + X^r)^p = \overline{g}(X)^p$$

$$\text{in } \mathbb{F}_p[X].$$

Thus, equation (12) can be rewritten as :

$$\overline{g}^p = \overline{f}.\overline{h}$$

and this shows that $\overline{f}$ and $\overline{g}$ are not relatively prime. Let then $\varphi(X) \in \mathbb{F}_p[X]$ be a non-constant common factor of $\overline{f}$ and $\overline{g}$. Equation (11) shows that φ^2 divides $X^n - 1$: let

$$X^n - 1 = \varphi^2.\psi \qquad \text{in } \mathbb{F}_p[X].$$

Then, tacking the derivatives of both sides, we get :

$$nX^{n-1} = \varphi(2\ \partial\varphi\cdot\Psi + \varphi\cdot\partial\Psi),$$

whence φ divides $X^n - 1$ and nX^{n-1} : this is impossible since $X^n - 1$ and nX^{n-1} are relatively prime in $\mathbb{F}_p[X]$. (It is here that the hypothesis that p does not divide n is needed). This contradiction shows that the hypothesis $f(\omega^p) \neq 0$ was absurd. □

41. THEOREM : For every integer $n \geq 1$, the cyclotomic polynomial Φ_n is irreducible over $\mathbb{Q}$.

Proof : Let f be a monic irreducible factor of Φ_n in $\mathbb{Q}[X]$. We shall prove that every root of Φ_n in $\mathbb{C}$ is a root of f. Since the roots of Φ_n are simple, it will then follow from proposition 5.15 that Φ_n divides f, hence that $\Phi_n = f$, since f and Φ_n divide each other and are both monic.
Let ζ be a root of f; then ζ is a root of Φ_n, which means that ζ is a primitive n-th root of unity. From proposition 7.15 we recall that any other primitive n-th root of unity has the form ζ^k, where k is an integer relatively prime to n between 0 and n. Factoring k into (not necessarily distinct) prime factors :

$$k = p_1 \cdots p_s,$$

we find, by successive applications of the preceding lemma:

$$f(\zeta) = 0 \Rightarrow f(\zeta^{p_1}) = 0 \Rightarrow f(\zeta^{p_1 p_2}) = 0 \Rightarrow \ldots \Rightarrow f(\zeta^{p_1 \cdots p_{s-1}}) = 0 \Rightarrow$$
$$f(\zeta^k) = 0.$$

Thus, f has as root every primitive n-th root of unity, i.e. every root of Φ_n. □

42. THEOREM : If m and n are relatively prime integers,

then Φ_n is irreducible over $\mathbb{Q}(\mu_m)$.

Proof : Let f be a monic irreducible factor of Φ_n in $\mathbb{Q}(\mu_m)[X]$ and let $\zeta \in \mathbb{C}$ be a root of f. Arguing as above, we see that it suffices to prove :

$$f(\zeta^k) = 0$$

for every integer k relatively prime to n between 0 and n.

Let η be a primitive m-th root of unity. As in n° 14, we have $\mathbb{Q}(\mu_m) = \mathbb{Q}(\eta)$, hence, by proposition 18, every coefficient of f is a polynomial expression in η with rational coefficients; therefore,

$$f(X) = \varphi(\eta, X)$$

for some polynomial $\varphi(Y,X) \in \mathbb{Q}[Y,X]$.

Let now $\rho = \zeta\eta$. Since m and n are relatively prime, it follows from proposition 7.13 that ρ is a primitive mn-th root of unity. Moreover, since m and n are relatively prime, theorem 7.11 shows that there exist integers r and s such that

$$mr + ns = 1.$$

Since $\zeta^n = 1$ and $\eta^m = 1$, this equation implies that

$$\zeta = \zeta^{mr} = \rho^{mr} \quad \text{and} \quad \eta = \eta^{ns} = \rho^{ns}.$$

Since $f(\zeta) = 0$, we have $\varphi(\eta, \zeta) = 0$, or :

$$\varphi(\rho^{ns}, \rho^{mr}) = 0.$$

Lemma 16 and the preceding theorem then show that $\Phi_{mn}(X)$ divides $\varphi(X^{ns}, X^{mr})$ and it follows that

$$\varphi(\omega^{ns}, \omega^{mr}) = 0 \qquad (13)$$

for every primitive mn-th root of unity ω.

For any integer k relatively prime to n between 0 and n, let

$$\ell = kmr + ns.$$

Since $mr + ns = 1$, we have $mr \equiv 1 \pmod{n}$ and $ns \equiv 1 \pmod{m}$, whence :

$$\ell \equiv k \pmod{n} \quad \text{and} \quad \ell \equiv 1 \pmod{m}. \qquad (14)$$

It follows that $\zeta^\ell = \zeta^k$ and $\eta^\ell = \eta$, and since we already observed that $\zeta = \rho^{mr}$ and $\eta = \rho^{ns}$, we have :

$$\rho^{\ell mr} = \zeta^k \quad \text{and} \quad \rho^{\ell ns} = \eta.$$

On the other hand, relations (14) also show that ℓ is relatively prime to mn. Therefore, ρ^ℓ is a primitive mn-th root of unity, and equation (13) yields :

$$\varphi(\rho^{\ell ns}, \rho^{\ell mr}) = 0,$$

i.e. $\varphi(\eta, \zeta^k) = 0,$

or $f(\zeta^k) = 0.$ □

43. <u>COROLLARY</u> : Let p be a prime number and let k be an integer which divides p-1. Let also $\zeta \in \mathbb{C}$ be a primitive p-th root of unity. Then every element in $\mathbb{Q}(\mu_k)(\mu_p)$ can be uniquely written in the form :

$$a_1 \zeta + a_2 \zeta^2 + \ldots + a_{p-1} \zeta^{p-1},$$

for some $a_1, \ldots, a_{p-1}$ in $\mathbb{Q}(\mu_k)$.

Proof : The hypothesis on k ensures that k is relatively prime to p, hence Φ_p is irreducible over $\mathbb{Q}(\mu_k)$, by the preceding theorem. The corollary then follows by the same arguments as in the proof of theorem 15. □

Appendix to chapter 12 : Ruler and compass construction of regular polygons

We aim to find a process to construct regular polygons in the plane, using ruler and compass only. It is clear that this construction should be possible whenever the center of the polygon (i.e. the center of the circumscribed circle) and one of its vertices are arbitrarily chosen. Therefore, we may regard the center O and one of the vertices A as given, and we have to determine the other vertices. From the two given points O and A, new points can be constructed by a (finite) sequence of operations of the following types :

1) draw a line through two points already determined.

2) draw a circle with center a point already determined and radius the distance between two points already determined.

New points are determined as intersection points of the lines or circles drawn according to (1) and (2). The points which can be thus determined are called : <u>constructible</u> points.

The problem is thus to decide for which values of n the vertices of the regular polygon with n sides, with center O and A as one of the vertices, are constructible. To solve this problem, we first give an algebraic characterization of the constructible points, via their coordinates in a suitable basis, which we construct as follows : we consider the perpendicular to OA through O and denote by B one of the intersection points of this perpendicular and the circle with center O and radius OA :

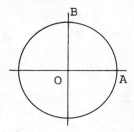

(Observe that the point B is constructible).

PROPOSITION : A point in the plane can be constructed by ruler and compass from O and A if and only if its coordinates in the basis (OA,OB) can be obtained from 0 and 1 by a (finite) sequence of operations of the following types :

i) rational operations.

ii) extraction of square roots.

Proof : First, we show that the points whose coordinates satisfy the condition above are constructible.
Since the perpendicular through a given point to a given line can be constructed by ruler and compass, a point with coordinates (a,b) is constructible if (and only if) the points (a,0) and (0,b) are constructible. Moreover, since (0,b) is the intersection of the axis OB with the circle with center O passing through (b,0), it suffices to consider points with coordinates (u,0). So, we have to prove that a point with coordinates (u,0) is constructible with ruler and compass if u is obtained from 0 and 1 by a sequence of operations (i) and (ii) above.
We argue inductively on the number of operations. Thus, we shall prove that if (u,0) and (v,0) are constructible, then (u+v,0), (u-v,0), (uv,0), (u/v,0) (assuming $v \neq 0$) and ($\sqrt{u}$,0) (assuming $u \geq 0$) are constructible.
This is clear for (u+v,0) and (u-v,0). In order to cons-

truct (uv,0) and (u/v,0), we consider the figure below :

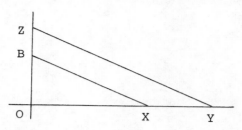

Since BX and YZ are parallel,

$$\frac{OX}{OB} = \frac{OY}{OZ}.$$

Since $B = (0,1)$, it follows that, denoting $X = (x,0)$, $Y = (y,0)$ and $Z = (0,z)$:

$$x = y/z, \quad \text{or} \quad y = xz.$$

Therefore, if we regard X and Z as given, we can construct Y by drawing the parallel to BX through Z; this construction yields (xz,0) from (x,0) and (0,z) (or, equivalently, (z,0)). On the other hand, if we regard Y and Z as given, then we can obtain X by drawing the parallel to YZ through B; this yields (y/z,0) from (y,0) and (0,z) (or (z,0)).

To complete the proof of the "if" part, it only remains to show that $(\sqrt{u},0)$ can be constructed from (u,0) (assuming $u \geq 0$). This can be done as follows :

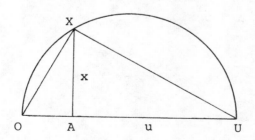

Let U be the point with coordinates (1+u,0) and let X be one of the intersection points of the perpendicular to OU

265

through A with the circle with diameter OU. We thus have
X = (1,x) for some x. Since the triangles OAX and XAU are
similar, we have

$$\frac{AX}{OA} = \frac{AU}{AX},$$

whence

$$\frac{x}{1} = \frac{u}{x}$$

or $x = \sqrt{u}$.

Since the point (x,0) can be easily determined from (1,x),
this construction yields $(\sqrt{u},0)$ from (u,0).
We have thus proved that the points whose coordinates are
obtained from 0 and 1 by rational operations and extraction
of square roots are constructible.

To prove the converse, we first observe that if a line
passes through two points (a_1,b_1) and (a_2,b_2), then its
equation has the form :

$$AX + BY = C$$

where A,B and C are rational expressions of a_1, a_2, b_1 and
b_2. (Specifically, $A = (b_2-b_1)$, $B = (a_1-a_2)$ and
$C = (b_1a_2-a_1b_2)$). Likewise, the equation of a circle with
center (a_1,b_1) and radius the distance between (a_2,b_2)
and (a_3,b_3) is :

$$(X-a_1)^2 + (Y-b_1)^2 = (a_2-a_3)^2 + (b_2-b_3)^2$$

hence it has the form :

$$X^2 + Y^2 = AX + BY + C$$

where A,B, C are rational expressions of a_1,a_2,a_3,b_1,b_2,b_3.

Now, direct calculations show that the coordinates of the intersection point of two lines

$$A_1 X + B_1 Y = C_1 \quad \text{and} \quad A_2 X + B_2 Y = C_2$$

are rational expressions of $A_1, B_1, C_1, A_2, B_2, C_2$. Thus, if a point is constructed as the intersection of two lines passing through given points, its coordinates are rational expressions of the coordinates of the given points. Similarly, it can be seen that the coordinates of the intersection point of a line and a circle :

$$A_1 X + B_1 Y = C_1 \quad \text{and} \quad X^2 + Y^2 = A_2 X + B_2 Y + C_2$$

are obtained by rational operations and extraction of a square root from $A_1, B_1, C_1, A_2, B_2, C_2$. Therefore, if a point is constructed as the intersection of a line through given points and a circle with given center and with radius the distance between given points, then its coordinates are obtained from the coordinates of the given points by rational operations and extraction of a square root. Finally, the intersection of two circles :

$$X^2 + Y^2 = A_1 X + B_1 Y + C_1 \quad \text{and} \quad X^2 + Y^2 = A_2 X + B_2 Y + C_2$$

can be obtained as the intersection of the circle

$$X^2 + Y^2 = A_1 X + B_1 Y + C_1$$

and the line

$$A_1 X + B_1 Y + C_1 = A_2 X + B_2 Y + C_2,$$

hence the same conclusion as for the preceding case holds. These arguments show that the coordinates of the constructible points are obtained by operations (i) and (ii) from the coordinates of O and A, i.e. from 0 and 1. □

This result seems to have been first published by Pierre Laurent Wantzel (1814-1848) in 1837, but it was undoubtedly known to Gauss (and presumably also to others) around 1796.

THEOREM : If p is a prime number of the form :

$$p = 2^m + 1 \qquad (m \in \mathbb{N})$$

then the regular polygon with p sides can be constructed with ruler and compass.

Proof : Since p-1 is a power of 2, the lattice of divisors of p-1 is a chain :

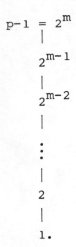

Therefore, the results of section 4 (see in particular n° 27) show that the periods of two terms can be determined by solving a sequence of quadratic equations. Since, by n° 24, the periods of two terms are the values $2\cos(2k\pi/p)$, for $k=1,\ldots,(p-1)/2$, and since the solution of a quadratic equation only requires rational operations and extraction of square roots, it follows that $\cos(2\pi/p)$ can be obtained from the integers (or even from 0 and 1) by rational operations and extraction of square roots. The preceding proposition then shows that the point with coordinates $(\cos(2\pi/p), 0)$ is constructible.

The point $P = (\cos(2\pi/p), \sin(2\pi/p))$ can then be obtained as the intersection of the circle with center O and radius OA with the perpendicular to OA through $(\cos(2\pi/p), 0)$. The point P is a vertex of the regular polygon with p sides. In fact, it is one of the two vertices which is closest to A, and the other vertices can be found by reproducing the distance AP on the circle. □

If a prime number p has the form $2^m + 1$, then it is easily seen that m is a power of 2 : indeed, if m is divisible by some odd integer k, then $2^m + 1$ is divisible by $2^{m/k} + 1$, as can be seen by letting $X = 2^{m/k}$ in the relation :

$$X^k + 1 = (X+1)(X^{k-1} + X^{k-2} + \ldots + X + 1).$$

Thus, the prime numbers which satisfy the hypothesis of the proposition are in fact of the form :

$$p = 2^{2^n} + 1$$

for some integer n. These prime numbers are called <u>Fermat primes</u>, after Pierre de Fermat, who conjectured that the number

$$F_n = 2^{2^n} + 1$$

is prime for every integer n. For n=0,1,2,3,4, this formula yields 3,5,17,257 and 65537, which are indeed prime, but in 1732 Euler showed that

$$F_5 = 641 \times 6700417.$$

Since then, the numbers F_n have been shown to be composite for various values of n, and no new Fermat prime has been found. Although it has not been proved that no other Fermat prime exists, it is at least known that there is no such prime between 65538 and 10^{39456} (i.e. F_n is not prime for $5 \leq n \leq 16$).

COROLLARY : The regular polygon with n sides can be constructed with ruler and compass if n is a product of distinct Fermat primes and of a power of 2.

Proof : Since the regular polygon with n sides is constructible when n is a power of 2 (by repeated bisections of angles) or when n is a Fermat prime (by the preceding theorem), it suffices to show that when n_1 and n_2 are relatively prime integers such that the regular polygons with n_1 and n_2 sides are constructible, then the regular polygon with $n_1 n_2$ sides is constructible.
If n_1 and n_2 are relatively prime, theorem 7.11 shows that there exist integers m_1 and m_2 such that

$$m_1 n_1 + m_2 n_2 = 1.$$

Multiplying both sides by $2\pi/n_1 n_2$, we get :

$$m_1 (2\pi/n_2) + m_2 (2\pi/n_1) = (2\pi/n_1 n_2).$$

Therefore, the arc $(2\pi/n_1 n_2)$ can be constructed by reproducing a certain number of times the arcs $(2\pi/n_1)$ and $(2\pi/n_2)$, and it readily follows that the regular polygon with $(n_1 n_2)$ sides can be constructed from the regular polygons with n_1 and n_2 sides. □

Remark : It can be proved that the converses of the theorem and of the corollary above also hold; thus, the regular polygon with n sides can be constructed with ruler and compass if and only if n is a product of distinct Fermat primes and of a power of 2.

This result is explicitly stated (without proof) by Gauss [Gau2, Art. 366], but a smooth proof of these converses requires a detailed analysis of field degrees, which would carry us too far afield. Therefore, we refer the interested reader to [Car, chap. 4] or [St, ch. 17] for a proof. We also refer to [HW, § 5.8] for an explicit geometric construction of the 17-gon with ruler and compass.

Exercises for chapter 12

1) Recall from exercise 7 of chapter 7 that for every integer $n \geq 2$, the number of integers which are relatively prime to n between 0 and n is denoted by $\varphi(n)$. Prove the following generalization (due to Euler) of Fermat's theorem 6 : $a^{\varphi(n)} \equiv 1 \pmod{n}$ for every integer $n \geq 2$ and every integer a relatively prime to n.

2) Show that Girard's theorem 6.1 readily follows from Gauss' lemma 12.

3) Prove that the periods with an even number of terms are real numbers.

4) Prove that the set of periods of f terms does not depend on the choice of a primitive root of p nor of a primitive p-th root of unity. More precisely, let ζ and $\zeta' \in \mathbb{C}$ be primitive p-th roots of unity, let $g, g' \in \mathbb{Z}$ be primitive roots of p and let $p-1 = e \cdot f$ for some positive integers e, f. Denote $\zeta_i = \zeta^{g^i}$ and $\zeta'_i = \zeta'^{g'^i}$ for $i = 0, \ldots, p-2$. Show that for any $i = 0, \ldots, f-1$, there is an integer j between 0 and f-1 such that

$$\zeta_i + \zeta_{e+i} + \ldots + \zeta_{e(f-1)+i} = \zeta'_j + \zeta'_{e+j} + \ldots + \zeta'_{e(f-1)+j}.$$

5) With the notations of n° 31, prove that if $K_g \subset K_f$, then f divides g. Moreover, show that in this case,

$$\dim_{K_g} K_f = g/f.$$

The following exercise provides complementary observations to the proof of corollary 39 :

6) Notations as in corollary 39. Show that $t(\omega) \neq 0$ and that $t(\omega^k)/t(\omega)^k$ has a rational expression in terms of ω, for $k \in \mathbb{Z}$. (Compare exercise 4 of chapter 10). Conclude that it suffices to extract a single $(n-1)$-th root to determine ζ_i.

7) By looking at the algebraic expression of 5-th roots of unity, find a construction of regular pentagons by ruler and compass. Find also a construction of regular 20-gons.

13 Ruffini and Abel on general equations

§ 1. Introduction

Lagrange's investigations were primarily aimed at the solution of 'general' equations, i.e. equations whose coefficients are letters :

$$x^n - s_1 x^{n-1} + s_2 x^{n-2} - \ldots + (-1)^n s_n = 0$$

(see definition 8.1). At about the same time when Gauss completed the solution of the class of particular equations which arise from the division of the circle (known as cyclotomic equations), Lagrange's line of investigation bore new fruits in the hands of Paolo Ruffini (1765-1822). In 1799, Ruffini published a massive two-volume treatise : "Teoria Generale delle Equazioni" [R, t. 1, pp. 1-324], in which he proves that the general equations of degree at least 5 are not solvable by radicals.

Ruffini's proof was received with scepticism by the mathematical community. Indeed, the proof was rather hard to follow through the 516 pages of his books. A few years after the publication, negative comments were made but, to Ruffini's dismay, no clear, focused objection was raised. Vague criticism was denying Ruffini the credit of having validly proved his claim. Negative reactions prompted Ruffini to simplify his proof, and he eventually came up with very neat arguments, but distrust of Ruffini's work did not subside. Typical in this respect is the following anecdote : in order to get a clear, motivated pronouncement from the French Academy of Sciences, Ruffini submitted a paper to the Academy in 1810. A year later, the referees (Lagrange, Lacroix and Legendre) had not yet given their conclusions. Ruffini then wrote to Delambre, who wa secretary of the Academy, to withdraw his paper. In his reply, Delambre explains the referees' attitude :

> "Whatever decision Your Referees would have reached, they needed a very considerable work either to motivate their approval or to refute Your proof. You know how precious is time to realize also how reluctant most geometers are to occupy themselves for a long time with the works of each other, and if they would have happened no to be of Your opinion, they would have had to be moved by a quite powerful motive to enter the lists against a geometer so learned and so skilful" [R, t. 3, p. 59].

At least, unconvincing as it was, Ruffini's proof seems to have completed the reversal of the current opinion towards general equations : while the works of Bezout and Euler around the middle of the eighteenth century were grounded on the opinion that general equations were solvable, and that finding the solution of the fifth degree equations wa only a matter of clever transformations, the opposite view became common in the beginning of the nineteenth century (see [Ay, p. 274]). Some comments of Gauss may also have been influential in this respect. In his proof of the fundamental theorem of algebra, [Gau1, § 9], Gauss writes :

> "After the works of many geometers left very little hope of ever arriving at the resolution of the general equation algebraically, it appears more and more likely that this resolution is impossible and contradictory"

and he voiced again the same scepticism in "Disquisitiones Arithmeticae" [Gau2, Art. 359].

Ruffini's credit also includes advances in the theory of permutations, which was crucial for his proof. Ruffini's results in this direction were soon generalized by Cauchy. Incidently, it is noteworthy that Cauchy was very appreciative of Ruffini's work and that he supported Ruffini's claim that his proof was valid [R, t. 3, pp. 88-89]. In fact, it now appears that Ruffini's proofs do have a significant gap, which we shall point out below.

In 1824, a new proof [A, n° 3] was found by Niels-Henrik Abel (1802-1829), independently of Ruffini's work. An expanded version of Abel's proof was published in 1826 in the first issue of Crelle's journal (the Journal für die reine und angewandte Mathematik) [A, n° 7]. This proof also contains some minor flaws (see [A, vol. 2, pp. 292-293]), but it essentially settled the issue of solvability of general equations.

Abel's approach is remarkably methodical. He explains it in some detail in the introduction to a subsequent paper : "Sur la résolution algébrique des équations" (1828) [A, n° 18] :

"To solve these equations [of degree at most 4], a uniform method has been found, and it was believed that it could be applied to equations of arbitrary degree; but in spite of the efforts of a Lagrange and other distinguished geometers, one was not able to reach this aim. This led to the presumption that the algebraic solution of general equations was impossible; but that could not be decided, since the method which was used could not lead to definite conclusions except in the case where the equations were solvable. Indeed, the purpose was to solve equations, without knowing whether this was possible. In this case, one could get the solution, although that was not sure at all; but if unfortunately the solution happened to be impossible, one could have sought it for ever without finding it. In order to obtain unfailingly something in this matter, it is therefore necessary to take another way. One has to cast the problem in such a form that it be always possible to solve, which can be done with any problem. Instead of seeking a relation of which it is not known whether it exists or not one has to seek whether such a relation is indeed possible. For instance, in the integral calculus, instead of trying by a kind of divination or by trial and error to integrate differential formulas, one has to look rather whether it is possible to integrate them in this or that way. When a problem is thus presented, the statement itself contains the seed of the solution and shows the **way that** is to be taken; and I think that there will be few cases where one could not reach more or less important propositions, even when one could not completely solve the question because the calculations would be too complicated".

The method which is thus advocated by Abel can be interpreted in the realm of algebraic equations as a kind of generic method : one has to find the most general form of the expected solution and work on it to investigate what kind of information can be obtained on this expression if it is a root of the general equation. Abel thus proves, by an intricate inductive argument, that if an expression by radicals is a root of the general equation of some degree, then every function of which it is composed is a rational expression of the roots (see theorem 13 for a precise statement). This fills a gap in Ruffini's proofs. Some delicate arguments involving the number of values of functions under permutations of the variables and, in particular, a theorem of Cauchy generalizing earlier results of Ruffini, complete the proof. This last part of the proof can be significantly streamlined by using arguments from the last of Ruffini's proofs, as Wantzel later noticed. In the following sections, we shall present this easy version, but we point out that this approach unfortunately downplays the advances in the theory of permutations (i.e. in the study of the symmetric group S_n) which were prompted by Ruffini's earlier work.

§ 2. Radical extensions

Abel's calculations with expressions by radicals, which we discuss in this section and the following as a first step in the proof that general equations of degree higher than 4 are not solvable, can be adequately cast into the vocabulary of field extensions. This point of view will be used throughout since it is probably more enlightening for the modern reader.

An expression by radicals is constructed from some quantities which are regarded as known (usually the coefficients of an equation, in this context) by the four usual operations of arithmetic and the extraction of roots : this means that any such expression lies in a field obtained from the field of rational expressions in the known quantities by successive adjunctions of roots of some orders. In fact, it is clearly sufficient to consider roots of prime order, since if

$$n = p_1 \ldots p_r$$

is the factorization of a positive integer n into prime factors, then

$$a^{1/n} = (\ldots((a^{1/p_1})^{1/p_2})\ldots)^{1/p_r}.$$

This shows that a n-th root of any element a can be obtained by extracting a p_1-th root a^{1/p_1} of a, next a p_2-th root of a^{1/p_1} and so on. Moreover, it obviously suffices to extract p-th roots of elements which are not p-th powers, otherwise the base field is not enlarged. We thus come to the notion of radical extensions of fields. Before spelling out this notion in mathematical terms, we note that, in order to avoid some technical difficulties, we shall restrict attention throughout the chapter to fields of characteristic zero; in other words, we shall assume that $1 + 1 + \ldots + 1 \neq 0$ or that every field under consideration contains (an isomorphic copy of) the field $\mathbb{Q}$ of rational numbers. This is of course the classical case, which was the only case considered by Ruffini and Abel.

1. DEFINITIONS : A field R containing a field F is called a <u>radical extension of height 1</u> of F if there exist a prime number p, an element $a \in F$ which is not a p-th power in F and an element $u \in R$ such that :

$$R = F(u) \quad \text{and} \quad u^p = a.$$

Such an element u is sometimes denoted by $a^{1/p}$ or $\sqrt[p]{a}$, and, accordingly, one sometimes writes :

$$R = F(a^{1/p}) \quad \text{or} \quad R = F(\sqrt[p]{a}).$$

This is in fact an abuse of notations, since the element u is not uniquely determined by a and p : there are indeed p different p-th roots of a. Worse still, the field R itself is in general not uniquely determined by F, a and p : for instance, there are three subfields of $\mathbb{C}$ which qualify as $\mathbb{Q}(2^{1/3})$. (See however exercises 4 and 5). Therefore, the notations above will be used with caution.

Radical extensions of height h, for any positive integer h, are defined inductively as radical extensions of height 1 of radical extensions of height h - 1. More precisely, a field R containing a field F is called a <u>radical extension of height h</u> of F if there is a field R_1 between R and F such that R is a radical extension of height 1 of R_1 and R_1 is a radical extension of height h - 1 of F. Thus, in this case we can find a tower of extensions between R and F :

$$R \supset R_1 \supset R_2 \supset \ldots \supset R_{h-1} \supset F$$

such that, letting $R = R_0$ and $F = R_h$, we have for $i=0,\ldots,h-1$:

$$R_i = R_{i+1}(a_i^{1/p_i})$$

for some prime number p_i and some element $a_i \in R_{i+1}$ which is not a p_i-th power in R_{i+1}.

We simply term <u>radical extension</u> any radical extension of some (finite) height and, for completeness, we say that any field is a <u>radical extension of height 0</u> of itself.

The definition above is quite convenient to translate into mathematically amenable terms the questions concerning expressions by radicals. For instance, to say that a complex number z has an expression by radicals means that there is a radical extension of the field $\mathbb{Q}$ of rational numbers containing z. More generally, we shall say that an element v of a field L <u>has an expression by radicals over some field F contained in L</u> if there is a radical extension of F containing v.

Likewise, we say that a polynomial equation $P(X) = 0$ over some field F is <u>solvable by radicals over F</u> if there is a radical extension of F containing a root of P. In the case of general equations :

$$P(X) = (X-x_1)\ldots(X-x_n) = X^n - s_1 X^{n-1} + s_2 X^{n-2} - \ldots + $$
$$+ (-1)^n s_n = 0,$$

we are concerned with radical expressions involving only the coefficients $s_1,\ldots,s_n$, so the base field F will be the field of rational fractions in $s_1,\ldots,s_n$ (which can be considered as independent indeterminates, according to remark 8.13(a)). To be more precise, we have to specify a field of reference in which the rational fractions are allowed to take their coefficients. A logical choice is of course the

field $\mathbb{Q}$ of rational numbers, but in fact, since we are aim-
ing at a negative result, the reference field can be chosen
arbitrarily large : indeed, we shall prove that if an equation is solvable by radicals over some field F, then it is
solvable by radicals over every field L containing F; therefore, if the general equation of degree n is not solvable
over $\mathbb{C}(s_1,\ldots,s_n)$, it is not solvable over $\mathbb{Q}(s_1,\ldots,s_n)$
either.

Of course, Ruffini and Abel did not address in these
terms the problem of assigning a reference field, but their
rather free use of roots of unity suggests that all the
roots of unity are at their disposal in the base field.
The choice $F = \mathbb{C}(s_1,\ldots,s_n)$ seems therefore closer in spirit to Ruffini's and Abel's work.

The hypothesis that the base field contains all the
roots of unity also has a technical advantage, in that it
allows more flexibility in the treatment of radical extensions, as the next result shows :

2. **PROPOSITION** : Let R be a field containing a field F. If
R has the form $R = F(u)$ for some element u such that $u^n \in F$
for some integer n, and if F contains a primitive n-th root
of unity (hence all the n-th roots of unity, since the
other roots are powers of this one), then R is a radical
extension of F.

In other words, in the definition of radical extensions, we
need not require that the exponent n be a prime number, nor
that u^n be not the n-th power of an element in F, provided
that F contains a primitive n-th root of unity.

Proof : We argue by induction on n. If $n = 1$, then $u \in F$,
hence $R = F$ and R is then a radical extension of height 0
of F. We may thus assume that $n \geq 2$ and that the proposition holds when the exponent of u is at most $n - 1$.

If n is not prime, let

$$n = rs$$

for some (positive) integers $r, s < n$. By the induction hypothesis, $F(u)$ is a radical extension of $F(u^r)$ and $F(u^r)$ is a radical extension of F, since u^r satisfies : $(u^r)^s \in F$. Therefore, $F(u)$ is a radical extension of F, since it is clear from the definition that the property of being radical is transitive, namely : in a tower of extensions $F \subset K \subset L$, if L is a radical extension of K and K is a radical extension of F, then L is a radical extension of F.

If n is prime, we consider two cases, according to whether u^n is or is not the n-th power of an element in F. If it is not, then R is a radical extension of F, by definition. If it is, let

$$u^n = b^n$$

for some $b \in F$. If $b = 0$, then $u = 0$ and $R = F$, a radical extension of height 0 of F. If $b \neq 0$, then the preceding equation yields :

$$(u/b)^n = 1,$$

hence u/b is a n-th root of unity. Since the n-th roots of unity are all in F, it follows that $u/b \in F$, hence $u \in F$ and again : $R = F$, a radical extension of height 0 of F. □

As an application, we have the following result, which will be useful later through its corollary :

3. PROPOSITION : Let R and L be subfields of a field K, both containing a subfield F. Assume F contains the field $\mathbb{C}$ of complex numbers, so that all the roots of unity are in F. If R is a radical extension of F, then there is a radical extension S of L containing R and contained in K.

Proof : We argue by induction on the height of R. If this height is zero, then R = F and we can choose S = L. We may thus let the height of R be h ⩾ 1 and assume that the proposition holds for radical extensions of height at most h - 1. By definition of radical extensions of height h, we can find inside R a radical extension R_1 of F of height h - 1 and an element u such that

$$R = R_1(u)$$

and $u^p \in R_1$

for some prime number p. By the induction hypothesis, there is a radical extension S_1 of L in K which contains R_1. Then $u^p \in S_1$ and proposition 2 shows that $S_1(u)$ is a radical extension of L. This extension is contained in K, since $u \in K$ and $S_1 \subset K$, and it contains R, since $R = R_1(u)$ and $R_1 \subset S_1$. It thus fulfills the required conditions.   □

4. COROLLARY : Let $v_1, \ldots, v_n$ be elements of a field K containing a field F. Assume that F contains $\mathbb{C}$ and that each of $v_1, \ldots, v_n$ lies in a radical extension of F contained in K. Then there is a single radical extension of F in K which contains all of $v_1, \ldots, v_n$.

Proof : We argue by induction on n. There is nothing to prove if n = 1, so we may assume that n ⩾ 2 and that the corollary holds for n - 1 elements. Hence, there is a radical extension L of F in K which contains $v_1, \ldots, v_{n-1}$. Let R be a radical extension of F in K containing v_n. The preceding proposition shows that there is a radical extension S of L in K containing R. Since S contains both L and R, it contains $v_1, \ldots, v_n$. Since moreover S is a radical extension of L, which is a radical extension of F, it is a radical extension of F.   □

So far, we have dealt only with the case where roots of unity are in the base field. In order to reduce more general situations to this case, we have to use Gauss' result that any root of unity has an expression by radicals. Since we now have a formal definition for 'expression by radicals', it seems worthwhile to spell out how Gauss' arguments actually fit in this framework.

<u>5. PROPOSITION</u> : For any integer n and any field F, the n-th roots of unity lie in a radical extension of F.

<u>Proof</u> : It suffices to show that a primitive n-th root of unity ζ lies in a radical extension of F, since the other n-th roots of unity are powers of ζ and lie therefore in the same radical extension as ζ.

We argue by induction on n. For n = 1, we have $\zeta = 1$, hence ζ lies in F, which is a radical extension of height 0 of itself. We may thus assume that $n \geq 2$ and that the proposition holds for roots of unity of exponent less than n.

If n is not prime, let

$$n = rs$$

for some (positive) integers r,s < n. Then ζ^r is a s-th root of unity. By the induction hypothesis, we can find a radical extension R_1 of F containing ζ^r. By the induction hypothesis again, we can find a radical extension R_2 of R_1 (hence also of F) which contains a primitive r-th root of unity. Then, since $\zeta^r \in R_2$, it follows from proposition 2 that $R_2(\zeta)$ is a radical extension of R_2, hence of F. The proposition is thus proved in this case.

If n is prime, then we have to resort to Gauss' results : first, we can find a radical extension R_1 of F which contains the (n-1)-th roots of unity, by the induction hypothesis. We then consider the Lagrange resolvents $t(\omega)$ as in the proof of corollary 12.39. By proposition 12.38,

$$t(\omega)^{n-1} \in R_1$$

for every (n-1)-th root of unity ω; hence proposition 2 shows that $R_1(t(\omega))$ is a radical extension of R_1. Adjoining successively all the Lagrange resolvents $t(\omega)$, we find a radical extension R_2 of R_1, whence of F, which contains $t(\omega)$ for all $\omega \in \mu_{n-1}$. From Lagrange's formula 10.2, it follows that ζ can be rationally calculated from the Lagrange resolvents, hence $\zeta \in R_2$ and the proof is complete. □

We now aim to prove the afore-mentioned fact that solvability of an equation by radicals over some field F implies solvability by radicals over any larger field L. This fact may seem obvious, since every expression by radicals involving elements of F is an expression by radicals involving elements of L; it needs however a careful justification : the point is that, in building radical extensions or expressions by radicals, we allow only extractions of p-th roots of elements which are not p-th powers in F, but these elements could become p-th powers in the larger field L.

6. LEMMA : Let L be a field containing a field F. For any radical extension R of F, there is a radical extension S of L such that R can be identified to a subfield of S.

Proof : We argue by induction on the height h of R. If $h = 0$, then $R = F$ and we can choose $S = L$.

If $h = 1$, let $R = F(u)$ where u is such that

$$u^p = a$$

for some element $a \in F$ which is not a p-th power in F. Let also K be a field containing L and over which the polynomial $x^p - a$ splits into a product of linear factors. (The

existence of such a field K follows from Girard's theorem
9.3). Since u is one of the roots of $X^p - a$, it can be
identified with an element in K, and every rational fraction in u with coefficients in F, i.e. every element in R,
is then identified with an element in K. We shall thus
henceforth assume that R is contained in K.

If a is not a p-th power in L, then L(u) is a radical
extension of height 1 of L, and this extension contains R
since it contains F and u : it thus fulfills the required
conditions.

If a is a p-th power in L, then let $b \in L$ be a p-th root
of a :

$$b^p = a.$$

Since the p-th powers of u and b are equal, it follows that

$$(u/b)^p = 1.$$

Therefore, u/b is a p-th root of unity, and proposition 5
shows that there is a radical extension S of L which contains u/b. Since $b \in L$, it follows that $u \in S$, hence
$R \subset S$ and the proof is complete in the case where the
height h of R is 1.

If $h \geq 2$, the lemma readily follows from the preceding
case and the induction hypothesis. Indeed, we can find in
R a subfield R_1 which is a radical extension of height
$h - 1$ of F and such that R is a radical extension of height
1 of R_1. By the induction hypothesis, we can assume that
R_1 is contained in a radical extension S_1 of L and, by the
case $h = 1$ already considered, R can be identified to a
subfield of a radical extension S of S_1. The field S is
then a radical extension of L and it satisfies the condition of the lemma. □

7. THEOREM : Let P be a polynomial with coefficients in a field F. If $P(X) = 0$ is solvable by radicals over F, then it is solvable by radicals over every field L containing F.

Proof : Let R be a radical extension of F containing a root r of P. The preceding lemma shows that we can assume R is contained in some radical extension S of L. The radical extension S then contains the root r, hence $P(X) = 0$ is solvable by radicals over L. □

The following special case of the theorem is particularly relevant for this chapter :

8. COROLLARY : If the general equation of degree n :

$$P(X) = (X-X_1)\ldots(X-X_n) = X^n - s_1 X^{n-1} + \ldots + (-1)^n s_n = 0$$

is not solvable by radicals over $\mathbb{C}(s_1,\ldots,s_n)$, then it is not solvable by radicals over $\mathbb{Q}(s_1,\ldots,s_n)$ either.

We may thus henceforth assume that the base field contains all the roots of unity.

§ 3. Abel's theorem on natural irrationalities

Any proof that the general equation of some degree is not solvable by radicals obviously proceeds <u>ad absurdum</u> : we thus assume on the contrary that there is a radical extension R of $\mathbb{C}(s_1,\ldots,s_n)$ containing a root X_i of the general equation

$$(X-X_1)\ldots(X-X_n) = X^n - s_1 X^{n-1} + s_2 X^{n-2} - \ldots + (-1)^n s_n = 0.$$

The first step in Abel's proof (which was missing in Ruffini's proofs) is to show that R can be supposed to lie inside $\mathbb{C}(X_1,\ldots,X_n)$: this means that the irrationalities which occur in an expression by radicals for a root of the general equation of degree n can be chosen to be <u>natural</u>, as opposed to <u>accessory</u> irrationalities, which denote the elements of extensions of $\mathbb{C}(s_1,\ldots,s_n)$ outside $\mathbb{C}(X_1,\ldots,X_n)$ [Ay, p. 268]. (The terms 'natural' and 'accessory' irrationalities were coined by Kronecker).

The aim of this section is to prove this result, following Abel's approach [A, n° 7, § 2].

9. LEMMA : Let p be a prime number and let a be an element of some field F, which is not a p-th power in F.
a) For $k = 1,\ldots,p-1$, the k-th power a^k is not a p-th power in F either.
b) The polynomial $X^p - a$ is irreducible over F.

Proof : a) If k is an integer between 1 and $p-1$, then it is relatively prime to p, whence by theorem 7.11 we can find integers ℓ and q such that

$$pq + k\ell = 1.$$

Then

$$a = (a^q)^p \cdot (a^k)^\ell.$$

Therefore, if

$$a^k = b^p$$

for some $b \in F$, then it follows that

$$a = (a^q b^\ell)^p,$$

in contradiction with the hypothesis that a is not a p-th power in F. This contradiction proves a).
b) : Let P and Q be polynomials in F[X] such that

$$X^p - a = PQ.$$

We may assume that P and Q are monic, and we have to prove that P or Q is the constant polynomial 1. Let K be an extension of F over which $X^p - a$ splits into a product of linear factors. (The existence of such a field follows from Girard's theorem 9.3). Since the roots of $X^p - a$ are the p-th roots of a, which are obtained from any of them by multiplication by the various p-th roots of unity (see § 7.3), we have in K[X] :

$$\prod_{\omega \in \mu_p} (X - \omega u) = PQ$$

where $u \in K$ is one of the p-th roots of a in K. This relation shows that P and Q split in K[X] into products of factors $X - \omega u$; more precisely, μ_p decomposes into a union of disjoint subsets I and J such that

$$P = \prod_{\omega \in I} (X - \omega u) \quad \text{and} \quad Q = \prod_{\omega \in J} (X - \omega u).$$

Consider then the independent term of P, which we denote by b. The above factorization of P shows that

$$b = (\prod_{\omega \in I} \omega) u^k$$

where k denotes the number of elements of I. Since $\omega^p = 1$ for any $\omega \in I$, we get by raising both sides of the preceding equality to the p-th power :

$$b^p = a^k.$$

Part a) of the lemma then shows that $k = 0$ or $k = p$. In the first case $P = 1$ and in the second $P = X^p - a$, whence $Q = 1$. □

Let now R be a radical extension of height 1 of some field F. By definition, this means that there exists an element $u \in R$ such that $R = F(u)$ and

$$u^p = a$$

for some element $a \in F$ which is not a p-th power in F. The preceding lemma allows us to give a standard form to the elements of R :

10. COROLLARY : Every element $v \in R$ can be written in a unique way as :

$$v = v_0 + v_1 u + v_2 u^2 + \ldots + v_{p-1} u^{p-1},$$

for some elements $v_0, v_1, \ldots, v_{p-1} \in F$.

Proof : This readily follows from proposition 12.18, by the preceding lemma. □

In fact, when $v \in R$ is given beforehand outside F, then the element u can be chosen in such a way that $v_1 = 1$ in the expression above, as we now show :

11. LEMMA : Let R be a radical extension of height 1 of some field F and let $v \in R$. If $v \notin F$, then the element $u \in R$ as above can be chosen in such a way that

$$v = v_0 + u + v_2 u^2 + \ldots + v_{p-1} u^{p-1}$$

for some $v_0, v_1, \ldots, v_{p-1} \in F$.

Proof : Let u' be an element of R such that $R = F(u')$ and

$$u'^p = a'$$

for some element $a' \in F$ which is not a p-th power in F. By corollary 10, we can write :

$$v = v_0' + v_1' u' + v_2' u'^2 + \ldots + v_{p-1}' u'^{p-1}, \quad (v_0', \ldots, v_{p-1}' \in F)$$

289

and $v'_1, \ldots, v'_{p-1}$ are not all zero since $v \notin F$. Let k be an index between 1 and p-1 such that $v'_k \neq 0$, and let

$$u = v'_k u'^k. \tag{1}$$

Raising both sides of this equation to the p-th power, we get :

$$u^p = v'^p_k a'^k.$$

This shows that u satisfies the equation :

$$u^p = a$$

with $a = v'^p_k a'^k \in F$.

If a is the p-th power of an element in F, then the preceding equation shows that a'^k also is a p-th power in F. But then, it follows from lemma 9(a) that a' itself is a p-th power in F, which contradicts the hypothesis on u'. Therefore, a is not a p-th power in F.

Since $u \in R$, we obviously have $F(u) \subset R$. In order to prove that $R = F(u)$, it thus suffices to show that every element in R has a rational expression in u with coefficients in F. We first show that the powers of u' have such expressions. For any $i = 0, \ldots, p-1$, we get by raising both sides of equation (1) to the power i :

$$u^i = v'^i_k u'^{ki}. \tag{2}$$

Now, recall the permutation σ_k of $\{0, 1, \ldots, p-1\}$ which maps every integer i between 0 and p-1 to the unique integer $\sigma_k(i)$ between 0 and p-1 such that :

$$\sigma_k(i) \equiv ik \pmod{p}$$

(see proposition 10.7). The last relation above shows that

$$ik - \sigma_k(i) = p.m$$

for some integer m (depending on i and k), whence

$$u'^{ik} = (u'^p)^m \, u'^{\sigma_k(i)}.$$

Therefore, recalling that $u'^p = a'$ and letting

$$b_i = (v_k'^i \, a'^m)^{-1} \qquad \text{for } i=0,\ldots,p-1,$$

we get from equation (2) :

$$b_i u^i = u'^{\sigma_k(i)} \qquad \text{for } i=0,\ldots,p-1.$$

Now, every $x \in R$ has an expression :

$$x = \Sigma_{i=0}^{p-1} x_i u'^i, \qquad (x_i \in F \text{ for } i=0,\ldots,p-1)$$

which can be alternatively written as :

$$x = \Sigma_{i=0}^{p-1} x_{\sigma_k(i)} \, u'^{\sigma_k(i)},$$

as σ_k is a permutation of $\{0,\ldots,p-1\}$.

Substituting $b_i u^i$ for $u'^{\sigma_k(i)}$, we obtain :

$$x = \Sigma_{i=0}^{p-1} x_{\sigma_k(i)} \, b_i u^i.$$

This shows that every element in R has a rational expression in u with coefficients in F, whence $R = F(u)$. For the given $v \in R$, the coefficient of u in this expression is 1, since taking $i = 1$ in the calculations above, we find $\sigma_k(1) = k$ and $m = 0$, whence $b_1 = v_k'^{-1}$. This completes the proof. □

12. LEMMA : We keep the same notations as in lemma 11 and assume moreover that F contains a primitive p-th root of unity ζ (whence all the p-th roots of unity, since the others are powers of ζ). If v is a root of an equation with coefficients in F, then R contains p roots of this equation, and $u, v_o, v_2, \ldots, v_{p-1}$ are rational expressions of these roots with coefficients in $\mathbb{Q}(\zeta)$.

Proof : Let $P \in F[X]$ be such that $P(v) = 0$. Using the expression of v as in lemma 11, we define from P another polynomial Q with coefficients in F :

$$Q(Y) = P(v_o + Y + v_2 Y^2 + \ldots + v_{p-1} Y^{p-1}) \in F[Y].$$

This definition is designed so that the equation $P(v) = 0$ yields :

$$Q(u) = 0.$$

On the other hand, u is also a root of the polynomial $Y^p - a$, which is irreducible by lemma 9(b). Therefore, lemma 12.16 shows that $Y^p - a$ divides $Q(Y)$, and it follows that every root of $Y^p - a$ is a root of $Q(Y)$. Since the roots of $Y^p - a$ are the p-th roots of a, which are of the form $\zeta^i u$, for $i=0, \ldots, p-1$, we have

$$Q(\zeta^i u) = 0 \qquad \text{for } i=0, \ldots, p-1.$$

Let then

$$z_i = v_o + \zeta^i u + v_2 \zeta^{2i} u^2 + \ldots + v_{p-1} \zeta^{(p-1)i} u^{p-1} \quad \text{for } i=0, \ldots, p-1;$$

the preceding equation yields

$$P(z_i) = 0 \qquad \text{for } i=0, \ldots, p-1,$$

which proves that R contains p roots of P. To complete

the proof, we now show that $u, v_0, v_2, \ldots, v_{p-1}$ are rational expressions of $z_0, \ldots, z_{p-1}$, by calculations which are reminiscent of Lagrange's formula 10.2. Grouping the terms which contain a given factor $v_j u^j$ in the sum of $\zeta^{-ik} z_i$, we have:

$$\Sigma_{i=0}^{p-1} \zeta^{-ik} z_i = \Sigma_{j=0}^{p-1} (\Sigma_{i=0}^{p-1} \zeta^{(j-k)i}) v_j u^j \quad \text{for } k=0, \ldots, p-1, \qquad (3)$$

where we have let $v_1 = 1$. If $j \neq k$, then ζ^{j-k} is a p-th root of unity other than 1, whence a root of

$$\Phi_p(X) = \Sigma_{i=0}^{p-1} X^i;$$

therefore,

$$\Sigma_{i=0}^{p-1} \zeta^{(j-k)i} = 0.$$

Hence, all the terms with index $j \neq k$ vanish in the right-hand side of (3), and it remains:

$$\Sigma_{i=0}^{p-1} \zeta^{-ik} z_i = p\, v_k u^k \quad \text{for } k=0, \ldots, p-1.$$

This proves that $v_k u^k$ is a rational expression (indeed a linear expression) of $z_0, \ldots, z_{p-1}$ with coefficients in $\mathbb{Q}(\zeta_p)$. In particular, for $k=1$, we see that u is such an expression, and since $v_k = (v_k u^k) u^{-k}$, it follows that $v_0, v_2, \ldots, v_{p-1}$ also are rational expressions of $z_0, \ldots, z_{p-1}$ with coefficients in $\mathbb{Q}(\zeta_p)$. □

Now, we let

$$K = \mathbb{C}(X_1, \ldots, X_n),$$

where $X_1, \ldots, X_n$ are independent indeterminates over $\mathbb{C}$, and we denote by F the subfield of symmetric fractions. By theorem 8.3, we have:

$$F = \mathbb{C}(s_1,\ldots,s_n),$$

where $s_1,\ldots,s_n$ are the elementary symmetric polynomials in $X_1,\ldots,X_n$.

13. THEOREM (of natural irrationalities) : If an element $v \in K$ lies in a radical extension of F, then there is inside K a radical extension of F containing v.

Proof : We argue by induction on the height of the radical extension R of F containing v, which is assumed to exist. There is nothing to prove if the height of R is 0 (i.e. if R = F) since in this case R lies inside K. We may thus assume the height of R is $h \geq 1$ and consider R as a radical extension of height 1 of some subfield R_1, which is a radical extension of F of height h-1.

If $v \in R_1$, then we are done by the induction hypothesis. For the rest of the proof, we may thus assume that v lies outside R_1. Lemma 11 then shows that

$$R = R_1(u)$$

for some element u such that $u^p \in R_1$ (for some prime p) and

$$v = v_0 + u + v_2 u^2 + \ldots + v_{p-1} u^{p-1} \qquad (4)$$

for some elements $v_0, v_2, \ldots, v_{p-1} \in R_1$. Now, proposition 10.1 (and its proof) show that every element in K is a root of a polynomial with coefficients in F, which splits into a product of linear factors over K (its roots are the various 'values' of the element under the permutations of $X_1,\ldots,X_n$). In particular, v is a root of an equation with coefficients in F (whence in R_1), whose roots all lie in K. Therefore, we can apply lemma 12 to conclude that $u, v_0, v_2, \ldots, v_{p-1} \in K$.

But $u^p, v_0, v_2, \ldots, v_{p-1}$ also lie in R_1, which is a radical extension of height h-1 of F. By the induction hypothesis, $u^p, v_0, v_2, \ldots, v_{p-1}$ all lie in radical extensions of F inside K and, by corollary 4, we can find a single radical extension R' of F inside K containing $u^p, v_0, v_2, \ldots, v_{p-1}$. Since $u^p \in R'$, the field R'(u) is a radical extension of R', hence a radical extension of F. Since moreover we have already observed that $u \in K$, we have $R'(u) \subset K$, and equation (4) shows that $v \in R'(u)$. This completes the proof. □

§ 4. Proof of the unsolvability of general equations of degree higher than 4

In order to prove that general equations of degree higher than 4 are not solvable by radicals, we have to show, according to definitions 1 above, that for $n \geq 5$ there is no radical extension of $\mathbb{C}(s_1, \ldots, s_n)$ containing a root X_i of the general equation of degree n :

$$(X-X_1) \ldots (X-X_n) = X^n - s_1 X^{n-1} + \ldots + (-1)^n s_n = 0.$$

The proof we give below is based upon Ruffini's last proof (1813) [R, vol. 2, pp. 162-170]. It is sometimes called the Wantzel modification of Abel's proof [Se, n° 516], [R, vol. 2, p. 505], although Wantzel was relying on Ruffini's papers [Ay, p. 270].

14. LEMMA : Let u and a be elements of $\mathbb{C}(X_1, \ldots, X_n)$ such that

$$u^p = a$$

for some prime number p, and assume $n \geq 5$. If a is invariant under the permutations

$$\sigma : X_1 \to X_2 \to X_3 \to X_1; \quad X_i \to X_i \quad \text{for } i > 3$$

and $\tau : X_3 \to X_4 \to X_5 \to X_3; \quad X_i \to X_i \quad \text{for } i=1,2 \text{ and } i > 5,$

then so is u.

Proof : Applying σ to both sides of the equation $u^p = a$, we get : $\sigma(u)^p = a$, hence

$$\sigma(u)^p = u^p.$$

Since the lemma is trivial if $u = 0$, we may assume $u \neq 0$ and divide both sides of the preceding equation by u^p :

$$(\sigma(u) u^{-1})^p = 1,$$

whence $\quad \sigma(u) = \omega_\sigma u$

for some p-th root of unity ω_σ. Applying σ to both sides of this last equation, we get : $\sigma^2(u) = \omega_\sigma^2 u$, next $\sigma^3(u) = \omega_\sigma^3 u$. Since σ^3 is the identity map, we have $\sigma^3(u) = u$, whence

$$\omega_\sigma^3 = 1. \tag{5}$$

Arguing similarly with τ instead of σ, we find :

$$\tau(u) = \omega_\tau u$$

with $\omega_\tau^3 = 1.$ (6)

From these relations, we also deduce :

$$\sigma\tau(u) = \omega_\sigma \omega_\tau u \quad \text{and} \quad \sigma^2\tau(u) = \omega_\sigma^2 \omega_\tau u.$$

However, since

$$\sigma\tau : X_1 \to X_2 \to X_3 \to X_4 \to X_5 \to X_1; \quad X_i \to X_i \text{ for } i > 5$$

and

$$\sigma^2\tau : X_1 \to X_3 \to X_4 \to X_5 \to X_2 \to X_1; \; X_i \to X_i \text{ for } i > 5,$$

we have $(\sigma\tau)^5 = (\sigma^2\tau)^5 = I$, whence the arguments above yield :

$$(\omega_\sigma \omega_\tau)^5 = (\omega_\sigma^2 \omega_\tau)^5 = 1. \tag{7}$$

Since

$$\omega_\sigma = \omega_\sigma^6 (\omega_\sigma \omega_\tau)^5 (\omega_\sigma^2 \omega_\tau)^{-5},$$

relations (5) and (7) yield :

$$\omega_\sigma = 1.$$

From (7), we then deduce : $\omega_\tau^5 = 1$, and since

$$\omega_\tau = \omega_\tau^6 \omega_\tau^{-5},$$

it follows from relation (6) that $\omega_\tau = 1$. This shows that u is invariant under σ and τ. □

15. COROLLARY : Let R be a radical extension of $\mathbb{C}(s_1,\ldots,s_n)$ in $\mathbb{C}(X_1,\ldots,X_n)$. If $n \geq 5$, then every element of R is invariant under the permutations σ and τ of lemma 14.

<u>Proof</u> : We argue by induction on the height of R, which we denote by h. If $h = 0$, then $R = \mathbb{C}(s_1,\ldots,s_n)$ and the corollary is obvious. If $h \geq 1$, then there is an element $u \in R$ and a radical extension R_1 of height (h-1) of $\mathbb{C}(s_1,\ldots,s_n)$ such that

$$R = R_1(u)$$

and $u^p \in R_1$

for some prime number p. By induction, we may assume that every element of R_1 is invariant under σ and τ. The lemma then shows that u is also invariant under σ and τ, and, since the elements in R are rational expressions of u, it readily follows that every element in R is invariant under σ and τ. □

We thus reach the conclusion :

16. THEOREM : If $n \geq 5$, the general equation of degree n :

$$P(X) = (X-X_1)\ldots(X-X_n) = X^n - s_1 X^{n-1} + \ldots + (-1)^n s_n = 0$$

is not solvable by radicals over $\mathbb{Q}(s_1,\ldots,s_n)$, nor over $\mathbb{C}(s_1,\ldots,s_n)$.

Proof : According to corollary 8, it suffices to show that $P(X) = 0$ is not solvable by radicals over $\mathbb{C}(s_1,\ldots,s_n)$. Assume on the contrary that there is a radical extension R of $\mathbb{C}(s_1,\ldots,s_n)$ containing a root X_i of P. Changing the numbering of $X_1,\ldots,X_n$ if necessary, we may assume that $i = 1$. Moreover, by theorem 13 (of natural irrationalities), this radical extension R can be assumed to lie within $\mathbb{C}(X_1,\ldots,X_n)$. Then, corollary 15 shows that every element of R is invariant under σ and τ. But $X_1 \in R$ and X_1 is not invariant under σ : this is a contradiction. □

Exercises for chapter 13

1) Show that over $\mathbb{R}$ and over $\mathbb{C}$, every equation of any degree is solvable by radicals.

2) Show that the general cubic equation

$$(X-X_1)(X-X_2)(X-X_3) = X^3 - s_1 X^2 + s_2 X - s_3 = 0$$

is solvable by radicals over $\mathbb{Q}(s_1,s_2,s_3)$. Construct explicitly a radical extension of $\mathbb{Q}(s_1,s_2,s_3)$ containing one of the roots of this cubic and show that this radical extension is not contained in $\mathbb{Q}(X_1,X_2,X_3)$: thus, the solution of the general cubic equation by radicals over $\mathbb{Q}(s_1,s_2,s_3)$ involves accessory irrationalities.

Same questions for the general equation of degree four.

3) Let ζ_7 (resp. ζ_3) be a primitive 7-th (resp. cube) root of unity. Show that $\mathbb{Q}(\zeta_7)$ is not a radical extension of $\mathbb{Q}$, but that $\mathbb{Q}(\zeta_7,\zeta_3)$ is a radical extension of $\mathbb{Q}$.

4) Let R be a radical extension of a field F, of the form $R = F(a^{1/p})$, for some $a \in F$ which is not a p-th power in F. Find an isomorphism leaving F elementwise invariant :

$$F[X]/(X^p - a) \xrightarrow{\sim} R.$$

Conclude that all the fields of the form $F(a^{1/p})$ are isomorphic, under isomorphisms leaving F invariant.

5) Show that there are three different subfields of $\mathbb{C}$ of the form $\mathbb{Q}(2^{1/3})$. Show that if F is a subfield of $\mathbb{C}$ containing a primitive p-th root of unity, then for any $a \in F$ which is not a p-th power in F there is only one subfield of $\mathbb{C}$ of the form $F(a^{1/p})$.

To make up partially for the lack of details on the early stages of the theory of groups in Ruffini's and Cauchy's works, the following exercise presents a result of Cauchy on the number of values of rational fractions under permutations of the indeterminates, which was used in Abel's proof that general equations are not solvable by radicals :

6) Let n be an integer, $n \geqslant 3$, let $\Delta = \Delta(X_1,\ldots,X_n)$ be the polynomial defined in n° 8.14 and let $I(\Delta) \subset S_n$ be the isotropy group of Δ, i.e. :

$$I(\Delta) = \{\sigma \in S_n \mid \sigma(\Delta) = \Delta\}.$$

(This subgroup of S_n is called the <u>alternating group</u> : see n° 14.34).

a) Show that any permutation of n elements is a composite of permutations which interchange two elements and leave the other elements invariant. (Permutations of this type are called <u>transpositions</u>).

b) Show that a permutation leaves Δ invariant if and only if it is a composite of an even number of transpositions.

c) Let p be an odd prime, $p \leq n$. Show that the cyclic permutations of length p :

$$i_1 \to i_2 \to \ldots \to i_p \to i_1$$

(where $i_1, \ldots, i_p \in \{1, \ldots, n\}$) generate $I(\Delta)$.

[Hint : by (b), it suffices to show that the composite of any two transpositions is a composite of cycles of length p].

d) Let again p be an odd prime, $p \leq n$, and let V be a rational fraction in $X_1, \ldots, X_n$ which takes strictly less than p values under the permutations of $X_1, \ldots, X_n$. Show that V has the form : $V = R + \Delta S$ where R and S are symmetric rational fractions, hence that the number of values of V is 1 or 2.

[Hint : show that V is invariant under the cyclic permutations of length p].

14 Galois

§ 1. Introduction

After Gauss, Ruffini and Abel, the two major classes of equations have been treated thoroughly, with divergent results : the cyclotomic equations of any degree are solvable by radicals, while the general equations of degree at least five are not. Thus, the next obvious question arises : which are the equations that are solvable by radicals ?

Abel himself addressed this question and returned several times to the theory of equations, which he called his "thème favori" [A, t. II, p. 260]. Following a clue from Gauss, he discovered a large class of solvable equations, which contains in particular the cyclotomic equations. In the introduction to the seventh chapter of Disquisitiones Arithmeticae, in which he discusses cyclotomic equations, Gauss had written :

> "Moreover, the principles of the theory that we are about to explain extend much farther than we let it see here. Indeed, they apply not only to circular functions, but also with the same success to numerous other transcendental functions, e.g. to these which depend on the integral $\int \frac{dx}{\sqrt{(1-x^4)}}$ and also to various kinds of congruences" [Gau2, Art. 335].

The integral $\int \frac{dx}{\sqrt{(1-x^4)}}$ occurs in the calculation of the length of an arc of the lemniscate, as the integral

$$\int \frac{dx}{\sqrt{(1-x^2)}} = \sin^{-1} x$$

occurs for the arc of the circle.

Following this clue, Abel realized that Gauss' method for cyclotomic equations could also be applied to the equations which arise from the division of the lemniscate. In complete analogy with Gauss' results on the constructibility of regular polygons with ruler and compass, Abel even proved that the lemniscate can be divided into $2^n + 1$ equal parts by ruler and compass, whenever $2^n + 1$ is a prime number (see [A, t. II, p. 261]). Pushing his investigations further, Abel eventually came up to the following grand generalization (published in 1829) :

THEOREM (Abel) : Let P be a polynomial with roots $r_1, \ldots, r_n$. If the roots $r_2, \ldots, r_n$ can be rationally expressed in terms of r_1, i.e. if there exist rational fractions $\theta_2, \ldots, \theta_n$ such that

$$r_i = \theta_i(r_1) \qquad i = 2, \ldots, n$$

and if moreover

$$\theta_i \theta_j (r_1) = \theta_j \theta_i (r_1) \qquad \text{for all } i, j$$

then the equation $P(X) = 0$ is solvable by radicals [A, t. I, p. 479].

This theorem applies in particular to cyclotomic equations:

$$\Phi_p(X) = X^{p-1} + X^{p-2} + \ldots + X + 1 = 0 \qquad (p \text{ prime}) :$$

indeed, the roots of Φ_p are the primitive p-th roots of unity and, denoting one of the roots by ζ, the other roots are powers of ζ; rational fractions $\theta_2, \ldots, \theta_{p-1}$ as above can thus be chosen to be :

$$\theta_i(X) = X^i \qquad i=2,\ldots,p-1$$

so the above condition obviously holds :

$$\theta_i\theta_j(\zeta) = \zeta^{i+j} = \theta_j\theta_i(\zeta).$$

Elaborating again on these results, Abel was closing in on general necessary and sufficient conditions for an equation to be solvable by radicals. He was working on a comprehensive memoir on this subject [A, t. II, n° 18], when he was prematurely carried off by tuberculosis in 1829.

The honour of finding a complete solution to the problem eventually fell to another young genius, Evariste Galois (1811-1832), who was only 18 in 1830 when he submitted to the Paris Academy of Sciences a memoir on the theory of equations. In this memoir, he described what is now known as the Galois group of an equation, and applied this new tool to derive conditions for an equation to be solvable by radicals. The referee was Jean-Baptiste Joseph Fourier (1768-1830), who died a few weeks later. Galois' memoir was then lost in Fourier's papers (see however [G, pp. 103-109]). The next year, a second memoir was submitted by Galois, but rejected by the Academy because it was not sufficiently developed. Galois died in a duel the following year, without having had the occasion to submit a more thorough (or rather, a less sketchy) exposition of his ideas. His "Mémoire sur les conditions de résolubilité des équations par radicaux" [G, pp. 43 ff] (see also [E, App. 1]) is indeed very concise and makes rather rough reading. Fortunately, Joseph Liouville (1809-1882) generously took the trouble to decipher Galois' memoir, and he published it in 1846 with some explanations of his own, thus rescuing Galois theory from complete oblivion.

The basic idea of Galois is to associate to any(*) equation a group of permutations of the roots. This group consists of all the permutations which preserve the relations among the roots; it thus shows to what extent the roots are interchangeable. Galois' brilliant insight was that this group provides an effective measure of the difficulty of an equation; in particular, the solvability of the equation by radicals can be translated in terms of the associated group. This is achieved by describing the behaviour of the group under extension of the base field. Of these fertile new ideas, Galois offers a single application, proving that irreducible equations of prime degree are solvable by radicals if and only if any of the roots can be rationally expressed by two of them.

This brief summary of Galois' memoir does not do justice to the novelty of the ideas it contains. Indeed, it is not clear at all how to characterize the permutations which preserve the relations among the roots in this general context. (For the particular case of cyclotomic equations, see §§ 11.3 and 12.4). This difficulty seems especially overwhelming if one avoids making use of the notion of field, which is the central notion in Galois theory, but which was not available at the time when Galois wrote his memoir. Galois solves the problem by using the irreducibility of polynomials with awesome virtuosity. The concept of field (and of extension of fields) becomes transparent in the first few lines of his memoir, where he emphasizes that in his discussion of irreducibility, the base field can be arbitrary :

> "Definitions. An equation is said to be reducible if it admits rational divisors; otherwise it is irreducible.
> It is necessary to explain what is meant by the word rational, because it will appear frequently.

(*) almost any, in fact : see 1 below.

When the equation has coefficients that are all numeric and rational, this means simply that the equation can be decomposed into factors which have coefficients that are numeric and rational.

But when the coefficients of an equation are not <u>all</u> numeric and rational, one must mean by a rational divisor a divisor whose coefficients can be expressed as rational functions of the coefficients of the proposed equation, and, more generally, by a rational quantity a quantity that can be expressed as a rational function of the coefficients of the proposed equation.

More than this : one can agree to regard as rational all rational functions of a certain number of determined quantities, supposed to be known <u>a priori</u>. For example, one can choose a particular root of a whole number and regard as rational every rational function of this radical.

When we agree to regard certain quantities as known in this manner, we shall say that we <u>adjoin</u> them to the equation to be resolved. We shall say that these quantities are <u>adjoined</u> to the equation.

With these conventions, we shall call <u>rational</u> any quantity which can be expressed as a rational function of the coefficients of the equation and of a certain number of <u>adjoined</u> quantities arbitrarily agreed upon. (...) One sees, moreover, that the properties and the difficulties of an equation can be altogether different, depending on what quantities are adjoined to it" [E, pp. 101-102].

Our discussion of Galois' memoir follows Galois' own order of propositions. We thus begin with the definition of the Galois group of an equation, next investigate the behaviour of the Galois group under extension of the base field, and deduce a necessary and sufficient condition for an equation to be solvable by radicals, in terms of its Galois group.

In the final section, the application of this condition to irreducible equations of prime degree will be described. In an appendix, we review Galois' notation for groups of permutations, which deviates from the modern notation.

§ 2. The Galois group of an equation

In this chapter, as in the preceding one, we consider only fields of characteristic zero, so that we are allowed to divide by non-zero integers unconcernedly. Another word of caution : we shall often have to replace in a rational fraction $f \in F(X_1,\ldots,X_n)$ the indeterminates $X_1,\ldots,X_n$ by elements $a_1,\ldots,a_n$ in F, yielding an element $f(a_1,\ldots,a_n)$ $\in$ F. Whenever this is done, it is implicitly assumed that the rational fraction f can be represented in the form :

$$f = P/Q$$

where P and Q are polynomials in $F[X_1,\ldots,X_n]$ such that $Q(a_1,\ldots,a_n) \neq 0$; we then set :

$$f(a_1,\ldots,a_n) = P(a_1,\ldots,a_n)/Q(a_1,\ldots,a_n).$$

<u>1</u>. For technical reasons which will be pointed out below (see lemma 9), Galois associates a group to equations with simple roots only. This is not a serious restriction, since Hudde's method transforms any equation into an equation with simple roots : see theorem 5.26 and remark 5.28. For the rest of this section, we shall thus consider a monic polynomial P(X) of degree n over a field F, which has n distinct roots in some field containing F (see Girard's theorem 9.3) :

$$P(X) = X^n - a_1 X^{n-1} + a_2 X^{n-2} - \ldots + (-1)^n a_n = (X-r_1)\ldots(X-r_n)$$

with $a_1,\ldots,a_n$ in F and $r_1,\ldots,r_n$ in some field containing F.

Extending slightly the notation of 12.17, we denote by $F(r_1,\ldots,r_n)$ the field of rational fractions in $r_1,\ldots,r_n$ with coefficients in F; thus

$$F(r_1,\ldots,r_n) = \{f(r_1,\ldots,r_n) \mid f \in F(X_1,\ldots,X_n)\}.$$

It is worth emphasizing that, since $r_1,\ldots,r_n$ are not independent indeterminates over F, an element in $F(r_1,\ldots,r_n)$ can be written in the form $f(r_1,\ldots,r_n)$ in more than one way : for instance, 0 can be written as $P(r_i)$ for any $i=1,\ldots,n$. This is very important in view of the fact that we shall consider permutations σ of $r_1,\ldots,r_n$; although $f(\sigma(r_1),\ldots,\sigma(r_n))$ is well-defined for any rational fraction $f \in F(X_1,\ldots,X_n)$ whose denominator does not vanish for $X_i = \sigma(r_i)$, defining $\sigma(f(r_1,\ldots,r_n))$ by :

$$\sigma(f(r_1,\ldots,r_n)) = f(\sigma(r_1),\ldots,\sigma(r_n)) \qquad (1)$$

requires caution, since it is not clear that the right-hand side depends on $f(r_1,\ldots,r_n)$ only (and not on the rational fraction $f(X_1,\ldots,X_n)$). More precisely, we have to check that if g is another rational fraction such that :

$$g(r_1,\ldots,r_n) = f(r_1,\ldots,r_n),$$

then

$$g(\sigma(r_1),\ldots,\sigma(r_n)) = f(\sigma(r_1),\ldots,\sigma(r_n)).$$

If this is not the case, then equation (1) does not make sense.

The distinction between the _form_ of an element $f(r_1,\ldots,r_n)$ of $F(r_1,\ldots,r_n)$ (i.e. the rational fraction $f(X_1,\ldots,X_n)$) and the _value_ of this element (in $F(r_1,\ldots,r_n)$) is emphasized by Galois himself :

"Here we call a function invariant not only if its form is unchanged by the substitutions of the roots, but also if its numerical value does not vary when these substitutions are applied" [G, p. 50] [E, p. 104].

<u>2</u>. In order to define the Galois group of the equation $P(X) = 0$, some preliminary results are needed. The proofs of these results will be given later, to avoid interrupting by lengthy proofs the course of reasoning.

<u>RESULT 1</u> : There is an element $V \in F(r_1,\ldots,r_n)$ such that

$$r_i \in F(V) \qquad \text{for } i=1,\ldots,n.$$

(The proof will be given in 10 below).
The elements V for which this condition holds are called <u>Galois resolvents</u> of the equation $P(X) = 0$ over the field F.
This terminology (which is of course not due to Galois) stems from the observation that in order to solve the equation $P(X) = 0$ it suffices to determine V since the roots $r_1,\ldots,r_n$ of P are rational fractions in V.

<u>RESULT 2</u> : For any element $u \in F(r_1,\ldots,r_n)$, there is a unique monic irreducible polynomial $\pi \in F[X]$ such that $\pi(u) = 0$. This polynomial π splits into a product of linear factors over $F(r_1,\ldots,r_n)$.

(The proof will be given in 11 below).
The polynomial π is called the <u>minimum polynomial of u over F</u>. (Compare remark 12.19).

The Galois group of the equation $P(X) = 0$ over F can now be described as follows : let V be a Galois resolvent, so that for $i=1,\ldots,n$,

$$r_i = f_i(V)$$

for some rational fraction $f_i(X) \in F(X)$. Let $V_1,\ldots,V_m \in F(r_1,\ldots,r_n)$ be the roots of the minimum polynomial of V over F (with $V = V_1$, say).

RESULT 3 : For any $j=1,\ldots,m$, the elements $f_1(V_j), f_2(V_j), \ldots, f_n(V_j)$ are the roots $r_1,\ldots,r_n$ of P, in some order.

(The proof will be given in 13 below).
From this result, it follows that for $j=1,\ldots,m$, the map :

$$\sigma_j : r_i \to f_i(V_j) \qquad i=1,\ldots,n$$

is a permutation of $\{r_1,\ldots,r_n\}$. The set $\{\sigma_1,\ldots,\sigma_m\}$ is called the <u>Galois group</u> of $P(X) = 0$ over F, and denoted by $\mathrm{Gal}(P/F)$. To justify this terminology, we shall prove :

RESULT 4 : The set $\mathrm{Gal}(P/F)$ is a subgroup of the group of all permutations of $\{r_1,\ldots,r_n\}$. It does not depend on the choice of the Galois resolvent V.

(The proof will be given in 16 and 17 below).
It is noteworthy that the order of the Galois group $\mathrm{Gal}(P/F)$, which is denoted above by m, is equal to the degree of the minimum polynomial π of a Galois resolvent V. Without further assumption on P, this order is not related in any way whatsoever to the degree n of P (see however exercise 1).

In the course of proving result 4 (which is not to be found explicitly in Galois' memoir), we shall establish the following major property of the Galois group, which is proposition 1 in Galois' memoir [G, p. 51], [E, p. 104] :

RESULT 5 : Let $f(X_1,\ldots,X_n)$ be a rational fraction in n indeterminates $X_1,\ldots,X_n$ with coefficients in F. Then

$$f(r_1,\ldots,r_n) \in F$$

if and only if

$$f(r_1,\ldots,r_n) = f(\sigma(r_1),\ldots,\sigma(r_n)) \text{ for all } \sigma \in \text{Gal}(P/F)$$

(The proof will be given in 14 below).
This result will enable us to prove moreover that formula (1) :

$$\sigma(f(r_1,\ldots,r_n)) = f(\sigma(r_1),\ldots,\sigma(r_n))$$

makes sense for $\sigma \in \text{Gal}(P/F)$ and defines an extension of σ to an automorphism of $F(r_1,\ldots,r_n)$ which leaves every element in F invariant.

<u>3</u>. To illustrate the steps which lead to the construction of the Galois group of an equation, we consider the following easy example : let

$$P(X) = (X-1)(X^2-2)(X^2-3) = (X-1)(X-\sqrt{2})(X+\sqrt{2})(X-\sqrt{3})(X+\sqrt{3}).$$

We denote the roots of P by

$$r_1 = 1, \quad r_2 = \sqrt{2}, \quad r_3 = -\sqrt{2}, \quad r_4 = \sqrt{3}, \quad r_5 = -\sqrt{3}.$$

In order to determine the Galois group of $P(X) = 0$ over the field $\mathbb{Q}$ of rational numbers, we first choose a Galois resolvent : we claim that the element

$$V = \sqrt{2} + \sqrt{3} = r_2 + r_4$$

fulfills the required conditions. To prove it, we square both sides of the equation

$$V - r_2 = r_4,$$

which yields :

$$V^2 - 2r_2V + 2 = 3 \tag{2}$$

and we obtain r_2, whence also r_3 since $r_3 = -r_2$, as rational expressions of V :

$$r_2 = (V^2-1)/2V \qquad\qquad r_3 = (1-V^2)/2V.$$

Similarly, from $V - r_4 = r_2$ we get :

$$r_4 = (V^2+1)/2V \qquad\qquad r_5 = -(V^2+1)/2V.$$

Since $r_1 = 1$ is a rational expression of r_1 in V (trivially), this shows that every root of P has a rational expression in V, hence V is a Galois resolvent of $P(X) = 0$ over $\mathbb{Q}$, as claimed.

The next step is to find the minimum polynomial of V over $\mathbb{Q}$. From equation (2), a rational equation in V can be obtained by isolating on one side the term containing r_2 and squaring both sides. We thus get :

$$V^4 - 10V^2 + 1 = 0,$$

hence V is a root of the polynomial $X^4 - 10X^2 + 1 \in \mathbb{Q}[X]$. It is easy to check that

$$X^4 - 10X^2 + 1 = (X-(\sqrt{2}+\sqrt{3}))(X-(\sqrt{2}-\sqrt{3}))(X-(-\sqrt{2}+\sqrt{3}))(X-(-\sqrt{2}-\sqrt{3}))$$

and that the factors on the right-hand side cannot be combined to yield a non-trivial divisor of $X^4 - 10X^2 + 1$ with rational coefficients. Therefore, $X^4 - 10X^2 + 1$ is irreducible, hence it is the minimum polynomial of V over $\mathbb{Q}$.

At the same time, we have found the roots of this polynomial :

$$V_1 = V = \sqrt{2}+\sqrt{3}, \quad V_2 = \sqrt{2}-\sqrt{3}, \quad V_3 = -\sqrt{2}+\sqrt{3}, \quad V_4 = -\sqrt{2}-\sqrt{3}.$$

The determination of the Galois group of $P(X) = 0$ is now only a matter of straightforward calculations : in the rational fractions $f_i(X)$ which are such that $r_i = f_i(V)$ for $i=1,\ldots,5$, i.e.

$$f_1(X) = 1, \quad f_2(X) = \frac{X^2-1}{2X}, \quad f_3(X) = -f_2(X), \quad f_4(X) = \frac{X^2+1}{2X},$$

$$f_5(X) = -f_4(X),$$

we substitute successively V_1, V_2, V_3, V_4 for X and we obtain the elements of $\text{Gal}(P/\mathbb{Q})$ as :

$$\sigma_j : r_i \to f_i(V_j) \qquad \text{for } j=1,\ldots,4.$$

Explicitly :

$$\sigma_1 = I \text{ (the identity)}$$

$$\sigma_2 : \begin{cases} r_1 \to r_1 \\ r_2 \to r_2 \\ r_3 \to r_3 \\ r_4 \to r_5 \\ r_5 \to r_4 \end{cases} \quad \sigma_3 : \begin{cases} r_1 \to r_1 \\ r_2 \to r_3 \\ r_3 \to r_2 \\ r_4 \to r_4 \\ r_5 \to r_5 \end{cases} \quad \sigma_4 : \begin{cases} r_1 \to r_1 \\ r_2 \to r_3 \\ r_3 \to r_2 \\ r_4 \to r_5 \\ r_5 \to r_4 \end{cases}$$

Thus, the Galois group of $P(X) = 0$ over $\mathbb{Q}$ consists of the permutations of $r_1,\ldots,r_5$ which leave r_1 invariant and which either leave invariant or interchange r_2 and r_3 on one side and r_4 and r_5 on the other side.

This was predictable from the heuristic point of view that the permutations in the Galois group are the permutations which preserve the relations among the roots; indeed, the roots $\sqrt{2}$ and $-\sqrt{2}$ play exactly the same role with respect to rational numbers : there is no way to distinguish one from the other with the aid of rational numbers; they can thus be interchanged by the Galois group. Similar arguments hold for $\sqrt{3}$ and $-\sqrt{3}$, but the roots of the various factors $X-1$, X^2-2 and X^2-3 of P cannot be interchanged, since for instance r_2 satisfies $r_2^2 - 2 = 0$ but r_4 does not. The permutations $\sigma_1, \sigma_2, \sigma_3, \sigma_4$ above are therefore the only permutations which preserve the relations among the roots.

With hindsight, it appears that the most tricky points in the determination of the Galois group of an equation are :

a) to find the roots of the given equation
b) to find a Galois resolvent
c) to determine its minimum polynomial
d) to find the roots of the minimum polynomial.

In fact, point (b) is not too much of a problem, since the proof of the existence of a Galois resolvent (which will be given in 10 below) is sufficiently explicit to provide a method to find one. Likewise, the proof of the existence of the minimum polynomial (see 11 below) yields a polynomial of which the Galois resolvent is a root. It thus "suffices" to find an irreducible factor of this polynomial which has the Galois resolvent as a root. This could be however a formidable task. Similarly, to find the roots of the given equation and of the minimum polynomial explicitly enough so that the subsequent calculations could be performed can prove to be a daunting problem.

Of course, Galois was well aware of these problems :

"If you now give me an equation that you have chosen at your pleasure, and if you want to know if it is or is not solvable by radicals, I could do nothing more than to indicate to you the means of answering your question, without wanting to give myself or anyone else the task of doing it. In a word, the calculations are impractical.

From that, it would seem that there is no fruit to derive from the solution that we propose. Indeed, it would be so if the question usually arose from this

point of view. But, most of the time, in the applications of the Algebraic Analysis, one is led to equations of which one knows beforehand all the properties : properties by means of which it will always be easy to answer the question by the rules we are going to explain.
(...) All, that makes this theory beautiful and at the same time difficult, is that one has always to indicate the course of analysis and to foresee its results without ever being able to perform [the calculations]" [G, pp. 39-40].

These last remarks will be clear from the following examples :

4. EXAMPLE : The Galois group of the general equation of degree n :

$$P(X) = X^n - s_1 X^{n-1} + \ldots + (-1)^n s_n = (X-X_1)\ldots(X-X_n) = 0$$

over the field F of rational fractions in $s_1,\ldots,s_n$ (over some reference field k) is the group of all permutations of $X_1,\ldots,X_n$. It can thus be identified to the full symmetric group S_n.

Indeed, if we assume on the contrary that Gal(P/F) is not the group of all permutations of $X_1,\ldots,X_n$, then by proposition 10.6 we can find a rational fraction $f(X_1,\ldots,X_n) \in k(X_1,\ldots,X_n)$ (= $F(X_1,\ldots,X_n)$) which is not symmetric (i.e. not in F) but such that

$$f(\sigma(X_1),\ldots,\sigma(X_n)) = f(X_1,\ldots,X_n) \text{ for all } \sigma \in \text{Gal}(P/F).$$

This contradicts result 5 above (in n° 2).

5. EXAMPLE : The Galois group over $\mathbb{Q}$ of the cyclotomic equation of prime index :

$$\Phi_p(X) = X^{p-1} + X^{p-2} + \ldots + X + 1 = 0 \qquad (p \text{ prime})$$

is a cyclic group of order $p-1$.

To prove this, we retrace the steps in the determination of the Galois group. Let ζ be any primitive p-th root of unity, i.e. any root of $\Phi_p(X)$. Since the other roots of $\Phi_p(X)$ are powers of ζ, we can choose ζ itself as a Galois resolvent of $\Phi_p(X)$. Since $\Phi_p(X)$ is irreducible, the minimum polynomial of ζ is $\Phi_p(X)$. Choosing a primitive root g of p, we denote:

$$\zeta_i = \zeta^{g^i} \qquad \text{for } i=0,\ldots,p-2,$$

as in 12.21. Thus, the roots of $\Phi_p(X)$ are $\zeta_0, \ldots, \zeta_{p-2}$, and the rational fractions f_i are now:

$$f_i(X) = X^{g^i}.$$

According to n° 2 above, the elements of $\mathrm{Gal}(\Phi_p/\mathbb{Q})$ are:

$$\sigma_j : \zeta_i \to f_i(\zeta_j) \qquad \text{for } j=0,\ldots,p-2.$$

Since

$$f_i(\zeta_j) = (\zeta^{g^j})^{g^i} = \zeta^{g^{i+j}},$$

and since, by Fermat's theorem 12.6,

$$g^{p-1} \equiv g^0 \pmod{p},$$

it follows that the above description of σ_j can be simplified to:

$$\sigma_j : \zeta_i \to \zeta_{i+j},$$

where the subscript $i+j$ is taken modulo $p-1$ (i.e. replaced

by the integer between 0 and p-2 congruent to i+j, if
i+j ⩾ p-1).
Therefore,

$$\sigma_j = \sigma_1^j \qquad \text{for } j=0,\ldots,p-2,$$

hence $\text{Gal}(\Phi_p/\mathbb{Q})$ is generated by the single element σ_1 (which was denoted by σ in 12.21). It is thus a cyclic group of order p-1.

6. EXAMPLE : Let P be a polynomial with simple roots $r_1, \ldots, r_n$ for which Abel's condition (in the theorem quoted in section 1) holds, i.e. there are rational fractions $\theta_i(X) \in F(X)$ such that

$$r_i = \theta_i(r_1) \qquad \text{for } i=2,\ldots,n$$

and

$$\theta_i \theta_j(r_1) = \theta_j \theta_i(r_1) \qquad \text{for all } i,j.$$

Then the Galois group of the equation $P(X) = 0$ over F is commutative. (This is why commutative groups are often called <u>abelian</u> groups).

Indeed, in this case one can choose r_1 as Galois resolvent. Since r_1 is a root of P, its minimum polynomial divides P, by lemma 12.16, hence the roots of the minimum polynomial of r_1 are among $r_1,\ldots,r_n$. Changing the numbering if necessary, we can assume that these roots are $r_1,\ldots,r_m$ (with m ⩽ n). According to n° 2 above, the elements of $\text{Gal}(P/F)$ are $\sigma_1,\ldots,\sigma_m$ where

$$\sigma_j : r_i \to \theta_i(r_j) \qquad \text{for } i=1,\ldots,n \text{ and } j=1,\ldots,m.$$

Since

$$\theta_i(r_j) = \theta_i\theta_j(r_1)$$

and since Abel's condition holds, we have :

$$\theta_i(r_j) = \theta_j\theta_i(r_1) = \theta_j(r_i),$$

hence

$$\sigma_j : r_i \to \theta_j(r_i) \quad \text{for } i=1,\ldots,n \text{ and } j=1,\ldots,m.$$

It then follows that, for all j,k between 1 and m,

$$\sigma_j \circ \sigma_k : r_i \to \theta_j\theta_k(r_i) = \theta_j\theta_k\theta_i(r_1)$$

and

$$\sigma_k \circ \sigma_j : r_i \to \theta_k\theta_j(r_i) = \theta_k\theta_j\theta_i(r_1).$$

Therefore, commutativity of Gal(P/F) readily follows from Abel's condition.

We now turn to the proofs of the results quoted above. First, we prove the following easy elaboration on lemma 12.16, which will be repeatedly used in the sequel :

7. LEMMA : Let $f \in F(X)$ be a rational fraction in one indeterminate over a field F and let V be a root of some irreducible polynomial $\pi \in F[X]$ (in some field containing F). If $f(V) = 0$, then $f(W) = 0$ for any root W of π.

<u>Proof</u> : Let $f = P/Q$ for some polynomials $P,Q \in F[X]$ such that $Q(V) \neq 0$ and $P(V) = 0$. By lemma 12.16, this last relation implies that π divides P, hence $P(W) = 0$ for any root W of π. On the other hand, if $Q(W) = 0$, then the same argument shows that $Q(V) = 0$, a contradiction. Therefore, $P(W) = 0$ and $Q(W) \neq 0$, hence $f(W) = 0$. □

(Compare lemma 1 in Galois' memoir [G, p. 47], [E, p. 102]).

8. LEMMA : Let g be a polynomial in n indeterminates $X_1,\ldots,X_n$ over some field K. If g is invariant under every permutation of $X_2,\ldots,X_n$, then it can be written as a polynomial in X_1 and the elementary symmetric polynomials $s_1,\ldots,s_{n-1}$ in $X_1,\ldots,X_n$.

Proof : We consider g as a polynomial in $X_2,\ldots,X_n$ with coefficients in $K[X_1]$. From theorem 8.4 and remark 8.13(b) it follows that g can be written as a polynomial in the elementary symmetric polynomials $s_1',\ldots,s_{n-1}'$ in $X_2,\ldots,X_n$, with coefficients in $K[X_1]$. Therefore, there exists a polynomial g' such that

$$g(X_1,\ldots,X_n) = g'(X_1,s_1',\ldots,s_{n-1}'), \qquad (3)$$

where

$$s_1' = X_2 + \ldots + X_n$$

$$s_2' = X_2 X_3 + \ldots + X_{n-1} X_n$$

$$\cdots\cdots$$

$$s_{n-1}' = X_2 X_3 \ldots X_n.$$

To complete the proof, it now suffices to observe that $s_1',\ldots,s_{n-1}'$ can be replaced by polynomials in X_1 and $s_1,\ldots,s_{n-1}$. A simple way to obtain explicit formulas for $s_1',\ldots,s_{n-1}'$ is to divide by $X - X_1$ the general polynomial

$$(X-X_1)\ldots(X-X_n) = X^n - s_1 X^{n-1} + \ldots + (-1)^n s_n$$

and to identify the result with

$$(X-X_2)\ldots(X-X_n) = X^{n-1} - s_1' X^{n-2} + \ldots + (-1)^{n-1} s_{n-1}'.$$

We thus get :

$$s_1' = s_1 - X_1$$

$$s_2' = s_2 - s_1 X_1 + X_1^2$$

$$s_3' = s_3 - s_2 X_1 + s_1 X_1^2 - X_1^3$$

......

$$s_{n-1}' = s_{n-1} - s_{n-2} X_1 + \ldots + (-1)^{n-1} X_1^{n-1},$$

hence, replacing $s_1', \ldots, s_{n-1}'$ in equation (3) we get :

$$g(X_1, \ldots, X_n) = g'(X_1, s_1 - X_1, \ldots, s_{n-1} - s_{n-2} X_1 + \ldots + (-1)^{n-1} X_1^{n-1}),$$

and the right-hand side is a polynomial in X_1 and $s_1, s_2, \ldots, s_{n-1}$. □

From this point on, we make use of the notations set in n° 1 above. Thus, P is a polynomial of degree n over some field F, with distinct roots $r_1, \ldots, r_n$ in some field containing F :

$$P(X) = X^n - a_1 X^{n-1} + \ldots + (-1)^n a_n = (X-r_1) \ldots (X-r_n).$$

9. LEMMA : There is a polynomial $f \in F[X_1, \ldots, X_n]$ such that the various elements in $F(r_1, \ldots, r_n)$ obtained from f by replacing the indeterminates $X_1, \ldots, X_n$ by $r_1, \ldots, r_n$ in all n! possible ways are all pairwise distinct.

<u>Proof</u> : Let $L(X_1, \ldots, X_n) = A_1 X_1 + \ldots + A_n X_n$, where $A_1, \ldots, A_n$ are indeterminates. The equality between two values of L obtained by replacing $X_1, \ldots, X_n$ by $r_1, \ldots, r_n$ in some ways is a linear equation in $A_1, \ldots, A_n$ (with coefficients in $F(r_1, \ldots, r_n)$). Writing down all the possible equalities yields a finite number ((n!)(n!-1)/2, in fact) of homogeneous linear equations in $A_1, \ldots, A_n$, none of which is tri-

vial, since $r_1, \ldots, r_n$ are pairwise distinct. The solutions of these equations in F^n form a union of proper (vector-) subspaces of F^n. Now, since F is infinite (as its characteristic is assumed to be zero), F^n is not a union of a finite number of proper subspaces, hence we can find a n-tuple $(a_1, \ldots, a_n)$ in F^n for which none of the equations in $A_1, \ldots, A_n$ holds. The resulting polynomial

$$f(X_1, \ldots, X_n) = a_1 X_1 + \ldots + a_n X_n$$

satisfies the condition of the lemma. □

This lemma is lemma 2 in Galois' memoir [G, p. 47], [E, p. 102]. It obviously does not hold if multiple roots are allowed, i.e. if $r_1, \ldots, r_n$ are not pairwise distinct. We can now prove result 1 of n° 2, which asserts the existence of Galois resolvents :

<u>10. PROPOSITION</u> : There is an element $V \in F(r_1, \ldots, r_n)$ such that

$$r_i \in F(V) \qquad \text{for } i=1, \ldots, n.$$

<u>Proof</u> : Let $f \in F[X_1, \ldots, X_n]$ be a polynomial satisfying the property in lemma 9 and let

$$V = f(r_1, \ldots, r_n) \in F(r_1, \ldots, r_n).$$

We are going to show that $r_1, \ldots, r_n$ are in $F(V)$. It is of course sufficient to spell out the arguments for one of the roots $r_1, \ldots, r_n$, say for r_1, since the same proof applies to any of them, by a simple change of numbering.

We consider the polynomial

$$g(X_1, \ldots, X_n) = \prod_\sigma (V - f(X_1, \sigma(X_2), \ldots, \sigma(X_n))) \in F(V)[X_1, \ldots, X_n],$$

where σ runs over all the permutations of $X_2,\ldots,X_n$. Since g is symmetric in $X_2,\ldots,X_n$, lemma 8 shows that g can be written as a polynomial in X_1 and the elementary symmetric polynomials $s_1,\ldots,s_{n-1}$ in $X_1,\ldots,X_n$:

$$g(X_1,X_2,\ldots,X_n) = h(X_1,s_1,\ldots,s_{n-1})$$

for some polynomial h with coefficients in $F(V)$. Therefore, replacing in various ways the indeterminates $X_1,\ldots,X_n$ by the roots $r_1,\ldots,r_n$ of P, which has the effect of replacing $s_1,\ldots,s_{n-1}$ by $a_1,\ldots,a_{n-1} \in F$, we get:

$$g(r_1,r_2,\ldots,r_n) = h(r_1,a_1,\ldots,a_{n-1}) \quad (4)$$

and

$$g(r_i,r_1,r_2,\ldots,r_{i-1},r_{i+1},\ldots,r_n) = h(r_i,a_1,\ldots,a_{n-1}). \quad (5)$$

Now, since f satisfies the property in lemma 9 and $V = f(r_1,\ldots,r_n)$, we have

$$V \neq f(r_i,\sigma(r_1),\sigma(r_2),\ldots,\sigma(r_{i-1}),\sigma(r_{i+1}),\ldots,\sigma(r_n))$$

for $i \neq 1$ and for any permutation σ of $\{r_1,\ldots,r_{i-1},r_{i+1},\ldots,r_n\}$. Therefore,

$$g(r_i,r_1,r_2,\ldots,r_{i-1},r_{i+1},\ldots,r_n) \neq 0 \quad \text{for } i \neq 1.$$

On the other hand, the definitions of g and V readily show that

$$g(r_1,\ldots,r_n) = 0.$$

In view of equations (4) and (5), these last relations show that the polynomial

$$h(X, a_1, \ldots, a_{n-1}) \in F(V)[X]$$

vanishes for $X = r_1$ but not for $X = r_i$ with $i \neq 1$; hence, it is divisible by $(X - r_1)$ but not by $(X - r_i)$ for $i \neq 1$.

We then consider the monic greatest common divisor $D(X)$ of $P(X)$ and $h(X, a_1, \ldots, a_{n-1})$ in $F(V)[X]$. Since in $F(r_1, \ldots, r_n)[X]$,

$$P(X) = (X-r_1) \ldots (X-r_n),$$

it follows that D splits over $F(r_1, \ldots, r_n)$ in a product of factors $X - r_i$. Since $X - r_1$ divides both $P(X)$ and $h(X, a_1, \ldots, a_{n-1})$, it divides D; on the other hand, $h(X, a_1, \ldots, a_{n-1})$ is not divisible by $X - r_i$ for $i \neq 1$, hence D has no other factor than $X - r_1$. Thus,

$$D = X - r_1$$

whence $r_1 \in F(V)$ since $D \in F(V)[X]$. □

This proposition is lemma 3 in Galois' memoir [G, p. 49], [E, p. 103]. We now turn to the proof of result 2, about the existence of minimum polynomials :

11. PROPOSITION :

a) Every element $u \in F(r_1, \ldots, r_n)$ has a polynomial expression in $r_1, \ldots, r_n$:

$$u = \varphi(r_1, \ldots, r_n)$$

for some polynomial $\varphi \in F[X_1, \ldots, X_n]$.

b) For every element $u \in F(r_1, \ldots, r_n)$, there is a unique monic irreducible polynomial $\pi \in F[X]$ such that $\pi(u) = 0$. This polynomial π splits into a product of linear factors over $F(r_1, \ldots, r_n)$.

Proof : The proofs of these two results are intertwined : we first establish b) for those elements u which have a polynomial expression in $r_1,\ldots,r_n$ and then deduce a); the proof of b) will then be complete.

STEP 1 : proof of b) for the elements which have a polynomial expression in $r_1,\ldots,r_n$:

Let $u \in F(r_1,\ldots,r_n)$ be such that

$$u = \varphi(r_1,\ldots,r_n)$$

for some polynomial $\varphi \in F[X_1,\ldots,X_n]$. According to 12.17 and remark 12.19, it suffices to show that u is a root of some polynomial with coefficients in F which splits into a product of linear factors over $F(r_1,\ldots,r_n)$. Let

$$\Theta(X,X_1,\ldots,X_n) = \Pi_\sigma (X-\varphi(\sigma(X_1),\ldots,\sigma(X_n)))$$

where σ runs over the set of all permutations of $X_1,\ldots,X_n$. Since Θ is symmetric in $X_1,\ldots,X_n$, we can write Θ as a polynomial in X and the elementary symmetric polynomials $s_1,\ldots,s_n$ in $X_1,\ldots,X_n$, by theorem 8.4 and remark 8.13(b) :

$$\Theta(X,X_1,\ldots,X_n) = \Psi(X,s_1,\ldots,s_n)$$

for some polynomial Ψ with coefficients in F. Replacing the indeterminates $X_1,\ldots,X_n$ by $r_1,\ldots,r_n$, we obtain :

$$\Theta(X,r_1,\ldots,r_n) = \Psi(X,a_1,\ldots,a_n) \in F[X]$$

and since, by definition of Θ,

$$\Theta(u,r_1,\ldots,r_n) = 0,$$

it follows that $\Psi(X,a_1,\ldots,a_n)$ is a polynomial in $F[X]$ which has u as a root. Moreover, since $\Theta(X,r_1,\ldots,r_n)$ is a product of linear factors, it follows that $\Psi(X,a_1,\ldots,a_n)$ splits into a product of linear factors over $F(r_1,\ldots,r_n)$.

(The point of taking for φ a polynomial is that $\Theta(X,X_1,\ldots,X_n)$ is then a polynomial, hence $\Theta(X,r_1,\ldots,r_n)$ is defined. Otherwise, it would not be clear that no denominator vanish when $X_1,\ldots,X_n$ are replaced by $r_1,\ldots,r_n$).

STEP 2 : proof of a) :

Let $V \in F(r_1,\ldots,r_n)$ be defined as in the proof of proposition 10. Since $r_1,\ldots,r_n$ have been shown to be rational fractions in V, it follows that u also is a rational fraction in V :

$$u \in F(V).$$

Since V has a polynomial expression in $r_1,\ldots,r_n$, we can apply step 1 and proposition 12.18 to derive that u can be expressed as a polynomial in V :

$$u = Q(V) \qquad \text{for some polynomial } Q \in F[X].$$

Replacing $V = f(r_1,\ldots,r_n)$, we get :

$$u = Q(f(r_1,\ldots,r_n))$$

which is a polynomial expression of u in $r_1,\ldots,r_n$, since Q and f are polynomials. □

12. COROLLARY : Let V be any Galois resolvent of $P(X) = 0$ over F and let $V_1,\ldots,V_m$ be the roots of its minimum polynomial over F (among which V lies). Then

$$F(r_1,\ldots,r_n) = F(V) = F(V_1,\ldots,V_m).$$

Proof : Since $r_1,\ldots,r_n$ are rational fractions of V, we have

$$F(r_1,\ldots,r_n) \subset F(V).$$

On the other hand, by the preceding proposition, the roots $V_1,\ldots,V_m$ of the minimum polynomial of V are in $F(r_1,\ldots,r_n)$, hence

$$F(V_1,\ldots,V_m) \subset F(r_1,\ldots,r_n).$$

Since the inclusion

$$F(V) \subset F(V_1,\ldots,V_m)$$

is obvious (as V lies among $V_1,\ldots,V_m$), the three inclusions above yield :

$$F(r_1,\ldots,r_n) = F(V) = F(V_1,\ldots,V_m). \qquad \square$$

We now come to the proof of result 3, which is lemma 4 in Galois' memoir [G, p. 49], [E, p. 104]. We let V be a Galois resolvent of $P(X) = 0$ and we let

$$r_i = f_i(V) \qquad \text{for } i=1,\ldots,n,$$

for some rational fraction $f_i(X) \in F(X)$. We denote by $V_1 = V, V_2,\ldots,V_m$ the roots of the minimum polynomial of V over F, which are in $F(r_1,\ldots,r_n)$, by result 2.

13. PROPOSITION : For $i=1,\ldots,n$ and $j=1,\ldots,m$, the element $f_i(V_j)$ is a root of P. Moreover, for any given $j=1,\ldots,m$, the roots $f_1(V_j),\ldots,f_n(V_j)$ are pairwise distinct, so that

$$\{f_1(V_j),\ldots,f_n(V_j)\} = \{r_1,\ldots,r_n\}.$$

<u>Proof</u> : Since $f_i(V_1) = r_i$ for $i=1,\ldots,n$, we have

$$P(f_i(V_1)) = 0,$$

hence

$$P(f_i(V_j)) = 0 \qquad \text{for } j=1,\ldots,m,$$

by lemma 7, applied to the rational fraction $P(f_i(X)) \in F(X)$. (From this argument, it follows at the same time that $f_i(V_j)$ is defined).
Moreover, if for some $i,k=1,\ldots,n$ and some $j=1,\ldots,m$,

$$f_i(V_j) = f_k(V_j),$$

then V_j is a root of the rational fraction $f_i - f_k$, hence by lemma 7 again :

$$f_i(V_1) = f_k(V_1).$$

This shows that

$$r_i = r_k,$$

whence $i=k$ since the roots $r_1,\ldots,r_n$ are assumed to be pairwise distinct. □

This proposition shows that the maps

$$\sigma_j : r_i = f_i(V_1) \to f_i(V_j) \qquad (i=1,\ldots,n),$$

where $j=1,\ldots,m$, are permutations of $\{r_1,\ldots,r_n\}$. We set :

$$\text{Gal}(P/F) = \{\sigma_1,\ldots,\sigma_m\}$$

(although it is not yet clear at this stage that this set does not depend on the choice of the Galois resolvent V), and we prove the following major property of Gal(P/F), which was announced as result 5 in n° 2 above :

14. THEOREM : Let $f(X_1,\ldots,X_n)$ be a rational fraction in n indeterminates $X_1,\ldots,X_n$, with coefficients in F. For $\sigma \in \text{Gal}(P/F)$, the element $f(\sigma(r_1),\ldots,\sigma(r_n)) \in F(r_1,\ldots,r_n)$ is defined whenever $f(r_1,\ldots,r_n)$ is defined. Moreover,

$$f(r_1,\ldots,r_n) \in F$$

if and only if

$$f(\sigma(r_1),\ldots,\sigma(r_n)) = f(r_1,\ldots,r_n) \quad \text{for all } \sigma \in \text{Gal}(P/F).$$

Proof : Let $f = \varphi/\psi$, where $\varphi,\psi \in F[X_1,\ldots,X_n]$. We have to prove first that if $\psi(r_1,\ldots,r_n) \neq 0$, then $\psi(\sigma(r_1),\ldots,\sigma(r_n)) \neq 0$ for every $\sigma \in \text{Gal}(P/F)$. Replacing $r_1,\ldots,r_n$ by their rational expression in V, we get :

$$\psi(r_1,\ldots,r_n) = \psi(f_1(V),\ldots,f_n(V)) = g(V)$$

for some rational fraction $g \in F(X)$.

Let now σ be a permutation in $\text{Gal}(P/F)$; if V' is the root of π such that

$$\sigma : r_i \to f_i(V'),$$

then

$$\psi(\sigma(r_1),\ldots,\sigma(r_n)) = \psi(f_1(V'),\ldots,f_n(V')) = g(V').$$

Therefore, if $\psi(\sigma(r_1),\ldots,\sigma(r_n)) = 0$, then V' is a root of g and by lemma 7 it follows that $g(V) = 0$, whence $\psi(r_1,\ldots,r_n) = 0$.

This shows that $f(\sigma(r_1),\ldots,\sigma(r_n))$ is defined when $f(r_1,\ldots,r_n)$ is defined. To prove the rest, we replace $r_1,\ldots,r_n$ by their rational expression in V in $f(r_1,\ldots,r_n)$:

$$f(r_1,\ldots,r_n) = f(f_1(V),\ldots,f_n(V)) = h(V)$$

where

$$h(X) = f(f_1(X),\ldots,f_n(X)) \in F(X).$$

If $f(r_1,\ldots,r_n) \in F$, then

$$h(X) - f(r_1,\ldots,r_n) \in F(X)$$

and since this rational fraction vanishes for $X = V$, it also vanishes for $X = V_1,\ldots,V_m$, by lemma 7. Therefore,

$$h(V_j) = f(f_1(V_j),\ldots,f_n(V_j)) = f(r_1,\ldots,r_n) \quad \text{for } j=1,\ldots,m$$

whence, using the definition of σ_j :

$$f(\sigma_j(r_1),\ldots,\sigma_j(r_n)) = f(r_1,\ldots,r_n) \quad \text{for } j=1,\ldots,m.$$

Conversely, if this last relation holds, then

$$f(r_1,\ldots,r_n) = h(V_j), \quad \text{for } j=1,\ldots,m$$

hence

$$f(r_1,\ldots,r_n) = \tfrac{1}{m}[h(V_1) + \ldots h(V_m)]. \tag{6}$$

Since the rational fraction $h(X_1) + \ldots + h(X_m)$ is clearly symmetric in the indeterminates $X_1,\ldots,X_m$, it can be expressed as a rational fraction in the elementary symmetric polynomials $s_1,\ldots,s_m$. Therefore, substituting $V_1,\ldots,V_m$ for $X_1,\ldots,X_m$, it follows that the right-hand side of (6) can be rationally calculated from the coefficients of the polynomial π which has as roots $V_1,\ldots,V_m$, and is thus an element in F. Hence, equation (6) shows that

$$f(r_1,\ldots,r_n) \in F. \qquad \square$$

15. COROLLARY : Each permutation in Gal(P/F) can be extended to an automorphism of $F(r_1,\ldots,r_n)$ which leaves every element in F invariant, by setting :

$$\sigma(f(r_1,\ldots,r_n)) = f(\sigma(r_1),\ldots,\sigma(r_n)) \text{ for } \sigma \in \text{Gal}(P/F)$$

for any rational fraction $f(X_1,\ldots,X_n) \in F(X_1,\ldots,X_n)$ for which $f(r_1,\ldots,r_n)$ is defined.

Proof : As pointed out before (in n° 1), we first have to prove that $\sigma(f(r_1,\ldots,r_n))$ is well-defined by the equation above, i.e. that it does not really depend on the rational fraction $f \in F(X_1,\ldots,X_n)$, but only on $f(r_1,\ldots,r_n)$. Assume thus :

$$f(r_1,\ldots,r_n) = g(r_1,\ldots,r_n)$$

for some rational fractions $f,g \in F(X_1,\ldots,X_n)$. Then the rational fraction $f-g$ vanishes for $X_i = r_i$ $(i=1,\ldots,n)$:

$$(f-g)(r_1,\ldots,r_n) = 0 \in F.$$

The preceding theorem shows that

$$(f-g)(\sigma(r_1),\ldots,\sigma(r_n)) = (f-g)(r_1,\ldots,r_n) = 0 \text{ for all } \sigma \in \text{Gal}(P/F),$$

hence

$$f(\sigma(r_1),\ldots,\sigma(r_n)) = g(\sigma(r_1),\ldots,\sigma(r_n)) \quad \text{for all } \sigma \in \text{Gal}(P/F).$$

This shows that $f(\sigma(r_1),\ldots,\sigma(r_n))$ depends only on the element $f(r_1,\ldots,r_n) \in F(r_1,\ldots,r_n)$, and not on the choice of the rational fraction $f(X_1,\ldots,X_n)$ which represents it.

Since σ is clearly bijective on $F(r_1,\ldots,r_n)$, the fact that it is an automorphism of $F(r_1,\ldots,r_n)$ readily follows from its definition, since

$$f(\sigma(r_1),\ldots,\sigma(r_n)) + g(\sigma(r_1),\ldots,\sigma(r_n)) = (f+g)(\sigma(r_1),\ldots,\sigma(r_n))$$

and

$$f(\sigma(r_1),\ldots,\sigma(r_n)) \cdot g(\sigma(r_1),\ldots,\sigma(r_n)) = (fg)(\sigma(r_1),\ldots,\sigma(r_n)). \quad \square$$

16. COROLLARY : The set Gal(P/F) does not depend on the choice of the Galois resolvent V.

<u>Proof</u> : Let $V' \in F(r_1,\ldots,r_n)$ be another Galois resolvent of $P(X) = 0$ and let π' be its minimum polynomial over F. Let also $f'_i \in F(X)$, for $i=1,\ldots,n$, be a rational fraction such that

$$r_i = f'_i(V') \qquad \text{for } i=1,\ldots,n. \tag{7}$$

We have to show that every element of Gal(P/F), as defined above with the aid of V, is also an element of Gal(P/F) as defined with respect to V'. The converse will then be clear, by interchanging V and V'.
Let thus $\sigma \in$ Gal(P/F). We have to show :

$$\sigma : r_i \to f'_i(W')$$

for some root W' of π'. In order to find a suitable W', we use the extension of σ to $F(r_1,\ldots,r_n)$; from equation (7), it follows by applying σ to both sides :

$$\sigma(r_i) = f'_i(\sigma(V')),$$

since σ leaves every element in F invariant. Similarly, since V' is a root of π', it follows that

$$\pi'(\sigma(V')) = 0,$$

i.e. $\sigma(V')$ is a root of π'. The element $W' = \sigma(V')$ thus satisfies the required conditions. □

We now complete the proof of result 4, and thus finish proving the results which were announced in n° 2 above :

17. COROLLARY : Gal(P/F) is a subgroup of the group of all permutations of $r_1,\ldots,r_n$.

Proof : That the identity map is in Gal(P/F) is clear, since this map is σ_1 in the notations of n° 2. It thus remains to show that Gal(P/F) is stable under composition of maps and under inversion.
Let $\sigma \in$ Gal(P/F); then by definition of Gal(P/F),

$$\sigma(r_i) = f_i(V_j)$$

for some $j=1,\ldots,m$. Proposition 13 shows that V_j is also a Galois resolvent of $P(X) = 0$, hence we can define Gal(P/F) with the aid of V_j instead of $V = V_1$. Since V_1 and V_j are roots of the same minimum polynomial π, it then follows that the map

$$f_i(V_j) \to f_i(V_1)$$

is also an element of Gal(P/F). This map is the inverse of σ, hence we have shown that Gal(P/F) is stable under inversion.
In order to prove that for any $\tau \in$ Gal(P/F) the composite map $\tau \circ \sigma$ also is in Gal(P/F), we consider again the definition of Gal(P/F) with respect to V_j; thus,

$$\tau : f_i(V_j) \to f_i(V_k) \qquad \text{for some } k=1,\ldots,m$$

and it follows that

$$\tau \circ \sigma : r_i \to f_i(V_k).$$

Therefore, $\tau \circ \sigma \in \text{Gal}(P/F)$. □

§ 3. The Galois group under field extension

In the definition of the Galois group of an equation, the base field F plays a rather inconspicuous, yet important, role. It is the purpose of this section to bring it into focus and to investigate what happens to the Galois group when the base field is enlarged by the adjunction of roots of auxiliary polynomials. In view of applications to solvability by radicals, the crucial case is the adjunction of p-th roots of elements, i.e. roots of auxiliary equations of the type $X^p - a$ (where p can be chosen to be prime : see the beginning of § 13.2).

As in the preceding section, we denote by P a monic polynomial of degree n over some field F, which has n distinct roots $r_1, \ldots, r_n$ in some field S containing F :

$$P(X) = X^n - a_1 X^{n-1} + a_2 X^{n-2} - \ldots + (-1)^n a_n = (X-r_1) \ldots (X-r_n).$$

The existence of such a field S follows from Girard's theorem 9.3. In fact, the field S can be chosen arbitrarily large, since only the subfield $F(r_1, \ldots, r_n)$ matters for the determination of the Galois group of $P(X) = 0$ over F. Therefore, if some field K containing F is given, we can assume that S contains K : it suffices to apply Girard's theorem with base field K instead of F. This allows us to mix elements in K and elements in $F(r_1, \ldots, r_n)$ in calculations and, in particular, to consider the field $K(r_1, \ldots, r_n)$ of rational fractions in $r_1, \ldots, r_n$ with coefficients in K. We can then determine the Galois group of $P(X) = 0$ over K as well as over F, by the method of n° 2 above. Here is how these Galois groups compare :

18. PROPOSITION : If K is a field containing F, then $\text{Gal}(P/K)$ is a subgroup of $\text{Gal}(P/F)$.

Proof : Let V be a Galois resolvent of $P(X) = 0$ over F. For $i=1,\ldots,n$,

$$r_i \in F(V)$$

and since $F(V) \subset K(V)$, every root r_i of P is a rational fraction of V with coefficients in K, hence V is also a Galois resolvent of $P(X) = 0$ over K. If, for $i=1,\ldots,n$,

$$r_i = f_i(V)$$

for some rational fraction $f_i \in F(X)$, then the same fraction f_i can be used to determine $\text{Gal}(P/K)$ and $\text{Gal}(P/F)$. The only difference is that the minimum polynomial π of V over F may not be irreducible over K; the minimum polynomial of V over K, which we denote by θ, is then different from π, but in any case θ divides π, by lemma 12.16. Therefore, the roots of θ are among those of π.

Since the permutations in $\text{Gal}(P/K)$ are of the form

$$\sigma : r_i \to f_i(V') \qquad (i=1,\ldots,n)$$

where V' is a root of θ, while the permutations in $\text{Gal}(P/F)$ have the same form, but with V' a root of π, it follows that $\text{Gal}(P/K) \subset \text{Gal}(P/F)$. □

19. Our aim in the rest of this section is to obtain additional information on the relations between $\text{Gal}(P/K)$ and $\text{Gal}(P/F)$, under certain assumptions on K. More precisely, we shall show that if K is obtained by adjoining a root of an irreducible auxiliary equation $T(X) = 0$, then the quotient $|\text{Gal}(P/F)|/|\text{Gal}(P/K)|$, i.e. the index[*] of $\text{Gal}(P/K)$

---

[*] Recall from 10.4 that the **index** of a subgroup H in a group G, denoted by (G:H), is the number of (left) cosets of H in G. If G is finite, the proof of Lagrange's theorem 10.4 shows that equivalently $(G:H) = |G|/|H|$.

in Gal(P/F), divides the degree of T. If on the other hand the field K is obtained by adjoining <u>all</u> the roots of the equation T(X) = 0, then the following property holds :

$$\text{for } \sigma \in \text{Gal}(P/F) \text{ and } \tau \in \text{Gal}(P/K), \ \sigma\tau\sigma^{-1} \in \text{Gal}(P/K).$$

This property is expressed by saying that Gal(P/K) is a <u>normal</u> subgroup of Gal(P/F).

<u>20. LEMMA</u> : Let π be an irreducible polynomial over a field F, and let K be a field containing F and such that π splits into a product of linear factors over K. Let also $f,g,h \in F[X,Y]$. If, for some root V of π in K,

$$f(X,V) = g(X,V)h(X,V) \qquad \text{in } K[X]$$

then

$$f(X,W) = g(X,W)h(X,W) \qquad \text{in } K[X]$$

for every root W of π.

<u>Proof</u> : Regarding f,g and h as polynomials in one indeterminate X over F[Y], we can write :

$$f(X,Y) - g(X,Y)h(X,Y) = c_r(Y)X^r + \ldots + c_0(Y)$$

for some polynomials $c_r(Y),\ldots,c_0(Y) \in F[Y]$. The hypothesis that $f(X,V) = g(X,V)h(X,V)$ implies that

$$c_i(V) = 0 \qquad\qquad \text{for } i=0,\ldots,r.$$

Therefore, lemma 7 implies :

$$c_i(W) = 0 \qquad\qquad \text{for } i=0,\ldots,r,$$

for every root W of π, hence

$$f(X,W) = g(X,W)h(X,W).$$ □

Henceforth, we denote by T an irreducible polynomial of degree t over F. From corollary 5.27 it follows that T has only simple roots in any field containing F. Thus, over a suitable field,

$$T(X) = (X-u_1)\ldots(X-u_t)$$

where $u_1,\ldots,u_t$ are pairwise distinct.

21. THEOREM : The index of $Gal(P/F(u_1))$ in $Gal(P/F)$ divides t.

Proof : Let V be a Galois resolvent of $P(X) = 0$ over F. As we have seen in the proof of proposition 18, V is also a Galois resolvent of $P(X) = 0$ over $F(u_1)$. We let θ (resp. π) denote its minimum polynomial over $F(u_1)$ (resp. F). Since the permutations in $Gal(P/F)$ are in 1-1 correspondence with the roots of π, and those in $Gal(P/F(u_1))$ with the roots of θ, we have to prove :

$$\deg \pi / \deg \theta \text{ divides } t.$$

From lemma 12.16, we know that θ divides π : let then

$$\pi = \theta \lambda \qquad (8)$$

for some polynomial $\lambda \in F(u_1)[X]$. Let also

$$\theta(X) = X^r + b_{r-1}X^{r-1} + \ldots + b_1X + b_0.$$

Since $b_0,\ldots,b_{r-1} \in F(u_1)$, these elements have polynomial expressions in u_1, by proposition 12.18 : let

$$b_i = \theta_i(u_1) \qquad \text{for } i=0,\ldots,r-1,$$

for some polynomial $\theta_i \in F[Y]$. Let then

$$\Theta(X,Y) = X^r + \theta_{r-1}(Y)X^{r-1} + \ldots + \theta_1(Y)X + \theta_0(Y) \in F[X,Y],$$

so that

$$\Theta(X,u_1) = \theta(X).$$

Acting similarly with λ, we construct a polynomial $\Lambda(X,Y) \in F[X,Y]$ such that

$$\Lambda(X,u_1) = \lambda(X).$$

Equation (8) can then be rewritten :

$$\pi(X) = \Theta(X,u_1) \, \Lambda(X,u_1)$$

and the preceding lemma yields :

$$\pi(X) = \Theta(X,u_i) \, \Lambda(X,u_i) \quad \text{for } i=1,\ldots,t.$$

Multiplying these equalities, we get :

$$\pi(X)^t = \Theta(X,u_1)\ldots\Theta(X,u_t) \, \Lambda(X,u_1)\ldots\Lambda(X,u_t) \qquad (9)$$

in $F(u_1,\ldots,u_t)[X]$.

We claim that in fact the product $\Theta(X,u_1)\ldots\Theta(X,u_t)$ is a polynomial with coefficients in F. Indeed, since the polynomial

$$\Theta(X,Y_1)\ldots\Theta(X,Y_t)$$

is clearly symmetric in the indeterminates $Y_1,\ldots,Y_t$, it can be expressed as a polynomial in X and the elementary symmetric polynomials in $Y_1,\ldots,Y_t$. Therefore, replacing $Y_1,\ldots,Y_t$ by $u_1,\ldots,u_t$ yields a polynomial in X whose coef-

ficients can be calculated from the coefficients of the
equation $T(Y) = 0$ which has $u_1,\ldots,u_t$ as roots. Since the
coefficients of T are in F, it follows that

$$\Theta(X,u_1)\ldots\Theta(X,u_t) \in F[X],$$

as claimed.

Relation (9) shows that this product divides $\pi(X)^t$. Since π is irreducible over F, it follows that

$$\Theta(X,u_1)\ldots\Theta(X,u_t) = \pi(X)^k \tag{10}$$

for some integer k between 1 and t. Comparing the degrees of both sides, we get :

$$tr = k \deg \pi$$

and since $r = \deg \theta$, it follows that

$$\deg \pi / \deg \theta \text{ divides } t. \qquad \square$$

With the same notations, we now prove the other property announced in n° 19 :

22. THEOREM : $\text{Gal}(P/F(u_1,\ldots,u_t))$ is a normal subgroup in $\text{Gal}(P/F)$, i.e. if $\sigma \in \text{Gal}(P/F)$ and $\tau \in \text{Gal}(P/F(u_1,\ldots,u_t))$, then

$$\sigma\tau\sigma^{-1} \in \text{Gal}(P/F(u_1,\ldots,u_t)).$$

<u>Proof</u> : Let V be a Galois resolvent of $P(X) = 0$ over F (whence also over $F(u_1,\ldots,u_t)$). We let φ (resp. π) denote the minimum polynomial of V over $F(u_1,\ldots,u_t)$ (resp. over F) and let $f_1,\ldots,f_n \in F(X)$ be rational fractions such that

$$r_i = f_i(V) \qquad \text{for } i=1,\ldots,n.$$

Any permutation $\tau \in \mathrm{Gal}(P/F(u_1,\ldots,u_t))$ then has the form :

$$\tau : r_i = f_i(V) \to f_i(V') \qquad (i=1,\ldots,n) \qquad (11)$$

where V' is a root of φ.

If $\sigma \in \mathrm{Gal}(P/F)$, then σ extends to an automorphism of $F(r_1,\ldots,r_n)$ which leaves every element of F invariant, by corollary 15; therefore, applying σ to both sides of the equation

$$\pi(V) = 0,$$

we get :

$$\pi(\sigma(V)) = 0.$$

This shows that $\sigma(V)$ is a root of π, hence, by proposition 13, every root $r_1,\ldots,r_n$ of P is a rational fraction in $\sigma(V)$. In other words, $\sigma(V)$ is a Galois resolvent of $P(X) = 0$ over F, hence also over $F(u_1,\ldots,u_t)$. Since $\mathrm{Gal}(P/F(u_1,\ldots,u_t))$ does not depend on the choice of a Galois resolvent (corollary 16), to describe its elements we can choose any Galois resolvent we find convenient. It turns out that $\sigma(V)$ is quite suitable to describe $\sigma\tau\sigma^{-1}$; indeed, (11) readily yields :

$$\sigma\tau\sigma^{-1} : f_i(\sigma(V)) \to f_i(\sigma(V')).$$

Therefore, in order to prove that $\sigma\tau\sigma^{-1} \in \mathrm{Gal}(P/F(u_1,\ldots,u_t))$, it suffices to prove that $\sigma(V')$ is a root of the minimum polynomial of $\sigma(V)$ over $F(u_1,\ldots,u_t)$.

Let W be a Galois resolvent of $T(X) = 0$ and let $W_1,\ldots,W_s$ be the roots of its minimum polynomial over F (among which W lies). By corollary 12, we have :

$$F(u_1,\ldots,u_t) = F(W) = F(W_1,\ldots,W_s).$$

In fact, since W can be any of $W_1,\ldots,W_s$, we also have

$$F(u_1,\ldots,u_t) = F(W_i) \qquad \text{for any } i=1,\ldots,s.$$

The extension $F(u_1,\ldots,u_t)$ can thus be regarded as an extension of F by a single element W_1. Duplicating the arguments in the proof of theorem 21 above, we produce a polynomial $\Phi(X,Y) \in F[X,Y]$ such that

$$\varphi(X) = \Phi(X,W_1) \in F(u_1,\ldots,u_t)[X]$$

and we obtain as in this proof an equation similar to (10) :

$$\Phi(X,W_1) \cdot \ldots \cdot \Phi(X,W_s) = \pi(X)^\ell$$

for some integer ℓ between 1 and s. Since $\sigma(V)$ is a root of π, this equality shows that $\sigma(V)$ is a root of some factor $\Phi(X,W_k)$.

In order to show that $\Phi(X,W_k)$ is the minimum polynomial of $\sigma(V)$ over $F(u_1,\ldots,u_t)$, it suffices to prove that this polynomial is irreducible. If it factors over $F(u_1,\ldots,u_t)$, then since $F(u_1,\ldots,u_t) = F(W_k)$, the factorization can be written in the form :

$$\Phi(X,W_k) = \Gamma(X,W_k) \Delta(X,W_k)$$

for some polynomials Γ,Δ with coefficients in F. Hence, by lemma 20 :

$$\Phi(X,W_1) = \Gamma(X,W_1) \Delta(X,W_1).$$

Since $\Phi(X,W_1) = \varphi(X)$ is irreducible, it follows that the above factorization is trivial, whence also the factorization of $\Phi(X,W_k)$. Therefore, $\Phi(X,W_k)$ is the minimum polynomial of $\sigma(V)$ over $F(u_1,\ldots,u_t)$.

Thus, what we have to prove is that $\sigma(V')$ is a root of

$\Phi(X,W_k)$, as $\sigma(V)$, assuming that V' is a root of $\Phi(X,W_1)$ $(= \varphi(X))$, as V.

Since by corollary 12, $F(r_1,\ldots,r_n) = F(V)$, we have

$$V' = g(V) \tag{12}$$

for some rational fraction $g(X) \in F(X)$. In fact, by proposition 12.18, we can choose g to be a polynomial in $F[X]$. Since $\Phi(V',W_1) = 0$, we have

$$\Phi(g(V),W_1) = 0$$

hence V is a root of the polynomial $\Phi(g(X),W_1) \in F(u_1,\ldots,u_t)[X]$ and, by lemma 12.16, $\Phi(X,W_1)$ divides $\Phi(g(X),W_1)$: let

$$\Phi(g(X),W_1) = \Phi(X,W_1) \Psi(X,W_1)$$

for some polynomial $\Psi \in F[X,Y]$. Lemma 20 then shows that

$$\Phi(g(X),W_k) = \Phi(X,W_k) \Psi(X,W_k)$$

and since $\sigma(V)$ is a root of $\Phi(X,W_k)$ it follows that

$$\Phi(g(\sigma(V)),W_k) = 0.$$

Now, applying σ to both sides of (12) yields :

$$\sigma(V') = g(\sigma(V)),$$

hence the preceding equation shows that $\sigma(V')$ is a root of $\Phi(X,W_k)$, as was to be shown. □

23. REMARK : Although this theorem does not yield the index of $\text{Gal}(P/F(u_1,\ldots,u_t))$ in $\text{Gal}(P/F)$, some information can be obtained from theorem 21; indeed, since u_2 is a root of the polynomial

$$T(X)/(X-u_1) \in F(u_1)[X],$$

which has degree $t-1$, it follows that the degree of the minimum polynomial of u_2 over $F(u_1)$ is at most $t-1$, hence by theorem 21 :

$$(\text{Gal}(P/F(u_1)) : \text{Gal}(P/F(u_1,u_2))) \leq t-1.$$

Likewise, since u_3 is a root of

$$T(X)/(X-u_1)(X-u_2) \in F(u_1,u_2)[X],$$

which has degree $t-2$, we have :

$$(\text{Gal}(P/F(u_1,u_2)) : \text{Gal}(P/F(u_1,u_2,u_3))) \leq t-2,$$

and so on.
Now, the index of $\text{Gal}(P/F(u_1,\ldots,u_t))$ in $\text{Gal}(P/F)$ can be calculated as :

$$\frac{|\text{Gal}(P/F)|}{|\text{Gal}(P/F(u_1,\ldots,u_t))|} = \frac{|\text{Gal}(P/F)|}{|\text{Gal}(P/F(u_1))|} \cdot \frac{|\text{Gal}(P/F(u_1))|}{|\text{Gal}(P/F(u_1,u_2))|} \cdot \ldots \cdot \frac{|\text{Gal}(P/F(u_1,\ldots,u_{t-1}))|}{|\text{Gal}(P/F(u_1,\ldots,u_t))|},$$

hence the above bounds and theorem 21 yield :

$$(\text{Gal}(P/F) : \text{Gal}(P/F(u_1,\ldots,u_t))) \leq t!.$$

24. EXAMPLE : As an illustration of theorems 21 and 22, we now show how the Galois group of the general equation of degree 4 is affected by the adjunction of one or all of the roots of a resolvent cubic. Let thus :

$$P(X) = (X-X_1)(X-X_2)(X-X_3)(X-X_4) = X^4 - s_1 X^3 + s_2 X^2 - s_3 X + s_4 = 0$$

be the general equation of degree 4, with base field the field of rational fractions in s_1, s_2, s_3, s_4 :

$$F = k(s_1, s_2, s_3, s_4)$$

(for some reference field k; for instance $k = \mathbb{Q}$ or $\mathbb{C}$). By n° 4, the Galois group Gal(P/F) is the group of all permutations of X_1, X_2, X_3, X_4, which can be identified to the symmetric group S_4 :

$$\text{Gal}(P/F) = S_4.$$

From Lagrange's discussion of the solution of equations of degree 4 (see § 10.2), it follows that the roots of Ferrari's resolvent cubic equation are :

$$u_1 = -(X_1+X_2)(X_3+X_4)/2$$

$$u_2 = -(X_1+X_3)(X_2+X_4)/2$$

$$u_3 = -(X_1+X_4)(X_2+X_3)/2.$$

Since none of these roots is in F, the resolvent cubic is irreducible over F, by corollary 5.18. We can therefore apply theorems 21 and 22 (and remark 23) to conclude that $\text{Gal}(P/F(u_1))$ is a subgroup of index 3 (whence of order 8) in Gal(P/F), and that $\text{Gal}(P/F(u_1, u_2, u_3))$ is a normal subgroup in Gal(P/F), of index at most 6.

In fact, it is not difficult to determine these groups explicitly, as we now show. By theorem 14, the permutations in $\text{Gal}(P/F(u_1))$ leave u_1 invariant; hence, denoting by $I(u_1)$ as in § 10.3 the subgroup of S_4 consisting in all the permutations which leave u_1 invariant,

$$\text{Gal}(P/F(u_1)) \subset I(u_1).$$

Since, by Lagrange's theorem 10.3, the index of $I(u_1)$ in S_4 is 3 (i.e. the number of values of u_1 under the permutations in S_4), it follows that

$$|I(u_1)| = |\text{Gal}(P/F(u_1))|,$$

hence

$$\text{Gal}(P/F(u_1)) = I(u_1).$$

More explicitly, the permutations in $I(u_1)$ are those which are induced on the vertices of the square :

```
1       3
 ┌─────┐
 │     │
 └─────┘
4       2
```

by the isometries which leave the square globally invariant. Such a group is called a <u>dihedral group</u> of order 8. The subgroups $I(u_1)$, $I(u_2)$ and $I(u_3)$ of S_4 correspond to the three inequivalent numberings of the vertices of a square.

The group $I(u_1)$ is not normal in S_4; indeed, if $\sigma \in S_4$ is a permutation which transforms u_1 into u_2, then

$$\sigma\, I(u_1)\sigma^{-1} = I(u_2) \quad (\neq I(u_1)).$$

On the contrary, the subgroup $\text{Gal}(P/F(u_1,u_2,u_3))$ is normal in S_4, and theorem 14 shows that u_1, u_2 and u_3 are all invariant under the permutations in this group. Therefore,

$$\text{Gal}(P/F(u_1,u_2,u_3)) \subset I(u_1) \cap I(u_2) \cap I(u_3).$$

The permutations in $I(u_1) \cap I(u_2) \cap I(u_3)$ are easy to find: they are : I (the identity), and the permutations which interchange 1,2,3,4 by pairs :

$$\sigma_1 : \begin{matrix} 1 \leftrightarrow 2 \\ 3 \leftrightarrow 4 \end{matrix} \qquad \sigma_2 : \begin{matrix} 1 \leftrightarrow 3 \\ 2 \leftrightarrow 4 \end{matrix} \qquad \sigma_3 : \begin{matrix} 1 \leftrightarrow 4 \\ 2 \leftrightarrow 3; \end{matrix}$$

thus,

$$|(\text{Gal}(P/F(u_1,u_2,u_3))| \leq |I(u_1) \cap I(u_2) \cap I(u_3)| = 4.$$

On the other hand, since the index of $\text{Gal}(P/F(u_1,u_2,u_3))$ in S_4 is at most 6, we have :

$$|\text{Gal}(P/F(u_1,u_2,u_3))| \geq 4;$$

therefore :

$$\text{Gal}(P/F(u_1,u_2,u_3)) = \{I,\sigma_1,\sigma_2,\sigma_3\}.$$

To finish this section, we record the following straightforward consequence of theorems 21 and 22 :

25. COROLLARY : Let K be a radical extension of height 1 of F :

$$K = F(u)$$

with

$$u^p = a$$

for some prime number p and some $a \in F$ which is not a p-th power in F. If F contains a primitive p-th root of unity, then $\text{Gal}(P/K)$ is a normal subgroup of index 1 or p in $\text{Gal}(P/F)$.

<u>Proof</u> : We apply the results above to the polynomial

$$T(X) = X^p - a,$$

which is irreducible, by lemma 13.9. Since F contains a
primitive p-th root of unity ζ, adjoining u, which is one
of the roots of T, amounts to adjoining all the roots of T,
since the other roots are $\zeta u, \zeta^2 u, \ldots, \zeta^{p-1} u$. Thus,

$$K = F(u) = F(u, \zeta u, \zeta^2 u, \ldots, \zeta^{p-1} u)$$

and therefore Gal(P/K) is a subgroup of Gal(P/F) which is
of index p by theorem 21 and is normal by theorem 22. □

Remark : The applications to the solvability of equations
by radicals use theorems 21 and 22 only through the above
corollary. In fact, only the special case of the corollary
was stated instead of theorem 22 in the original version of
Galois' memoir, with a sketch of proof. It was replaced
by the general statement of theorem 22 at a much later stage, presumably on the eve of the duel, with the comment :
"one will find the proof". The above proof comes from [E].

§ 4. Solvability by radicals

The solvability of an equation by radicals can now be
translated into a condition on the Galois group of the
equation. However, the notion of solvability by radicals
in Galois' memoir is slightly different from that of 13.1,
in that Galois requires all the roots of the equation (instead of one of them) to have an expression by radicals.
To distinguish this condition from that of 13.1, we say
that a polynomial equation with coefficients in a field F
is completely solvable (by radicals) over F if there is a
radical extension of F containing all the roots of the
equation.

This distinction is significant when dealing with arbitrary equations, and more specifically with equations
P(X) = 0 in which the left-hand side is a reducible polynomial; in this case, solving the equation amounts to finding a root of one of the factors of P, and the difficulty

of finding such a root can be completely different from factor to factor. For instance, over $\mathbb{C}(s_1,\ldots,s_n)$ the equation

$$(X-1)(X^n - s_1 X^{n-1} + s_2 X^{n-2} - \ldots + (-1)^n s_n) = 0$$

is solvable by radicals, since $X-1 = 0$ is solvable, but it is not completely solvable by radicals if $n \geq 5$, since general equations of degree at least 5 are not solvable by radicals (theorem 13.16).

We shall prove however that if the polynomial P is irreducible over F, then the equation $P(X) = 0$ is solvable by radicals over F if and only if it is completely solvable by radicals over F.

In his memoir, Galois considers only the complete solvability of equations without multiple roots; since the crucial case is the solution of irreducible equations (to which one is led by factoring the given polynomial) and since in this case both notions are equivalent, it turns out that Galois' results are actually sufficient to investigate the solvability of equations by radicals.

The central result of this section is the following :

26. THEOREM : Let P be a polynomial over a field F, and assume P has only simple roots in any field containing F. The equation $P(X) = 0$ is completely solvable over F if and only if its Galois group Gal(P/F) contains a sequence of subgroups :

$$\mathrm{Gal}(P/F) = G_0 \supset G_1 \supset G_2 \supset \ldots \supset G_t = \{I\}$$

such that, for $i=1,\ldots,t$, the subgroup G_i is normal of prime index in G_{i-1}. [G, Proposition 5, pp. 57 ff], [E, pp. 108-109].

Accordingly, a finite group G is said to be __solvable__ if it satisfies the condition of the theorem, i.e. if it contains a sequence of subgroups starting with G and ending with {I}, such that each subgroup is normal of prime index in the preceding one.

The proof of theorem 26 can actually be adapted to yield a necessary and sufficient condition for one of the roots r of P to have an expression by radicals : the condition is the same except that G_t is not required to be reduced to the identity alone, but is instead required to contain only permutations which leave r invariant.

Although the "only if" part follows relatively easily from the preceding results, the "if" part requires some preparation. More specifically, we need some results on group theory. First, we recall from 10.4 that (left) cosets of a subgroup H in a group G are the subsets of G of the form:

$$\sigma H = \{\sigma \xi \mid \xi \in H\}, \qquad \text{for } \sigma \in G,$$

and that the number of distinct cosets of H in G is the index (G:H), which is equal to the quotient $|G|/|H|$ if G is finite, by Lagrange's theorem 10.4.

27. LEMMA : $\sigma H = \tau H$ if and only if $\sigma^{-1}\tau \in H$.

Proof : If $\sigma H = \tau H$, then, in particular, $\tau.1 \in \sigma H$, hence

$$\tau = \sigma \xi$$

for some $\xi \in H$, and

$$\sigma^{-1}\tau = \xi \in H.$$

Conversely, if $\sigma^{-1}\tau \in H$, then the relation

$$\sigma \xi = \tau[(\sigma^{-1}\tau)^{-1}\xi] \qquad \text{for } \xi \in H$$

shows that $\sigma H \subset \tau H$, while the relation

$$\tau\xi = \sigma[(\sigma^{-1}\tau)\xi] \qquad \text{for } \xi \in H$$

shows that $\tau H \subset \sigma H$. □

28. LEMMA : Let $G_1 \supset G_2 \supset G_3$ be a chain of subgroups. If G_1 is finite, then

$$(G_1:G_3) = (G_1:G_2)(G_2:G_3).$$

In particular, if G_3 is a subgroup of prime index in G_1, then either $G_2 = G_1$ or $G_2 = G_3$.

<u>Proof</u> : This is clear if the index of G_3 in G_1 is calculated as :

$$\frac{|G_1|}{|G_3|} = \frac{|G_1|}{|G_2|} \cdot \frac{|G_2|}{|G_3|}.$$

□

<u>Remark</u> : The lemma also holds without the hypothesis that G_1 is finite, but the proof is more delicate. This more general case will not be needed.

29. PROPOSITION : Let H and N be subgroups of a group G, and define a subset H.N of G by :

$$H.N = \{\xi\nu \mid \xi \in H,\ \nu \in N\}.$$

If N is normal in G, then H.N is a subgroup of G and H ∩ N is a normal subgroup of H. If moreover N has prime index in G, and G is finite, then either H.N = N or else H.N = G. If H.N = N, then H ⊂ N, hence H ∩ N = H.
If H.N = G, then the index of H ∩ N in H is equal to the index of N in G :

$$(H:H \cap N) = (G:N).$$

Moreover, in this case, every coset of N in G has the form ξN for some $\xi \in H$.

Proof : The normality of $H \cap N$ in H readily follows from that of N in G, and showing that H.N is a subgroup of G is a straightforward verification. First, the unit element 1 in G is in H.N since it can be written as the product of the element $1 \in H$ by the element $1 \in N$.
Next, H.N is stable under products, since for $\xi_1, \xi_2 \in H$ and $\nu_1, \nu_2 \in N$:

$$(\xi_1 \nu_1)(\xi_2 \nu_2) = (\xi_1 \xi_2)[(\xi_2^{-1} \nu_1 \xi_2) \nu_2]$$

where $\xi_2^{-1} \nu_1 \xi_2 \in N$ since N is normal.
Finally, H.N contains the reciprocal of each of its elements, since for $\xi \in H$ and $\nu \in N$:

$$(\xi \nu)^{-1} = \xi^{-1}(\xi \nu^{-1} \xi^{-1}).$$

We thus have inclusions of subgroups :

$$G \supset H.N \supset N.$$

If the index of N in G is prime, then it follows from lemma 28 that either H.N = N or H.N = G. In the first case, we have $H \subset N$ since H is obviously contained in H.N. In the latter case, we can find, for every element $\sigma \in G$, elements $\xi \in H$ and $\nu \in N$ such that

$$\sigma = \xi \nu.$$

From lemma 27, it then follows that

$$\sigma N = \xi N,$$

hence every coset of N in G has the form ξN for some $\xi \in H$.

To prove the rest, we define a bijection between the set of cosets of $H \cap N$ in H and the set of cosets of N in G, by :

$$\xi(H \cap N) \to \xi N \qquad \text{for } \xi \in H.$$

That this map is onto follows from the last observation, that every coset of N has the form ξN for some $\xi \in H$. To prove the injectivity, we assume ξ_1 and $\xi_2 \in H$ are such that

$$\xi_1 N = \xi_2 N.$$

Lemma 27 then shows that

$$\xi_1^{-1} \xi_2 \in N,$$

hence

$$\xi_1^{-1} \xi_2 \in H \cap N$$

since ξ_1 and ξ_2 are both in H. Applying lemma 27 again, we obtain :

$$\xi_1(H \cap N) = \xi_2(H \cap N).$$

We have thus proved that

$$(G:N) = (H:H \cap N). \qquad \square$$

Remark : The results in this proposition can be put in a somewhat better perspective by making use of the notion of <u>factor group</u> of a group by a normal subgroup : essentially, the proposition asserts that the inclusion of H in $H.N$ induces an isomorphism of factor groups :

$$\frac{H}{H \cap N} \to \frac{H \cdot N}{N}.$$

This result is valid even when G is infinite. We have avoided this presentation, however, since the notion of factor group does not appear in Galois' papers and may have been unknown to Galois.

30. COROLLARY : Let N be a normal subgroup of prime index p in a finite group G. If σ is an element of G outside N, then $\sigma^p \in N$.

Proof : Consider the p+1 cosets :

$$N, \sigma N, \ldots, \sigma^p N.$$

Since the index of N in G is p, these cosets cannot be pairwise distinct, hence we can find integers m,n between 0 and p, with m < n, such that

$$\sigma^m N = \sigma^n N.$$

Lemma 27 then yields :

$$\sigma^{n-m} \in N.$$

We may thus consider the smallest integer k > 0 such that

$$\sigma^k \in N.$$

The preceding argument shows that $k \leqslant p$; to complete the proof, it thus suffices to show that $k \geqslant p$. In order to do this, we are going to show that every coset of N in G is one of the following :

$$N, \sigma N, \ldots, \sigma^{k-1} N;$$

it then readily follows that the index p of N in G is at most equal to k. Let $H = \{\sigma^i \mid i \in \mathbb{Z}\}$. This is clearly a subgroup of G and since $\sigma \notin N$, it follows that $H.N \neq N$. Therefore, the preceding proposition shows that $H.N = G$ and that any coset of N in G has the form $\sigma^i N$ for some $i \in \mathbb{Z}$. Dividing i by k, we get :

$$i = kq + r$$

for some integers q and r, with $0 \leq r < k$. Then

$$\sigma^{i-r} = (\sigma^k)^q \in N$$

hence by lemma 27 :

$$\sigma^i N = \sigma^r N.$$

This proves the claim, since r is between 0 and (p-1). □

As a further consequence of proposition 29, we record the following result, which will be quite useful in proofs by induction :

31. COROLLARY : Every subgroup H of a (finite) solvable group G is solvable.

Proof : Let

$$G = G_0 \supset G_1 \supset \ldots \supset G_t = \{1\}$$

be a sequence of subgroups in G, each of which is normal of prime index in the preceding one. We have to find a similar sequence in H. Consider :

$$H = H \cap G_0 \supseteq H \cap G_1 \supseteq \ldots \supseteq H \cap G_t = \{1\}. \qquad (13)$$

Applying proposition 29 with G_i instead of G, with G_{i+1} instead of N and $H \cap G_i$ instead of H, we deduce that $H \cap G_{i+1}$ is a normal subgroup of $H \cap G_i$, and that either

$$H \cap G_{i+1} = H \cap G_i$$

or

$$(H \cap G_i : H \cap G_{i+1}) = (G_i : G_{i+1}).$$

Therefore, after deleting repetitions in the sequence (13), we get a sequence of subgroups in H with the required properties. □

After all these preliminaries on group theory, we now come back to the solution of equations by radicals. We use the same notations as in the preceding sections : we consider a polynomial

$$P(X) = X^n - a_1 X^{n-1} + a_2 X^{n-2} - \ldots + (-1)^n a_n = (X-r_1)\ldots(X-r_n)$$

with coefficients $a_1, \ldots, a_n$ in a field F and pairwise distinct roots $r_1, \ldots, r_n$ in some field containing F. Our next result is a kind of converse of corollary 25 :

32. <u>LEMMA</u> : Let N be a normal subgroup of prime index p in Gal(P/F). If F contains a primitive p-th root of unity, then there exists a radical extension K of F in $F(r_1, \ldots, r_n)$, of the form :

$$K = F(a^{1/p})$$

for some $a \in F$, such that

$$\text{Gal}(P/K) = N.$$

Proof : We proceed by several steps. First, we pick a permutation σ in Gal(P/F) but not in N.

STEP 1 : Let $x \in F(r_1,\ldots,r_n)$ be such that $\nu(x) = x$ for all $\nu \in N$.
If $\sigma(x) = x$, then $x \in F$.
If $\sigma(x) \neq x$, and if $\tau \in \text{Gal}(P/F)$ is such that $\tau(x) = x$, then $\tau \in N$.

Let X be the set of permutations in Gal(P/F) which leave x invariant :

$$X = \{\tau \in \text{Gal}(P/F) \mid \tau(x) = x\}.$$

This set is obviously a group, which contains N, by hypothesis. From the inclusions :

$$\text{Gal}(P/F) \supset X \supset N$$

and from the hypothesis that the index of N in Gal(P/F) is prime, we deduce by lemma 28 :

$$X = N \quad \text{or} \quad X = \text{Gal}(P/F).$$

If $\sigma(x) = x$, then $\sigma \in X$ and therefore $X \neq N$ since $\sigma \notin N$. It then follows that $X = \text{Gal}(P/F)$, hence theorem 14 shows that $x \in F$.
If $\sigma(x) \neq x$, then $\sigma \notin X$, hence $X \neq \text{Gal}(P/F)$; therefore, $X = N$, which means that every permutation in Gal(P/F) which leaves x invariant is in N.

STEP 2 : There is an element $v \in F(r_1,\ldots,r_n)$ which is invariant under every permutation in N but is not in F.

Let $f(X_1,\ldots,X_n)$ be a polynomial in $F[X_1,\ldots,X_n]$ which has the property of lemma 9, i.e. that the n! elements of

$F(r_1,\ldots,r_n)$ obtained by replacing the indeterminates by $r_1,\ldots,r_n$ in all possible ways are pairwise different, and let $V = f(r_1,\ldots,r_n)$.

The proof of proposition 10 shows that V is a Galois resolvent of $P(X) = 0$ over F, hence the degree of its minimum polynomial over F is equal to $|\mathrm{Gal}(P/F)|$. Consider then the polynomial :

$$\Pi_{\nu \in N}(X - \nu(V)).$$

The coefficients of this polynomial are clearly invariant under N. If they were all in F, then V would be a root of a polynomial of degree $|N|$ over F; this is impossible since the minimum polynomial of V over F has degree larger than $|N|$. Therefore, at least one of the coefficients is invariant under N but is not in F. This coefficient can be chosen for v.

For every p-th root of unity $\omega \in F$, we define a kind of Lagrange resolvent :

$$t(\omega) = v + \omega\sigma(v) + \ldots + \omega^{p-1}\sigma^{p-1}(v).$$

<u>STEP 3</u> : $\sigma(t(\omega)) = \omega^{-1}t(\omega)$, and $\nu(t(\omega)) = t(\omega)$ for all $\nu \in N$.

The powers of ω are in F and are therefore invariant under every permutation in $\mathrm{Gal}(P/F)$, by theorem 14. Thus,

$$\sigma(t(\omega)) = \sigma(v) + \omega\sigma^2(v) + \ldots + \omega^{p-1}\sigma^p(v)$$

or, equivalently,

$$\sigma(t(\omega)) = \omega^{-1}(\sigma^p(v) + \omega\sigma(v) + \ldots + \omega^{p-1}\sigma^{p-1}(v)),$$

and

$$\nu(t(\omega)) = \nu(v) + \omega\nu\sigma(v) + \ldots + \omega^{p-1}\nu\sigma^{p-1}(v),$$

for every $\nu \in N$.
Corollary 30 shows that $\sigma^p \in N$, hence $\sigma^p(v) = v$ and it readily follows that

$$\sigma(t(\omega)) = \omega^{-1}t(\omega).$$

On the other hand, since N is a normal subgroup of $Gal(P/F)$, we have

$$\sigma^{-i}\nu\sigma^i \in N \quad \text{for every } \nu \in N \text{ and every } i=0,\ldots,p-1$$

whence

$$\sigma^{-i}\nu\sigma^i(v) = v \quad \text{for every } \nu \in N \text{ and every } i=0,\ldots,p-1.$$

Applying σ^i to both sides of this equation, we get :

$$\nu\sigma^i(v) = \sigma^i(v) \quad \text{for every } \nu \in N \text{ and every } i=0,\ldots,p-1,$$

hence

$$\nu(t(\omega)) = t(\omega) \quad \text{for every } \nu \in N.$$

<u>STEP 4</u> : $t(\omega)^p \in F$ for every p-th root of unity ω, and there is a p-th root of unity $\omega \neq 1$ such that $t(\omega) \neq 0$.

From step 3 it follows that $t(\omega)^p$ is invariant under σ and under every permutation in N. Step 1 shows that $t(\omega)^p$ lies therefore in F.
If we assume $t(\omega) = 0$ for every p-th root of unity $\omega \neq 1$, then Lagrange's formula 10.2

$$v = (1/p)[\Sigma_\omega t(\omega)]$$

yields :

$$v = (1/p)\, t(1)$$

and this equality shows, by step 3, that v is invariant under σ. Since it is also invariant under N, it follows by step 1 that v is in F; this is a contradiction, since v has been chosen outside F in step 2.

Let thus ω be a p-th root of unity such that $\omega \neq 1$ and $t(\omega) \neq 0$, and let

$$K = F(t(\omega)).$$

Step 4 and proposition 13.2 show that K is a radical extension of F, of the form $F(a^{1/p})$. To complete the proof, it now suffices to show :

STEP 5 : Gal(P/K) = N.

Since $t(\omega) \neq 0$ and $\omega \neq 1$, step 3 shows that $t(\omega)$ is not invariant under σ.
Therefore, $K \neq F$, hence K is a radical extension of height 1 of F. From corollary 25, it then follows that Gal(P/K) is a subgroup of index p in Gal(P/F), whence

$$|\mathrm{Gal}(P/K)| = |N|.$$

Moreover, since $\sigma(t(\omega)) \neq t(\omega)$, step 1 shows that every permutation in Gal(P/F) which leaves $t(\omega)$ invariant is in N. As $t(\omega) \in K$, the permutations in Gal(P/K) leave $t(\omega)$ invariant, by theorem 14, hence

$$\mathrm{Gal}(P/K) \subset N.$$

Since these groups have the same order, N cannot be strictly larger than Gal(P/K) :

$$\mathrm{Gal}(P/K) = N. \qquad \square$$

Proof of theorem 26 : We first prove, by induction on
$|\text{Gal}(P/F)|$, that if $P(X) = 0$ is completely solvable over F,
then Gal(P/F) is solvable.
If $|\text{Gal}(P/F)| = 1$, then Gal(P/F) = {I}, and this group is
trivially solvable. We may thus assume that completely
solvable equations with Galois group of order less than
that of $P(X) = 0$ over F have solvable Galois groups.
Let R be a radical extension of F which contains all the
roots of P. From theorem 14 it follows that every element
in R is invariant under Gal(P/R); therefore, every root of
P is invariant under the permutations in Gal(P/R), which
means that Gal(P/R) = {I}. This shows that there exist
radical extensions K of F such that

$$|\text{Gal}(P/K)| < |\text{Gal}(P/F)|.$$

We may thus consider the smallest prime p for which the extraction of a p-th root begins to decrease the order of the Galois group of P. We let p be the smallest prime number for which there exists a radical extension L of F such that

$$\text{Gal}(P/L) = \text{Gal}(P/F)$$

and

$$|\text{Gal}(P/L(a^{1/p}))| < |\text{Gal}(P/F)|$$

for some $a \in L$ which is not a p-th power in L.

By proposition 13.5, there is a radical extension of L which contains a primitive p-th root of unity. Moreover, inspection of the proof of this proposition shows that there is such an extension R' which is obtained from L by extractions of q-th roots for prime numbers $q < p$. Therefore, by definition of p, we have :

$$\text{Gal}(P/R') = \text{Gal}(P/L) = \text{Gal}(P/F).$$

Moreover, by proposition 18,

$$\text{Gal}(P/R'(a^{1/p})) \subset \text{Gal}(P/L(a^{1/p})),$$

hence

$$|\text{Gal}(P/R'(a^{1/p}))| < |\text{Gal}(P/F)|.$$

Since R' contains a primitive p-th root of unity, corollary 25 shows that $\text{Gal}(P/R'(a^{1/p}))$ is a normal subgroup of index p in $\text{Gal}(P/R') = \text{Gal}(P/F)$. Since $P(X) = 0$ is completely solvable over F, it is also completely solvable over $R'(a^{1/p})$, by theorem 13. (Note that the proof of this theorem is valid word for word for complete solvability instead of solvability). Therefore, by the induction hypothesis, we can find a sequence of subgroups :

$$\text{Gal}(P/R'(a^{1/p})) \supset G_2 \supset \ldots \supset G_t = \{I\}$$

such that each subgroup is normal of prime index in the preceding one. The sequence

$$\text{Gal}(P/F) \supset \text{Gal}(P/R'(a^{1/p})) \supset G_2 \supset \ldots \supset G_t = \{I\}$$

then shows that $\text{Gal}(P/F)$ is solvable.

We now prove that, conversely, solvability of $\text{Gal}(P/F)$ implies complete solvability of $P(X) = 0$ by radicals over F. We argue again by induction on $|\text{Gal}(P/F)|$. If $|\text{Gal}(P/F)| = 1$, then the only permutation in $\text{Gal}(P/F)$ is the identity, which leaves every root invariant. Therefore, by theorem 14, all the roots of P are in F, which is a radical extension of height 0 of itself.
We may thus assume, by induction, that equations with solvable Galois group of order less than that of P over F are completely solvable.
Since $\text{Gal}(P/F)$ is solvable, it contains a normal subgroup

N of prime index. Let p = (Gal(P/F) : N). By proposition 13.5, there exists a radical extension R of F which contains all the p-th roots of unity.
If

$$|Gal(P/R)| < |Gal(P/F)|,$$

then we can resort to the induction hypothesis, since by corollary 31, Gal(P/R) is a solvable group. The equation P(X) = 0 is thus completely solvable over R, hence there exists a radical extension R' of R which contains all the roots of P. Since the field R' is also a radical extension of F, the proof is complete in this case.
If on the contrary

$$Gal(P/R) = Gal(P/F),$$

then we resort to lemma 32, which shows that there is a radical extension R" of R such that Gal(P/R") = N. Since then

$$|Gal(P/R")| < |Gal(P/F)|,$$

we conclude as above by the induction hypothesis. □

33. REMARK : Assume all the roots of unity are in the base field F. The last part of the proof above then shows that if the Galois group Gal(P/F) is solvable :

$$Gal(P/F) = G_0 \supset G_1 \supset \ldots \supset G_t = \{I\},$$

with G_i normal of prime index in G_{i-1} for i=1,...,t, then a radical extension of F containing all the roots of P can be obtained by t extractions of roots : first, the extraction of a $(G_0:G_1)$-th root, which reduces the Galois group to G_1, next the extraction of a $(G_1:G_2)$-th root, which reduces the Galois group to G_2, and so on.

34. EXAMPLE : Theorem 26 illuminates the solution of equations of degree 3 and 4 by radicals, as we now show. First, we define for any integer $n \geq 1$ a subgroup A_n of the symmetric group S_n : it is the group of all permutations in S_n which leave invariant the polynomial $\Delta(X_1,\ldots,X_n)$ used in the definition of the discriminant (see 8.14) :

$$\Delta(X_1,\ldots,X_n) = \Pi_{1 \leq i < j \leq n} (X_i - X_j).$$

Thus, with the notations of § 10.2 :

$$A_n = I(\Delta).$$

The group A_n is called the <u>alternating group</u> on $\{1,\ldots,n\}$.

As noted in 8.14, any permutation of the indeterminates either leaves Δ invariant or transforms it into its opposite $-\Delta$. Therefore, by Lagrange's theorem 10.3, the subgroup A_n has index 2 in S_n :

$$|A_n| = n!/2.$$

Moreover, it is easily seen that A_n is normal in S_n : one has to see that for all $\sigma \in S_n$ and for all $\tau \in A_n$, $\sigma\tau\sigma^{-1} \in A_n$; this is clear if $\sigma \in A_n$, since A_n is stable under products; if on the contrary $\sigma \notin A_n$, then $\sigma^{-1} \notin A_n$, whence

$$\sigma(\Delta) = \sigma^{-1}(\Delta) = -\Delta$$

and thus

$$\sigma\tau\sigma^{-1}(\Delta) = -\sigma\tau(\Delta) = -\sigma(\Delta) = \Delta,$$

which shows that $\sigma\tau\sigma^{-1} \in A_n$, as claimed.

Consider now the general equation of degree n :

$$P(X) = (X-X_1)\ldots(X-X_n) = X^n - s_1 X^{n-1} + s_2 X^{n-2} - \ldots + (-1)^n s_n = 0$$

over $F = \mathbb{C}(s_1,\ldots,s_n)$. (The reference field is chosen to be $\mathbb{C}$ so that all the roots of unity are in F). By example 4, the Galois group Gal(P/F) can be identified to S_n. Since Δ^2 is the discriminant :

$$\Delta^2 = D(s_1,\ldots,s_n) \in F,$$

it follows that Δ is a root of the polynomial

$$X^2 - D(s_1,\ldots,s_n) \in F[X].$$

Hence, by theorem 21, Gal(P/F(Δ)) is a subgroup of index 2 in Gal(P/F) = S_n. Since the elements in Gal(P/F(Δ)) leave Δ invariant (theorem 14), we have

$$\text{Gal}(P/F(\Delta)) \subset A_n$$

whence

$$\text{Gal}(P/F(\Delta)) = A_n$$

since these groups both have index 2 in S_n and have therefore the same number of elements. (The fact that A_n is normal in S_n then also follows from theorem 22, since adjoining one of the root of a quadratic polynomial amounts to adjoining both roots).

Let now n = 3. The sequence of subgroups :

$$S_3 \supset A_3 \supset \{I\}$$

shows that S_3 is solvable, since the index of A_3 in S_3 is 2 and that of $\{I\}$ in A_3 is $|A_3| = 3$. Therefore, the general equation of degree 3 is completely solvable by radicals. Moreover, a radical extension of F containing all the roots can be obtained by two extractions of roots : first, the extraction of a square root, which reduces the Galois group from S_3 to A_3, and then the extraction of a cube root, which reduces the Galois group to $\{I\}$. More precisely, the above discussion shows that the square root which has to be extracted first is that of the discriminant $D(s_1,s_2,s_3)$, since indeed $Gal(P/F(\Delta)) = A_3$. (In Cardano's formula, the radicand of the square root is indeed the discriminant : compare 8.15).

Now, consider $n = 4$. In example 24, we have seen that the adjunction to F of all the roots of Ferrari's resolvent cubic reduces the Galois group to

$$V = \{I,\sigma_1,\sigma_2,\sigma_3\},$$

where σ_1,σ_2 and σ_3 commute and satisfy : $\sigma_1^2 = \sigma_2^2 = \sigma_3^2 = I$. Therefore, $\{I,\sigma_1\}$, $\{I,\sigma_2\}$ and $\{I,\sigma_3\}$ are normal subgroups of index 2 in V. Moreover, a direct verification shows that σ_1,σ_2 and σ_3 leave Δ invariant, whence

$$V \subset A_4.$$

Counting elements, we get :

$$(A_4 : V) = 3.$$

Moreover, since V is normal in S_4 (see n° 24), it is <u>a fortiori</u> normal in A_4. The sequence :

$$S_4 \supset A_4 \supset V \supset \{I,\sigma_1\} \supset \{I\}$$

then shows that S_4 is solvable. Consequently, the general equation of degree 4 is completely solvable by radicals over F. The above sequence of subgroups shows moreover

that a radical extension of F containing all the roots is obtained by :
1) the extraction of a square root of the discriminant $D(s_1,s_2,s_3,s_4)$, which reduces the Galois group to A_4;
2) the extraction of a cube root, which reduces the Galois group to V;
3) and 4) the successive extraction of two square roots, which reduce the Galois group to $\{I,\sigma_1\}$ and then to $\{I\}$.

In fact, the first two steps are achieved by the solution of Ferrari's resolvent cubic, which requires first the extraction of a square root of its discriminant. Therefore, extracting a square root of $D(s_1,s_2,s_3,s_4)$ amounts to extracting a square root of the discriminant of the resolvent cubic. Indeed, it is readily verified by a direct computation that these two discriminants are equal. (See exercise 3 of chapter 8).

From the preceding discussion, it follows at the same time that every equation of degree 3 or 4 is completely solvable by radicals, since the Galois group of such an equation is a group of permutation of the roots, which can be identified, via a numbering of the roots, to a subgroup of S_3 or S_4 and is therefore solvable, by corollary 31.

To finish this section, we turn to the (not necessarily complete) solvability of equations by radicals :

35. THEOREM : Let r be a root of a polynomial P over some field F, and assume P has only simple roots in any field containing F. The root r has an expression by radicals over F if and only if Gal(P/F) contains a sequence of subgroups:

$$\text{Gal}(P/F) = G_0 \supset G_1 \supset G_2 \supset \ldots \supset G_t$$

such that G_i is normal of prime index in G_{i-1} for $i=1,\ldots,t$, and r is invariant under every permutation in G_t.

The proof is the same as for theorem 26, except that one has to use induction on other integers than $|Gal(P/F)|$: for the "only if" part, use induction on the number of elements in the set

$$\{\sigma(r) \mid \sigma \in Gal(P/F)\}$$

(which is called the <u>orbit</u> of r under Gal(P/F)); for the "if" part, use induction on the length t of the sequence of subgroups of Gal(P/F). Details are left to the reader.

Using this theorem and theorem 26, we are now able to show that, as claimed in the introduction to this section, the two notions of solvability by radicals are equivalent for irreducible polynomials. We shall need the following characterization of irreducible equations through their Galois group :

36. PROPOSITION : A non-constant polynomial $P \in F[X]$ without multiple root in any extension of F is irreducible over F if and only if the Galois group Gal(P/F) is transitive on the roots of P, i.e. : for any roots r_i, r_j of P, there is a permutation $\sigma \in Gal(P/F)$ such that

$$\sigma(r_i) = r_j.$$

Proof : If P is not irreducible, let

$$P = P_1 P_2$$

for some non-constant polynomials $P_1, P_2 \in F[X]$. If r_1 is a root of P_1, then

$$P_1(r_1) = 0,$$

whence, applying $\sigma \in Gal(P/F)$ to both sides of this equation :

$$P_1(\sigma(r_1)) = 0 \quad \text{for all } \sigma \in \text{Gal}(P/F).$$

Thus, no permutation in Gal(P/F) carries r_1 to a root of P_2; therefore, Gal(P/F) is not transitive on the roots of P. Conversely, assume Gal(P/F) is not transitive on the roots of P. Then, there are roots r_i, r_j of P such that r_i is not mapped onto r_j by any permutation in Gal(P/F). Let

$$R = \{\sigma(r_i) \mid \sigma \in \text{Gal}(P/F)\}$$

and let

$$P_1 = \Pi_{r \in R} (X-r).$$

The coefficients of P_1 are in F, by theorem 14, since they are invariant under Gal(P/F). Moreover, the elements in R are roots of P, hence P_1 divides P. On the other hand, the hypothesis on r_i, r_j implies that $r_j \notin R$; there is thus a root of P which is not a root of P_1. Therefore, the degree of P_1 is strictly smaller than that of P, and it follows that P is not irreducible in F[X]. □

37. PROPOSITION : Let P be an irreducible polynomial over some field F. The equation P(X) = 0 is completely solvable by radicals over F if and only if it is solvable by radicals over F.

Proof : It obviously suffices to prove the "if" part. We first note that, by theorem 5.26, irreducible polynomials have no multiple root in any extension of the base field. We may thus use theorems 26 and 35, together with proposition 36, to translate the proposition into a purely group-theoretical statement :

CLAIM : Let G be a transitive group of permutations of a set $\{r_1, \ldots, r_n\}$. If G contains a sequence of subgroups

$$G = G_0 \supset G_1 \supset \ldots \supset G_t \tag{S}$$

such that G_i is normal of prime index in G_{i-1} for $i=1,\ldots,t$, and such that every permutation in the last subgroup G_t leaves one of the elements (r_1, say) invariant, then G is solvable.

Since G is transitive, we can find, for $i=2,\ldots,n$, a permutation $\sigma_i \in G$ such that

$$\sigma_i(r_1) = r_i.$$

We then use the inner automorphism of $G: \tau \to \sigma_i \tau \sigma_i^{-1}$, which transforms the given sequence of subgroups into:

$$G = G_0 \supset \sigma_i G_1 \sigma_i^{-1} \supset \ldots \supset \sigma_i G_t \sigma_i^{-1}, \tag{S_i}$$

a sequence in which each subgroup is normal of prime index in the preceding one, and the last subgroup $\sigma_i G_t \sigma_i^{-1}$ leaves r_i invariant.

Intersecting with G_t all the subgroups in the sequence (S_2), we get:

$$G_t \supset (G_t \cap \sigma_2 G_1 \sigma_2^{-1}) \supset (G_t \cap \sigma_2 G_2 \sigma_2^{-1}) \supset \ldots \supset (G_t \cap \sigma_2 G_t \sigma_2^{-1}). \tag{S_2'}$$

By proposition 29 (see also the proof of corollary 31), each subgroup is normal of index 1 or a prime number in the preceding subgroup. The given sequence (S) can thus be continued up to $(G_t \cap \sigma_2 G_t \sigma_2^{-1})$, which leaves invariant both r_1 and r_2.

Intersecting with $(G_t \cap \sigma_2 G_t \sigma_2^{-1})$ all the subgroups in the sequence S_3, we get a sequence similar to (S_2'), beginning with $(G_t \cap \sigma_2 G_t \sigma_2^{-1})$ and ending with

$$G_t \cap \sigma_2 G_t \sigma_2^{-1} \cap \sigma_3 G_t \sigma_3^{-1},$$

a subgroup which leaves invariant r_1, r_2 and r_3. This sequence can be used to extend the sequence previously constructed.

Continuing in the same way (and deleting the possible repetitions), we construct a sequence of subgroups of G, each of which is normal of prime index in the preceding one, which ends with

$$G_t \cap \sigma_2 G_t \sigma_2^{-1} \cap \ldots \cap \sigma_n G_t \sigma_n^{-1}.$$

Since this group leaves invariant all of $r_1, \ldots, r_n$, it is reduced to {I}, and the sequence of subgroups thus constructed shows that G is solvable. □

§ 5. Applications

We present two applications of Galois' theory : the first one is due to Galois himself, and deals with irreducible equations of prime degree; the second one proves the theorem of Abel stated in section 1, namely that equations which satisfy Abel's condition are solvable by radicals.

In the last part of his memoir, Galois determines the Galois groups of irreducible equations of prime degree which are solvable by radicals (completely or not; these properties are equivalent, by proposition 37). In view of theorem 26 and proposition 36, this amounts to the determination of the solvable transitive groups of permutations of p elements, for p prime, i.e. of the solvable transitive subgroups of S_p.

Before discussing Galois' result, we review some basic observations about groups of permutations, which will play a crucial role in the sequel.

38. DEFINITIONS : Let G be a group of permutations of a set E. For any $a \in E$, we define the <u>orbit</u> of a under G as the set of elements in E where a can be mapped by a permutation in G :

$$G(a) = \{\sigma(a) \mid \sigma \in G\}.$$

We also define the <u>isotropy subgroup</u> of a in G as the set of permutations in G which leave a invariant :

$$I_G(a) = \{\sigma \in G \mid \sigma(a) = a\}$$

(compare § 10.3).
The same arguments as in the proof of Lagrange's theorem 10.3 yield the following result :

<u>39. THEOREM</u> : If G is finite, then for any $a \in E$:

$$|G| = |G(a)| \cdot |I_G(a)|.$$

<u>Proof</u> : For any $b \in E$, let

$$I_G(a \to b) = \{\sigma \in G \mid \sigma(a) = b\}.$$

Arranging the elements of G according to where they map a, we obtain a decomposition of G into disjoint subsets :

$$G = \cup_{b \in G(a)} I_G(a \to b),$$

whence

$$|G| = \Sigma_{b \in G(a)} |I_G(a \to b)|.$$

Now, if $b \in G(a)$ then $b = \sigma(a)$ for some $\sigma \in G$ and it is readily checked that

$$I_G(a \to b) = \sigma \, I_G(a).$$

Therefore, all the sets $I_G(a \to b)$ for $b \in G(a)$ have the same number of elements as $I_G(a)$, and the preceding equality yields :

$$|G| = |G(a)| \cdot |I_G(a)|.$$ □

The following observations on the orbits under G will often be useful in the sequel :

40. PROPOSITION : With the same notations as in 38 :
a) For any $x \in G(a)$, we have $G(x) = G(a)$.
b) Any two orbits under G are either disjoint or identical.
c) The set E decomposes into a union of disjoint orbits under G.

Proof : a) Let $x = \sigma(a)$ for some $\sigma \in G$. Then

$$G(x) = G\sigma(a).$$

Since multiplication on the right by σ is a bijection from G onto G, it follows that $G\sigma = G$, hence

$$G(x) = G(a).$$

b) Let $a,b \in E$ be such that $G(a) \cap G(b)$ is not empty; there is then an element $x \in G(a) \cap G(b)$. By a), we have $G(x) = G(a)$ and $G(x) = G(b)$, hence $G(a)$ and $G(b)$ coincide.
c) readily follows from b) : if A is a subset of E containing one element from each orbit under G, then

$$E = \bigcup_{a \in A} G(a),$$

and the orbits $G(a)$ are pairwise disjoint.   □

We thus obtain a first result on transitive subgroups of S_p :

41. COROLLARY : Let G be a transitive group of permutations of a set E with p elements (p **prime**), and let N be a normal subgroup of G. If $N \neq \{I\}$, then N is transitive on E.

Proof : Decompose E into a union of disjoint orbits under N :

$$E = \bigcup_{a \in A} N(a).$$

Counting elements, it follows that

$$p = \sum_{a \in A} |N(a)|.$$

To complete the proof, it now suffices to show that any two orbits have the same number of elements; indeed, denoting this number by n, the equality above yields :

$$p = n \cdot |A|$$

and since p is prime, there are only two possibilities : either $n = 1$, which means that N leaves every element of E invariant, hence that $N = \{I\}$; or $n = p$ and $|A| = 1$, which means that N is transitive on E. Thus, it only remains to prove :

CLAIM : $|N(a)| = |N(b)|$ for any $a,b \in E$.

Since G is transitive on E, there is a permutation $\sigma \in G$ such that $\sigma(a) = b$. Then

$$\sigma N \sigma^{-1}(b) = \sigma N(a);$$

but $\sigma N \sigma^{-1} = N$ since N is normal in G, hence

$$N(b) = \sigma N(a).$$

Therefore, the permutation σ induces a bijection from $N(a)$ onto $N(b)$. This proves the claim.   □

In order to state Galois' classification of the solvable

transitive subgroups of S_p, we recall the group GA(p) defined before theorem 10.8. For notational convenience, we consider S_p as the group of permutations of the set $\{0,1,\ldots,p-1\}$ (instead of $\{1,\ldots,p\}$), and we define :

$$\tau : 0 \to 1 \to \ldots \to p-1 \to 0.$$

Using the congruence relation modulo p, we can recast the definition of τ as :

$$\tau : x \to x+1 \quad (\bmod\ p)$$

for $x \in \{0,\ldots,p-1\}$, since $(p-1)+1 \equiv 0 \pmod{p}$. For $i=1,\ldots,p-1$, we also define permutations σ_i by :

$$\sigma_i : x \to ix \quad (\bmod\ p)$$

(compare the definition before theorem 10.8). The group GA(p) is the subgroup of S_p generated by $\sigma_1,\ldots,\sigma_{p-1}$ and τ. It is readily checked that the elements of GA(p) are the permutations of the form :

$$x \to ax+b \quad (\bmod\ p)$$

where $a \in \{1,\ldots,p-1\}$ and $b \in \{0,\ldots,p-1\}$. (This shows that $|GA(p)| = p(p-1)$, as claimed in section 10.3).

Galois' result (in proposition 7 of his memoir [G, pp. 65 ff], [E, pp. 111-112]) is that GA(p) is essentially the only solvable transitive subgroup of S_p :

42. THEOREM : Every solvable transitive subgroup of S_p is conjugate to a subgroup of GA(p), i.e. is of the form $\alpha H \alpha^{-1}$ for some $\alpha \in S_p$ and some subgroup H of GA(p). In particular, the order of such a group divides $p(p-1)$. Conversely, every subgroup of S_p which is conjugate to a subgroup of GA(p) is solvable.

An interpretation of this theorem, in the light of Lagrange's investigations, is that no reduction can be carried out beyond the equation of degree $(p-2)!$ of theorem 10.8. The following property of $GA(p)$ is crucial for the proof of the theorem :

43. LEMMA : Let $\tau \in GA(p)$ be defined as above by :

$$\tau(x) = x+1 \pmod{p} \qquad \text{for } x \in \{0,\ldots,p-1\}$$

and let $\theta \in S_p$. If $\theta\tau\theta^{-1} \in GA(p)$, then $\theta \in GA(p)$.

Proof : We first show that $\theta\tau\theta^{-1}$ is a power of τ. Assume on the contrary :

$$\theta\tau\theta^{-1} : x \to ax+b \pmod{p}$$

for some $a \not\equiv 0,1 \pmod{p}$. Then it is easily seen by induction on i that

$$(\theta\tau\theta^{-1})^i : x \to a^i x + (a^{i-1} + \ldots + a + 1)b \pmod{p}.$$

In particular,

$$(\theta\tau\theta^{-1})^{p-1} : x \to a^{p-1} x + (a^{p-2} + \ldots + a + 1)b \pmod{p}.$$

Now, since $a \not\equiv 0 \pmod{p}$, we have $a^{p-1} \equiv 1 \pmod{p}$ by Fermat's theorem 12.6. Moreover, multiplying the coefficient of b by $a-1$ we get :

$$(a-1)(a^{p-2} + \ldots + a + 1) = a^{p-1} - 1 \equiv 0 \pmod{p}.$$

Since $a-1 \not\equiv 0 \pmod{p}$, it follows that the coefficient of b is $0 \pmod{p}$, hence :

$$(\theta\tau\theta^{-1})^{p-1} = I.$$

Since

$$(\theta\tau\theta^{-1})^{p-1} = \theta\tau^{p-1}\theta^{-1},$$

this last equality yields :

$$\tau^{p-1} = I,$$

a contradiction. Therefore, as claimed,

$$\theta\tau\theta^{-1} = \tau^i$$

for some integer i between 1 and p-1.
Multiplying both sides by θ on the right, we get :

$$\theta\tau = \tau^i\theta$$

which means that for all $x=0,\ldots,p-1$,

$$\theta(x+1) = \theta(x) + i \pmod{p}.$$

Arguing by induction on x, it follows that

$$\theta(x) = ix + \theta(o) \pmod{p}$$

for all $x=0,\ldots,p-1$. This shows that $\theta = \tau^{\theta(o)}\sigma_i$, whence $\theta \in GA(p)$. □

<u>Proof of theorem 42</u> : Let G be a solvable transitive subgroup of S_p and let

$$G = G_o \supset G_1 \supset \ldots \supset G_t = \{I\} \qquad (14)$$

be a sequence of subgroups, each of which is normal of prime index in the preceding one. Since G is transitive, corollary 41 implies that G_1 is transitive (unless G_1 =

{I}, i.e. t = 1), whence also that G_2 is transitive (unless G_2 = {I}, i.e. t = 2), and so on up to G_{t-1}.
By theorem 39, the number of elements in the orbit of any element in {0,...,p-1} divides $|G_{t-1}|$. Since G_{t-1} is transitive on {0,...,p-1}, it follows that p divides $|G_{t-1}|$; but the order of G_{t-1}, which is the index of G_t in G_{t-1}, is prime, hence

$$|G_{t-1}| = p.$$

The subgroup of G_{t-1} generated by any element other than I is then equal to G_{t-1}, since its order is a divisor of p. It follows that G_{t-1} is generated by a single permutation, which must be a cycle of length p, i.e. a permutation

$$\gamma : i_1 \to i_2 \to \ldots \to i_p \to i_1$$

where $i_1, i_2, \ldots, i_p$ are $0, 1, \ldots, p-1$ in some order.
Let $\alpha \in S_p$ be the permutation :

$$\alpha : \begin{cases} 0 \to i_1 \\ 1 \to i_2 \\ \vdots \\ p-1 \to i_p. \end{cases}$$

Then

$$\alpha^{-1} \gamma \alpha : 0 \to 1 \to 2 \to \ldots \to p-1 \to 0,$$

i.e. with the same notations as above :

$$\alpha^{-1} \gamma \alpha = \tau.$$

For i=0,1,...,t, let G_i' be the image of G_i under the inner automorphism of $S_p : \xi \to \alpha^{-1} \xi \alpha$:

375

$$G'_i = \alpha^{-1} G_i \alpha.$$

Transforming the given sequence (14) by the inner automorphism, we get a similar sequence:

$$G'_0 \supset G'_1 \supset \ldots \supset G'_t = \{I\}$$

in which each subgroup is normal of prime index in the preceding subgroup. Moreover, since G_{t-1} is generated by γ, the next-to-last subgroup G'_{t-1} is generated by τ; thus, $G'_{t-1} \subset GA(p)$.
The subgroup G'_{t-1} is normal in G'_{t-2}, hence for any $\theta \in G'_{t-2}$,

$$\theta \tau \theta^{-1} \in G'_{t-1}$$

and consequently

$$\theta \tau \theta^{-1} \in GA(p) \qquad \text{for all } \theta \in G'_{t-2},$$

since $G'_{t-1} \subset GA(p)$. Lemma 43 then shows that $G'_{t-2} \subset GA(p)$. We can now repeat the same arguments with G'_{t-2} and G'_{t-3} instead of G'_{t-1} and G'_{t-2}, to conclude that $G'_{t-3} \subset GA(p)$. Repeating the same arguments as many times as needed, we eventually obtain $G'_0 \subset GA(p)$. Since $G = \alpha G'_0 \alpha^{-1}$, it follows that G is conjugate to the subgroup G'_0 of $GA(p)$.
In order to prove that, conversely, every subgroup of S_p which is conjugate to a subgroup of $GA(p)$ is solvable, it suffices, by corollary 31, to prove that $GA(p)$ itself is solvable. To this end, we choose a primitive root g of p (see theorem 12.2) and, for any factor e of $p-1$, we define a subset H_e of $GA(p)$ by:

$$H_e = \{x \to g^{ei}x + c \pmod{p} \mid c=0,\ldots,p-1 \text{ and } i=0,\ldots,(p-1)e^{-1}-1\}.$$

A straightforward verification shows that this set is a

normal subgroup of GA(p). Moreover, we clearly have :

$$|H_e| = p(p-1)/e,$$

and if e,e' are factors of $p-1$ such that e divides e', then

$$H_e \supset H_{e'},$$

and by comparing the orders of these groups we see that the index of $H_{e'}$ in H_e is e'/e. Let now

$$p-1 = q_1 \ldots q_r$$

be the decomposition of $p-1$ into a product of (not necessarily distinct) primes, and let

$$e_0 = 1, e_1 = q_1, e_2 = q_1 q_2, \ldots, e_{r-1} = q_1 \ldots q_{r-1}, e_r = (p-1),$$

so that e_{i-1} divides e_i for $i=1,\ldots,r$, with a quotient e_i/e_{i-1} which is a prime number. The sequence of subgroups

$$GA(p) = H_{e_0} \supset H_{e_1} \supset \ldots \supset H_{e_r} \supset \{I\}$$

then shows that GA(p) is solvable. □

Another characterization of the solvable transitive subgroups of S_p can be derived from the preceding theorem :

44. THEOREM : A transitive subgroup of S_p is solvable if and only if no permutation in G leaves two elements of $\{0,\ldots,p-1\}$ invariant, except the identity.

Proof : Assume first that G is solvable. By theorem 42, there is a permutation $\alpha \in S_p$ such that

$$\alpha^{-1} G \alpha \subset GA(p).$$

If $\theta \in G$ leaves two elements u,v invariant, then $\alpha^{-1}\theta\alpha \in GA(p)$ leaves $\alpha^{-1}(u)$ and $\alpha^{-1}(v)$ invariant. But it is readily verified that no permutation of the form

$$x \to ax+b \qquad (a \in \{1,\ldots,p-1\},\ b \in \{0,\ldots,p-1\})$$

(i.e. no permutation in $GA(p)$) except the identity, leaves two elements invariant.

Therefore, $\alpha^{-1}\theta\alpha = I$, hence $\theta = I$.

Conversely, if no permutation in G, besides the identity, leaves two elements invariant, then for $u,v \in \{0,\ldots,p-1\}$ with $u \neq v$:

$$I_G(u) \cap I_G(v) = \{I\}.$$

Therefore, the set of permutations in G, other than I, which leave an element invariant decomposes into a union of disjoint sets :

$$\cup_{u \in \{0,\ldots,p-1\}} (I_G(u) - \{I\}). \tag{15}$$

In order to calculate the number of permutations in this set, we observe that, since G is transitive :

$$G(u) = \{0,\ldots,p-1\} \qquad \text{for all } u \in \{0,\ldots,p-1\};$$

hence, by theorem 39,

$$p \cdot |I_G(u)| = |G| \qquad \text{for all } u \in \{0,\ldots,p-1\}.$$

Denoting

$$q = |I_G(u)| \qquad \text{for any } u \in \{0,\ldots,p-1\},$$

we have

$$|G| = p \cdot q$$

and, using the decomposition (15), it follows that the number of elements in G, other than I, which leave an element invariant is p.(q-1). There are therefore p-1 permutations in G which leave no element invariant. Let θ be such a permutation.

CLAIM : θ is a cycle of length p :

$$\theta : i_1 \to i_2 \to \ldots \to i_p \to i_1$$

where $i_1, \ldots, i_p$ are $0, 1, \ldots, p-1$ in some order, and the elements of G which leave no element invariant are $\theta, \theta^2, \ldots, \theta^{p-1}$.

Let T be the subgroup of G generated by θ :

$$T = \{\theta^k \mid k \in \mathbb{Z}\}.$$

We first show that $I_T(u) = \{I\}$ for every $u \in \{0, \ldots, p-1\}$; indeed, if

$$\theta^k(u) = u,$$

then, applying θ to both sides :

$$\theta^k(\theta(u)) = \theta(u),$$

hence every permutation in $I_T(u)$ leaves invariant the two elements u, θ(u). From the hypothesis on G, it follows that $I_T(u)$ is reduced to {I}.
By theorem 39, this result implies that

$$|T(u)| = |T| \qquad \text{for any } u \in \{0, \ldots, p-1\}.$$

Considering then the decomposition of $\{0, \ldots, p-1\}$ into a union of disjoint orbits under T :

$$\{0,\ldots,p-1\} = \cup_{u \in U} T(u),$$

we get by counting elements :

$$p = n \cdot |T|$$

where n is the number of distinct orbits. Since p is prime and $|T| > 1$, it follows that

$$|T| = p \quad \text{and} \quad n = 1;$$

hence, θ is a cycle of length p. Then, $\theta, \theta^2, \ldots, \theta^{p-1}$ leave no element invariant and lie in G; there is no other permutation in G with this property, since their total number is p-1, as previously noted. This proves the claim.

Let then $\alpha \in S_p$ be defined by

$$\alpha : \begin{cases} 0 \to i_1 \\ 1 \to i_2 \\ \vdots \\ p-1 \to i_p \end{cases}$$

so that, with the same notations as above :

$$\alpha^{-1} \theta \alpha = \tau.$$

Let ρ be any element in G. Since $\rho \theta \rho^{-1}$ is an element of G which leaves no element of E invariant, it is some power of θ : let

$$\rho \theta \rho^{-1} = \theta^k$$

for some k between 1 and (p-1). Transforming this equation by the inner automorphism : $\xi \to \alpha^{-1} \xi \alpha$ of S_p, we get :

$$(\alpha^{-1} \rho \alpha) \tau (\alpha^{-1} \rho \alpha)^{-1} = \tau^k.$$

Lemma 43 then shows that $\alpha^{-1}\rho\alpha \in GA(p)$. We have thus proved :

$$\alpha^{-1} G \alpha \subset GA(p),$$

hence G is conjugate to a subgroup of GA(p), and is therefore solvable. □

Of course, in order to justify the introduction of groups and to demonstrate the power and usefulness of this new tool, one has to come up with some new results which do not refer to groups in their statement but require some group theory in their proof. Only a couple of such results are quoted by Galois. The following is proposition 8 in his memoir [G, p. 69], [E, p. 113] :

45. COROLLARY : Let P be an irreducible polynomial of prime degree over a field F. The equation $P(X) = 0$ is solvable by radicals over F if and only if all the roots of P can be rationally expressed over F from any two of them.

<u>Proof</u> : Denoting by $r_1,\ldots,r_p$ the roots of P (in some extension of F) and transforming the condition on solvability of $P(X) = 0$ into a condition on groups by theorem 26 (and proposition 37), we have to prove : Gal(P/F) is solvable if and only if

$$r_1,\ldots,r_p \in F(r_i,r_j) \quad \text{for any } i,j=1,\ldots,p \text{ with } i \neq j.$$

First, we note that the irreducibility hypothesis on P implies that Gal(P/F) is transitive on $r_1,\ldots,r_p$, by proposition 36; we will thus be able to apply the preceding results on transitive subgroups of S_p.
If $r_1,\ldots,r_p$ have rational expressions in r_i, r_j over F, then every permutation in Gal(P/F) which leaves invariant r_i and r_j must leave invariant $r_1,\ldots,r_p$; it is thus the

identity. From the characterization of solvable transitive groups of permutations of p elements in theorem 44, it then follows that Gal(P/F) is solvable.

Conversely, if Gal(P/F) is solvable, then by the same characterization,

$$\text{Gal}(P/F(r_i, r_j)) = \{I\} \quad \text{for any } i,j=1,\ldots,p \text{ with } i \neq j,$$

since theorem 14 shows that $\text{Gal}(P/F(r_i,r_j))$ only contains permutations which leave r_i and r_j invariant. Therefore, $r_1,\ldots,r_p$ are invariant under $\text{Gal}(P/F(r_i,r_j))$ whence

$$r_1,\ldots,r_p \in F(r_i, r_j),$$

by theorem 14. □

This corollary can be effectively used to produce examples of non-solvable equations over the field $\mathbb{Q}$ of rational numbers. We have for instance :

46. COROLLARY : Let P be an irreducible polynomial of prime degree over $\mathbb{Q}$. If at least two roots of P, but not all, are real, then P(X) = 0 is not solvable by radicals over $\mathbb{Q}$.

Proof : Let $r_1,\ldots,r_p$ be the roots of P, and assume $r_1, r_2 \in \mathbb{R}$ and $r_p \notin \mathbb{R}$. Then $\mathbb{Q}(r_1, r_2) \subset \mathbb{R}$, hence $r_p \notin \mathbb{Q}(r_1, r_2)$ and the preceding corollary shows that P(X) = 0 is not solvable over $\mathbb{Q}$. □

The above condition on the roots of P is not very hard to check in specific examples. If the degree of P is a prime congruent to 1 modulo 4, it can even be done purely arithmetically with the aid of the discriminant :

47. COROLLARY : Let P be a monic irreducible polynomial of prime degree p over $\mathbb{Q}$. Assume $p \equiv 1 \pmod 4$. If the discriminant of P is negative, then $P(X) = 0$ is not solvable by radicals over $\mathbb{Q}$.

Proof : By exercise 4 of chapter 8, the condition on the discriminant readily implies that the number of real roots of P is not 1 nor p. □

As a specific example, equations

$$X^5 - pq\, X + p = 0$$

where p is prime and q is an integer, $q \geq 2$ (or $q \geq 1$ and $p \geq 13$) are not solvable by radicals over $\mathbb{Q}$, since $X^5 - pq\, X + p$ is irreducible over $\mathbb{Q}$, as Eisenstein's criterion 12.13 readily shows, and its discriminant is negative (see exercise 1 of chapter 8 for the calculation of this discriminant).

As a last application of Galois' investigations of equations of prime degree, we now give another proof of the Ruffini-Abel theorem 13.16 :

48. COROLLARY : For $n \geq 5$, the general equation of degree n is not solvable by radicals (over $k(s_1, \ldots, s_n)$).

Proof : We have seen in example 4 that the Galois group of the general equation of degree n is the group of all permutations of the roots, which can be identified to S_n. Since this group is obviously transitive on the roots, it follows from proposition 36 that the general equation of degree n is irreducible (over $k(s_1, \ldots, s_n)$), hence that solvability of this equation implies its complete solvability (by proposition 37), which implies the solvability of S_n (by theorem 26).
Consider first the case $n = 5$: in this case we can apply theorem 42 to conclude that S_5 is not solvable, since $|S_5| > 4.5$.

It then follows from corollary 31 that S_n is not solvable for $n \geq 5$, since S_5 can be identified to the subgroup of S_n which leaves all the elements of $\{1,\ldots,n\}$ invariant, except $\{1,\ldots,5\}$. □

We now turn to the theorem of Abel quoted in section 1 :

49. THEOREM : Let P be a polynomial of degree n over some field F, with roots $r_1,\ldots,r_n$ in some field containing F. If there exist rational fractions $\theta_2,\ldots,\theta_n \in F(X)$ such that

$$r_i = \theta_i(r_1) \qquad \text{for } i=2,\ldots,n$$

and $\quad \theta_i(\theta_j(r_1)) = \theta_j(\theta_i(r_1)) \qquad$ for all i,j,

then the equation $P(X) = 0$ is completely solvable by radicals over F.

<u>Proof</u> : Using Hudde's trick in theorem 5.26, we can assume without loss of generality that the roots $r_1,\ldots,r_n$ are pairwise distinct. By example 6, the Galois group $Gal(P/F)$ is abelian. Therefore, by theorem 26, it suffices to prove the following group-theoretical statement :

50. PROPOSITION : Every (finite) abelian group is solvable.

<u>Proof</u> : Since every subgroup of an abelian group is normal, it suffices to prove that every finite abelian group $G \neq \{I\}$ contains a subgroup G_1 of prime index. Arguing by induction on the order of G, we then construct a sequence of subgroups :

$$G \supset G_1 \supset G_2 \supset \ldots \supset G_r = \{I\}$$

each of which is normal of prime index in the preceding

one. This sequence shows that G is solvable.
It thus only remains to prove the existence of a subgroup
of prime index in each finite non-trivial abelian group.
This is a special case (H = {1}) of the following result :

51. LEMMA : Let H be a subgroup of a finite abelian group G. If
H $\neq$ G, then there exists in G a subgroup G_1 of prime index
which contains H.

Proof : We argue by induction on the index (G:H), which is
assumed to be at least 2. If (G:H) = 2,3 or any other
prime number, then G_1 = H fulfills the required conditions.
Assume then (G:H) is not prime. Pick σ in G but not in H
and consider the minimal exponent e > 0 for which $\sigma^e \in$ H.
Let also p be a prime factor of e and

$$\rho = \sigma^{e/p}.$$

Then $\rho \notin$ H (otherwise e would not be minimal), and $\rho^p \in$ H.
Consider then

$$H' = \{\rho^i \mu \mid i=0,\ldots,p-1; \mu \in H\}.$$

It is easily checked that H' is a subgroup of G containing
H, and that the cosets of H in H' are : $H', \rho H', \rho^2 H', \ldots,$
$\rho^{p-1} H'$ (compare the proof of corollary 30), so that

$$(H':H) = p.$$

Therefore, H' $\neq$ G, since by hypothesis (G:H) is not prime,
and

$$(G:H') < (G:H).$$

From the induction hypothesis, it follows that there exists
in G a subgroup G_1 of prime index which contains H', hence

also H.

Remark : Proposition 50 can be used to show that the definition of solvability given in n° 26 in the case of finite groups has other equivalent formulations, which make sense for infinite groups as well.
For each group G, we define a derived subgroup G' : it is the subgroup of G generated by commutators $\sigma\tau\sigma^{-1}\tau^{-1}$, for $\sigma,\tau \in G$. For any positive integer $n \geq 2$, the n-th derived group $G^{(n)}$ is inductively defined as the derived subgroup of $G^{(n-1)}$.

PROPOSITION : The following conditions on a group G are equivalent :

a) there exists an integer n such that $G^{(n)} = \{1\}$
b) G contains a sequence of subgroups :

$$G = G_0 \supset G_1 \supset \ldots \supset G_t = \{1\}$$

such that each subgroup G_i is normal in the preceding subgroup G_{i-1}, with abelian factor group G_{i-1}/G_i, for i=1,..., r.
Moreover, if G is finite, these conditions are also equivalent to :

c) G contains a sequence of subgroups :

$$G = G_0 \supset G_1 \supset \ldots \supset G_r = \{1\}$$

such that each subgroup is normal of prime index in the preceding one.

Proof : a) $\Rightarrow$ b) : The sequence defined by : $G_i = G^{(i)}$ satisfies the required conditions.
b) $\Rightarrow$ a) : Since G_{i-1}/G_i is abelian, we have : $G'_{i-1} \subset G_i$.

By induction, it follows that $G^{(t)} \subset G_t$, hence $G^{(t)} = \{1\}$.

c) $\Rightarrow$ b) : Each factor group G_{i-1}/G_i for $i=1,\ldots,r$ has prime order and is therefore abelian, since it is generated by any single element (except the identity).

b) $\Rightarrow$ c) if G is finite : by proposition 50, each factor group G_{i-1}/G_i contains a sequence of subgroups :

$$G_{i-1}/G_i \supset H_{i1} \supset \ldots \supset H_{ir_i} = \{1\}$$

in which each subgroup has prime index in the preceding subgroup. Taking inverse images of $H_{i1},\ldots,H_{ir_i}$ under the canonical projection $\pi : G_{i-1} \to G_{i-1}/G_i$, we obtain a sequence of subgroups :

$$G_{i-1} \supset \pi^{-1}(H_{i1}) \supset \ldots \supset \pi^{-1}(H_{ir_i}) = G_i$$

in which each subgroup is normal of prime index in the preceding subgroup. The sequences thus obtained can be joined end to end to produce a sequence of subgroups starting at G and ending with $\{1\}$, in which each subgroup is normal of prime index in the preceding one. This shows c). □

Appendix to chapter 14 : Galois' description of groups of permutations

Although groups of permutations have been widely used in the preceding discussion of Galois' results, it should be observed that the notion of group in Galois' papers is slightly different from the modern one. Indeed, in Galois' approach to groups of permutations of a set E, the central role is played by the <u>arrangements</u>(*) of the elements of E, which are the various ways of ranging the elements of E in

───────────

(*) Galois uses the term "permutation" for what is called an "arrangement" here, and "substitution" for what is usually called "permutation" nowadays. Because of possible confusions, we avoid using the term "permutation" as far as possible in the appendix, and use "arrangement" and "substitution" instead.

a row, while nowadays the fundamental objects are the <u>substitutions</u> (or permutations), i.e. the 1-1 mappings from E onto itself.

The purpose of this appendix is to present Galois' description of groups and to point out how Galois' definitions are related to modern ones.

Let E be a finite set and let Ω be the set of arrangements of the elements of E. Thus, if for instance E = {a,b,c}, then

$$\Omega = \{abc, acb, bac, bca, cab, cba\}.$$

We denote by Sym(E) the set of substitutions of E (which have been up to here called permutations of E). The substitutions of E induce substitutions of Ω in an obvious way: a substitution σ transforms an arrangement α = abc... into $\sigma(\alpha) = \sigma(a)\sigma(b)\sigma(c)...$. This action of Sym(E) onto Ω has the following remarkable, yet obvious, property : for any $\alpha, \beta \in \Omega$, there is one and only one substitution $\sigma \in$ Sym(E) such that $\sigma(\alpha) = \beta$. (This property is sometimes expressed as follows : Ω is a <u>principal homogeneous set</u> under Sym(E)).

<u>DEFINITION</u> : A <u>group of arrangements</u> of E is a non-empty subset A of Ω which has the following property : for any $\xi, \eta, \zeta \in A$, the substitution which transforms ξ into η transforms ζ into an arrangement which also belongs to A. In other words, if $\sigma \in$ Sym(E) is such that

$\sigma(\xi) \in A$ for some $\xi \in A$,
then $\sigma(\zeta) \in A$ for all $\zeta \in A$,

i.e. $\sigma(A) \subset A$

(and in fact $\sigma(A) = A$ since the number of arrangements in $\sigma(A)$ is the same as in A). A <u>group of substitutions</u> of E is (as usual) a subgroup of Sym(E).

PROPOSITION : There is a 1-1 correspondence between groups of substitutions of E and groups of arrangements of E which contain a given arrangement α; this correspondence associates to any group of substitutions G the orbit $G(\alpha) \subset \Omega$, and to any group of arrangements A the set $\{\sigma \in \text{Sym}(E) \mid \sigma(\alpha) \in A\}$.

Proof : First, we show that $G(\alpha)$ is a group of arrangements of E. Let $\sigma \in \text{Sym}(E)$ be such that $\sigma(\xi) \in G(\alpha)$ for some $\xi \in G(\alpha)$; we have to prove : $\sigma G(\alpha) = G(\alpha)$. The hypotheses that ξ and $\sigma(\xi)$ are in G yield :

$$\xi = \tau(\alpha) \quad \text{and} \quad \sigma(\xi) = \theta(\alpha) \quad \text{for some } \tau, \theta \in G.$$

Hence,

$$\sigma\tau(\alpha) = \theta(\alpha)$$

and therefore

$$\sigma\tau = \theta.$$

This equality shows that $\sigma \in G$, whence $\sigma G = G$ and $\sigma G(\alpha) = G(\alpha)$.

Next, we prove that if A is a group of arrangements containing α, then the set

$$S_\alpha(A) = \{\sigma \in \text{Sym}(E) \mid \sigma(\alpha) \in A\}$$

is a subgroup of $\text{Sym}(E)$. This set contains the identity, since

$$I(\alpha) = \alpha \in A.$$

If $\sigma, \tau \in S_\alpha(A)$, then the property of groups of arrangements, applied with $\xi = \alpha$, $\eta = \sigma(\alpha)$ and $\zeta = \tau(\alpha)$, yields :

$$\sigma\tau(\alpha) \in A,$$

hence

$$\sigma\tau \in S_\alpha(A).$$

Likewise, if $\sigma \in S_\alpha(A)$, the same property applied with $\xi = \sigma(\alpha)$, $\eta = \alpha$ and $\zeta = \alpha$ yields :

$$\sigma^{-1}(\alpha) \in A,$$

hence

$$\sigma^{-1} \in S_\alpha(A).$$

This shows that $S_\alpha(A)$ is a group of substitutions.

To complete the proof, it remains to see that the maps

$$G \to G(\alpha) \quad \text{and} \quad A \to S_\alpha(A)$$

are reciprocal bijections, i.e. that

$$S_\alpha(G(\alpha)) = G \tag{16}$$

and

$$S_\alpha(A)(\alpha) = A; \tag{17}$$

but these equalities both readily follow from the definitions (and the fact that Ω is a principal homogeneous set under $Sym(E)$). □

COROLLARY : If α and β both belong to a group of arrangements A, then

$$\{\sigma \in Sym(E) \mid \sigma(\alpha) \in A\} = \{\sigma \in Sym(E) \mid \sigma(\beta) \in A\},$$

i.e., with the notations of the preceding proof:

$$S_\alpha(A) = S_\beta(A).$$

Proof: Equality (17) yields:

$$\beta \in S_\alpha(A)(\alpha),$$

which means that β is in the orbit of α under the group $S_\alpha(A)$; therefore, by proposition 40(a):

$$S_\alpha(A)(\alpha) = S_\alpha(A)(\beta)$$

whence, by equation (17):

$$A = S_\alpha(A)(\beta).$$

Taking the images of both sides under S_β and applying (16) (with β instead of α and $S_\alpha(A)$ instead of G), we get:

$$S_\beta(A) = S_\alpha(A). \qquad \square$$

This corollary shows that the group of substitutions $S_\alpha(A)$ which corresponds to a given group of arrangements A does not depend on the choice of a particular reference arrangement α in A. On the contrary, the choice of a reference arrangement α plays an important role for the passage from groups of substitutions to groups of arrangements: different groups of arrangements may correspond to the same group of substitutions. For instance, if $E = \{a,b,c\}$ as above and if $G = \{I,\tau\}$, where τ interchanges a and b and leaves c invariant, then choosing as reference α the arrangement abc we get the group

$$\{abc, bac\}$$

while taking bca as reference we get

{bca,acb}.

This shows that groups of substitutions are more natural than groups of arrangements, in that they do not depend on the choice of a reference arrangement. Certain passages in his memoir leave no doubt that Galois was aware of the fact that the basic notion was ultimately that of substitution, instead of arrangement; however, he seems to have settled for arrangements because of their more concrete, tangible nature :

(From the introductory principles :)

"The initial permutation one uses to describe substitutions is entirely arbitrary when one is dealing with functions, because there is no reason, in a function of several letters, for a letter to occupy one position rather than another. Nonetheless, since one can hardly comprehend the idea of a substitution without that of a permutation, we shall frequently speak of permutations, and we shall consider substitutions only as the passage from one permutation to another" [G, p. 47], [E, p. 102].

(After proposition 1, i.e. theorem 14 above :)

"Scholium. Clearly in the group of permutations under discussion the disposition of the letters is of no importance, but only the <u>substitutions</u> of the letters by which one passes from one permutation to the other" [G, p. 53], [E, p. 106].

A positive point in Galois' description with groups of arrangements is that the notion of subgroup, and particularly of normal subgroup, arises in a fairly natural way, as we now show.

If H is a subgroup of a group of substitutions G, then the corresponding group of arrangements $H(\alpha)$ (obtained from

a reference arrangement α) is clearly a subgroup of $G(\alpha)$.
Moreover, the decomposition of G into left cosets of H :

$$G = \bigcup_{\sigma \in R} \sigma H$$

where R is a set of representatives of the cosets of H in
G, i.e. a subset of G containing one and only one element
from each coset, yields a decomposition of the group of
arrangements $G(\alpha)$:

$$G(\alpha) = \bigcup_{\sigma \in R} \sigma H(\alpha).$$

The subsets $\sigma H(\alpha)$, for $\sigma \in R$, are pairwise disjoint and are
in fact subgroups of $G(\alpha)$, since the equality

$$\sigma H(\alpha) = (\sigma H \sigma^{-1})(\sigma(\alpha))$$

shows that $\sigma H(\alpha)$ is the orbit of $\sigma(\alpha)$ under the group of
substitutions $\sigma H \sigma^{-1}$. The set $\sigma H(\alpha)$ is therefore the group
of arrangements containing $\sigma(\alpha)$ and associated to the group
of substitutions $\sigma H \sigma^{-1}$.

The normality of H in G translates as follows : the groups
of substitutions of each $\sigma H(\alpha)$ are all equal to H. Indeed,
this condition amounts to :

$$\sigma H \sigma^{-1} = H \qquad \text{for all } \sigma \in R$$

and since every element in G has the form $\sigma \tau$ for some
$\sigma \in R$ and some $\tau \in H$, it follows that

$$\rho H \rho^{-1} = H \qquad \text{for all } \rho \in G.$$

For instance, let $E = \{a,b,c\}$, let $G = \text{Sym } E$ and $H = \{I,\tau\}$
where τ interchanges a and b and leaves c invariant.
Choose $\alpha = abc$ as reference arrangement. Then $G(\alpha)$ is the
group of all arrangements of a,b,c, which decomposes into

three subgroups : H(α) and two other subgroups of the form σH(α), which are obtained by applying a single substitution to all the arrangements of H(α) :

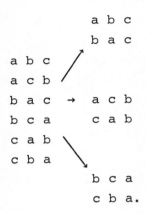

One passes from the first subgroup of arrangements (which is H(α)) to the second one by applying on all the arrangements the substitution b ↔ c (or, equivalently, the substitution a → c → b → a), and to the third one by applying a → b → c → a (or, equivalently, a ↔ c).

That H is not normal in G is reflected by the fact that the three groups of arrangements do not have the same group of substitutions. Indeed, the first group of substitutions is H, the second is {I, a ↔ c} and the third is {I, b ↔ c}

If we choose instead of H the group N = {I, a → b → c → a, a → c → b → a} (which is the alternating group on E), then the corresponding decomposition of G(α) is :

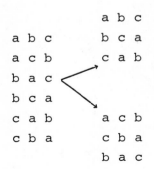

The second subgroup of arrangements is obtained from the first by applying the substitution b ↔ c, and the two subgroups both have N as group of substitutions.

Exercises for chapter 14

1) Let P be a polynomial over some field F. Show that if P is irreducible over F, then $|Gal(P/F)|$ is divisible by the degree of P.

2) Let V be a Galois resolvent of an equation $P(X) = 0$ over a field F. Show that for any $\sigma \in Gal(P/F)$, the function $\sigma(V)$ also is a Galois resolvent of $P(X) = 0$.

3) Show that an equation $P(X) = 0$ is completely solvable by radicals over a field F if and only if for each irreducible factor Q of P, the equation $Q(X) = 0$ is solvable by radicals over F.

4) Let $P(X) = (X-X_1)\ldots(X-X_n) = X^n - s_1 X^{n-1} + s_2 X^{n-2} - \ldots + (-1)^n s_n$ be the general polynomial of degree n over some reference field k, and let $F = k(s_1,\ldots,s_n)$. Let $u \in k(X_1,\ldots,X_n)$ and let $u_1,\ldots,u_r$ be the various (distinct) values of u under the permutations of $X_1,\ldots,X_n$ (with $u = u_1$, say). Show that the polynomial

$$(X-u_1)\ldots(X-u_r)$$

is irreducible over F. Deduce as in example 34 that $Gal(P/F(u)) = I(u)$.

5) Let G be a group of substitutions of a finite set E, let H be a subgroup of G and let α be an arrangement of the elements of E. Show that the group of arrangements G(α) can be decomposed into subgroups which have H as group of substitutions. Show that this decomposition is identical to $G(\alpha) = \bigcup_{\sigma \in R} \sigma H(\alpha)$ (where R is a set of representatives of the left cosets of H in G) if and only if H is normal in G.

Epilogue

Although Galois' memoir is nowadays regarded as the climax of several decades of research on algebraic equations, the first reactions to Galois' theory were negative. It was rejected by the referees, because the arguments were "not clear enough nor developed enough" [Ta, p. 121], but also for another, deeper motive : it did not yield any workable criterion to determine whether an equation is solvable by radicals. In that respect, even the application to equations of prime degree indicated by Galois (see corollary 14.45) is hardly useful, as the referees pointed out :

> "However, one should observe that [the memoir] does not contain, as [its] title promised, the condition of solvability of equations by radicals; indeed, assuming as true M. Galois' proposition, one would not derive from it any good way of deciding whether a given equation of prime degree is solvable or not by radicals, since one would have first to verify whether this equation is irreducible and next whether any of its roots can be expressed as a rational fraction of two others. The condition for solvability, if it exists, ought to have an external character which can be verified by inspecting the coefficients of a given equation or, at most, by solving other equations of degrees lower than that of the proposed equation" [Ta, p. 121].

Galois' criterion (see theorem 14.26) was very far from being external; indeed, Galois always worked with the roots of the proposed equation, never with its coefficients[*]. Thus, Galois' theory did not correspond to what was expected, it was too novel to be readily accepted.

After the publication of Galois' memoir by Liouville, its importance dawned upon the mathematical world, and it was eventually realized that Galois had discovered a mathematical gem much more valuable than any hypothetical external characterization of solvable equations. After all, the problem of solving equations by radicals was utterly artificial. It had focused the efforts of several generations of brilliant mathematicians because it displayed some strange, puzzling phenomena. It contained something mysterious, profoundly appealing. Galois had taken the pith out of the problem, by showing that the difficulty of an equation was related to the ambiguity of its roots and pointing out how this ambiguity could be measured by means of a group. He had thus set the theory of equations and, indeed, the whole subject of algebra, on a completely different track.

"Now, I think that the simplifications produced by the elegance of calculations (intellectual simplifications, I mean; there is no material simplification) are limited; I think the moment will come where the algebraic transformations foreseen by the speculations of analysts will not find nor the time nor the place to occur any more; so that one will have to be content with having foreseen them (...).
Jump above calculations; group the operations, classify them according to their complexities rather than their appearances; this, I believe, is the mission of future mathematicians; this is the road on which I am embarking in this work" [G, p. 9].

---

[*] It is telling that the proposed equation is nowhere displayed in Galois' memoir.

Thereafter, the theory of equations slowly disappeared, while new subjects emerged, such as the theory of groups and of various algebraic structures. This final stage in the evolution of a mathematical theory has been beautifully described by A. Weil :

> "Nothing is more fruitful, as all mathematicians know, than these dim analogies, these foggy glimpses from one theory to the other, these stealthy caresses, these inexplicable jumbles; nothing also gives more pleasure to the research worker. A day comes when the illusion dissipates; the vagueness changes into certainty; the twin theories disclose their common fount before vanishing; as the Gītā teaches, one reaches knowledge and indifference at the same time. Metaphysics has become mathematics, ready to make the substance of a treatise whose cold beauty could not move us any more" [W1, p. 52].

The subsequent developments arising from Galois theory do not fall within the scope of these lectures, so we refer to the papers by Kiernan [Ki] and by Van der Waerden [VW3] and to the book by Nový [No] for detailed accounts. There is however one major trend in this evolution that we want to point out : the gradual elimination of polynomials and equations from the foundations of Galois theory. Indeed, it is revealing of the profoundness of Galois' ideas to see, through the various textbook expositions, how this theory initially designed to answer a question about equations progressively outgrew its original context.

The first step in this direction is the emergence of the notion of field, through the works of Kronecker and Dedekind. Their approaches were quite different but complementary. Kronecker's point of view was constructivist; to define a field according to this point of view is to describe a process by which the elements of the field can be constructed. On the contrary, Dedekind's approach was set-

theoretic : he does not hesitate to define the field generated by a set P of complex numbers as the intersection of all the fields which contain P; this definition is hardly useful for determining whether a given complex number belongs to the field thus defined. Although Dedekind's approach has become the usual point of view nowadays, Kronecker's constructivism also led to important results, such as the algebraic construction of fields in which polynomials split into linear factors : see section 9.2.
The next step is the observation by Dedekind, around the end of the nineteenth century, that the permutations in the Galois group of an equation can be considered as automorphisms of the field of rational fractions of the roots (see corollary 14.15). Moreover, the newly developed linear algebra was brought to bear on the theory of fields, as the larger field in an extension can be regarded as a vector space over the smaller field.

These ideas came to fruition in the first decades of the twentieth century, as witnessed by the famous treatise of B.L. Van der Waerden : "Moderne Algebra" (1930) (of which [VW1] is the seventh edition). The treatment of Galois theory in this book is based on lectures by E. Artin. It states as its "fundamental theorem" a 1-1 correspondence between subfields of certain extensions (those which are obtained by adjoining all the roots of a polynomial without multiple root), nowadays called <u>Galois</u> extensions, and the subgroups of the associated Galois group. This correspondence is not quite explicit in Galois' memoir. It can be observed in the dual statements : corollary 14.25 and lemma 14.32 and in the proof of the criterion for solvability of an equation by radicals (theorem 14.26), but in this proof it is obscured by the fact that roots of unity are not assumed to be in the base field, while they are needed for the application of corollary 14.25 or lemma 14.32.
In Van der Waerden's book, the treatment of Galois theory

clearly emphasizes fields and groups, while polynomials and equations play a secondary role. They are used as tools in the proofs, but the main theorems do not involve polynomials in their statement.

A few years later, the exposition of Galois theory further evolved under the influence of Emil Artin, who once wrote :

> "Since my mathematical youth, I have been under the spell of the classical theory of Galois. This charm has forced me to return to it again and again, and to try to find new ways to prove these fundamental theorems" [Ar2, p. 380].

In his book "Galois Theory" (1942) [Ar1], Artin proposes a new, highly original, definition of Galois extension. The extension is looked at from the point of view of the larger field instead of the smaller; an extension of fields is then called Galois if the smaller field is the field of invariants under a (finite) group of automorphisms of the larger. This definition and some improvements in the proofs enabled Artin to further reduce the role of polynomials in the basic results of Galois theory, so that the fundamental theorem can now be proved without ever mentioning polynomials (see the appendix).

Artin's exposition has nowadays become the classical treatment of Galois theory from an elementary point of view. However, several other expositions have been proposed in more recent times, inspired by the applications of Galois theory in related areas. For instance, the Jacobson-Bourbaki correspondence [J1, p. 22] yields a uniform treatment of both the classical Galois theory and the Galois theory for purely inseparable field extensions of height 1, where restricted p-Lie algebras replace groups. In another direction, the Galois theory of commutative rings [CHR] due to Chase, Harrison and Rosenberg, has inspired new expositions which stress the analogy between extensions of fields and coverings of locally compact topological spaces : see [Dou] (compare also the new version of Bourbaki's treatise [Bou2]).

Through its applications in various areas and as a source of inspiration for new investigations, Galois theory is far from being a closed issue.

Appendix : The fundamental theorem of Galois theory

To conclude these lectures, we now give an account of the 1-1 correspondence which is now regarded as the fundamental theorem of Galois theory, after Artin's classical exposition in [Ar1].

1. DEFINITIONS : Let K be a field containing a subfield F. The dimension of K, regarded as a vector space over F, is called the rank of K over F, and is denoted by [K:F] :

$$[K:F] = \dim_F K.$$

The group of (field-)automorphisms of K which leave F elementwise invariant is called the Galois group of K over F, and is denoted by Gal(K/F) :

$$Gal(K/F) = Aut_F K.$$

The extension K/F is called a Galois extension if F is the field of all elements which are invariant under some finite group of automorphisms of K. In other words, denoting by K^G the field of invariants under a group G of automorphisms of K, i.e.

$$K^G = \{x \in K \mid \sigma(x) = x \text{ for all } \sigma \in G\},$$

the extension K/F is Galois if and only if there exists a finite group G of automorphisms of K such that $F = K^G$.

For instance, if F is a field of characteristic zero and if $r_1,\ldots,r_n$ are the roots of a polynomial with coefficients in F, then theorem 14.14 (and corollary 14.15) show that $F(r_1,\ldots,r_n)$ is a Galois extension of F.

2. THEOREM (Fundamental theorem of Galois theory) : If $F = K^G$ for some finite group G, then

$$[K:F] = |G| \quad \text{and} \quad G = \text{Gal}(K/F).$$

The field K is then a Galois extension of every subfield containing F. Moreover, there is a 1-1 correspondence between the subfields of K containing F and the subgroups of G, which associates to any subfield L the Galois group $\text{Gal}(K/L) \subset G$ and to any subgroup $H \subset G$ its field of invariants K^H.
Under this correspondence, the rank over F of a subfield of K corresponds to the index in G of the associated subgroup:

$$[L:F] = (G:\text{Gal}(K/L)) \qquad (G:H) = [K^H:F].$$

Furthermore, a subfield L of K is Galois over F if and only if the corresponding subgroup $\text{Gal}(K/L)$ is normal in G. The Galois group $\text{Gal}(L/F)$ is obtained by restricting to L the automorphisms in G, and the restriction homomorphism induces an isomorphism : $\dfrac{G}{\text{Gal}(K/L)} \xrightarrow{\sim} \text{Gal}(L/F)$.

Thus, it follows from theorem 14.14 that the group $\text{Gal}(P/F)$ defined in section 14.2 is the Galois group of $F(r_1,\ldots,r_n)$ over F, provided that its elements are considered as field-automorphisms of $F(r_1,\ldots,r_n)$ instead of permutations of $r_1,\ldots,r_n$.

The proof of this theorem requires some preparation. We start with a very simple observation which parallels lemma 14.28 :

3. **LEMMA** : Let $K \supset L \supset F$ be a tower of fields. Then

$$[K:F] = [K:L][L:F].$$

<u>Proof</u> : Let $(k_i)_{i \in I}$ be a basis of K over L and $(\ell_j)_{j \in J}$ be a basis of L over F. If we prove $(k_i \ell_j)_{(i,j) \in I \times J}$ is a basis of K over F, the lemma readily follows.

This family spans K since every element $x \in K$ can be written

$$x = \Sigma_{i \in I} \, k_i x_i$$

for some $x_i \in L$, and decomposing

$$x_i = \Sigma_{j \in J} \, \ell_j y_{ij}$$

with $y_{ij} \in F$, we end up with :

$$x = \Sigma_{\substack{i \in I \\ j \in J}} (k_i \ell_j) y_{ij}.$$

To show that the family $(k_i \ell_j)$ is linearly independent over F, consider :

$$\Sigma_{\substack{i \in I \\ j \in J}} k_i \ell_j y_{ij} = 0$$

for some $y_{ij} \in F$. Collecting terms which have the same index i, we get :

$$\Sigma_{i \in I} \, k_i (\Sigma_{j \in J} \, \ell_j y_{ij}) = 0$$

hence

$$\Sigma_{j \in J} \, \ell_j y_{ij} = 0 \qquad \text{for all } i \in I,$$

since (k_i) is linearly independent over L, hence also

$$y_{ij} = 0 \qquad \text{for all } i \in I,\ j \in J,$$

since (ℓ_j) is linearly independent over F. □

The basic observation which lies at the heart of the proof of the fundamental theorem is due to Artin; it generalizes an earlier result of Dedekind :

4. **LEMMA** (of linear independence of homomorphisms. Dedekind-Artin) : Let $\sigma_1,\ldots,\sigma_n$ be n distinct homomorphisms of a field L into a field K. Then $\sigma_1,\ldots,\sigma_n$, viewed as elements of the K-vector space $\mathcal{F}(L,K)$ of all maps from L to K, are linearly independent over K : if $a_1,\ldots,a_n \in K$ are such that

$$a_1\sigma_1(x) + \ldots + a_n\sigma_n(x) = 0 \qquad \text{for all } x \in L,$$

then $a_1 = \ldots = a_n = 0$.

Proof : Assume on the contrary that $\sigma_1,\ldots,\sigma_n$ are not independent, and choose $a_1,\ldots,a_n \in K$ such that

$$a_1\sigma_1(x) + \ldots + a_n\sigma_n(x) = 0 \qquad \text{for all } x \in L \qquad (1)$$

with $a_1,\ldots,a_n$ not all zero, but such that the number of $a_i \neq 0$ be minimal. This number is at least equal to 2, otherwise one of the σ_i would map L to $\{0\}$; this is impossible since, by definition of homomorphisms of fields, $\sigma_i(1) = 1$ for all i. Changing the numbering of $\sigma_1,\ldots,\sigma_n$ if necessary, we may thus assume without loss of generality that $a_1 \neq 0$ and $a_2 \neq 0$.
Choose $\ell \in L$ such that $\sigma_1(\ell) \neq \sigma_2(\ell)$. (This is possible since $\sigma_1 \neq \sigma_2$).
Multiplying both sides of (1) by $\sigma_1(\ell)$, we get :

$$a_1\sigma_1(\ell)\sigma_1(x) + \ldots + a_n\sigma_1(\ell)\sigma_n(x) = 0 \text{ for all } x \in L. (2)$$

On the other hand, applying relation (1) with ℓx instead of x and using the multiplicative property of σ_i, we get:

$$a_1 \sigma_1(\ell) \sigma_1(x) + \ldots + a_n \sigma_n(\ell) \sigma_n(x) = 0 \quad \text{for all } x \in L. \quad (3)$$

Subtracting (3) from (2), the first terms of each relation cancel out and we obtain:

$$a_2(\sigma_1(\ell) - \sigma_2(\ell)) \sigma_2(x) + \ldots + a_n(\sigma_1(\ell) - \sigma_n(\ell)) \sigma_n(x) = 0$$

$$\text{for all } x \in L.$$

The coefficients are not all zero since $\sigma_1(\ell) \neq \sigma_2(\ell)$, but this linear combination has fewer non-zero terms than (1), a contradiction. □

Remark: Only the multiplicative property of $\sigma_1, \ldots, \sigma_n$ has been used. The same proof thus shows the linear independence of distinct homomorphisms from any group to the multiplicative group of a field.

5. COROLLARY : Let $\sigma_1, \ldots, \sigma_n$ be as in lemma 4 and let

$$F = \{x \in L \mid \sigma_1(x) = \ldots = \sigma_n(x)\}.$$

Then $[L:F] \geq n$.

Proof: Let $[L:F] = m$ and assume $m < n$. Let $\ell_1, \ldots, \ell_m$ be a basis of L over F and consider the matrix $(\sigma_i(\ell_j))_{\substack{1 \leq i \leq n \\ 1 \leq j \leq m}}$ with entries in K.
The rank of this matrix is at most m, since the number of its columns is m, hence its rows are linearly dependent over K. We can thus find elements $a_1, \ldots, a_n \in K$, not all zero, such that

$$a_1 \sigma_1(\ell_j) + \ldots + a_n \sigma_n(\ell_j) = 0 \quad \text{for } j=1, \ldots, m. \quad (4)$$

Now, any x ∈ L can be expressed as :

$$x = \sum_{j=1}^{m} \ell_j x_j$$

for some $x_j \in F$. Multiplying relations (4) by $\sigma_1(x_j)$ (which is equal to $\sigma_i(x_j)$ for all i, since $x_j \in F$) and adding the relations thus obtained, we get :

$$a_1 \sigma_1(x) + \ldots + a_n \sigma_n(x) = 0.$$

This contradicts lemma 4. □

We are now ready for the proof of theorem 2. We let $F = K^G$ for some finite group G of automorphisms of the field K.

STEP 1 : $[K:F] = |G|$.

Applying corollary 5 with L = K and $\{\sigma_1, \ldots, \sigma_n\} = G$, we already have :

$$[K:F] \geq |G|.$$

If $[K:F] > |G|$, then for some m > n (= $|G|$) we can find a sequence $k_1, \ldots, k_m$ of elements of K which are linearly independent over F. The matrix

$$(\sigma_i(k_j))_{\substack{1 \leq i \leq n \\ 1 \leq j \leq m}}$$

has rank at most n, hence its colums are linearly dependent over K; let $a_1, \ldots, a_m \in K$, not all zero, such that

$$\sigma_i(k_1) a_1 + \ldots + \sigma_i(k_m) a_m = 0 \quad \text{for } i=1,\ldots,n. \quad (5)$$

Changing the numbering of $k_1, \ldots, k_m$ if necessary, we may assume $a_1 \neq 0$. Moreover, multiplying $a_1, \ldots, a_m$ by a common non-zero element of K, we can transform a_1 into any other non-zero element of K, and we may therefore assume that a_1 has the following property :

$$\sigma_1^{-1}(a_1) + \ldots + \sigma_n^{-1}(a_1) \neq 0.$$

(The existence of elements in K which have this property follows from lemma 4, which implies that $\sigma_1^{-1}, \ldots, \sigma_n^{-1}$ are linearly independent over K).

Applying σ_i^{-1} to relations (5) and adding up the relations thus obtained, we get :

$$k_1(\sigma_1^{-1}(a_1) + \ldots + \sigma_n^{-1}(a_1)) + \ldots + k_m(\sigma_1^{-1}(a_m) + \ldots + \sigma_n^{-1}(a_m)) = 0$$

i.e., since $G = \{\sigma_1, \ldots, \sigma_n\} = \{\sigma_1^{-1}, \ldots, \sigma_n^{-1}\}$:

$$k_1(\Sigma_{\sigma \in G} \sigma(a_1)) + \ldots + k_m(\Sigma_{\sigma \in G} \sigma(a_m)) = 0.$$

The coefficients of $k_1, \ldots, k_m$ are invariant under G and are not all zero, by the hypothesis on a_1. Therefore, this relation is in contradiction with the hypothesis that $k_1, \ldots, k_m$ are linearly independent over F. Thus, $[K:F] = |G|$.

STEP 2 : $G = \text{Gal}(K/F)$.

Since the inclusion $G \subset \text{Gal}(K/F)$ is obvious, it suffices to show :

$$|G| = |\text{Gal}(K/F)|.$$

To this end, we first observe that from the inclusion $G \subset \text{Gal}(K/F)$ it follows that every element of K invariant under $\text{Gal}(K/F)$ is also invariant under G :

$$K^G \supset K^{\text{Gal}(K/F)}.$$

Now, F is obviously invariant under $Gal(K/F)$; since moreover $F = K^G$, we have :

$$F = K^G \supset K^{Gal(K/F)} \supset F,$$

hence $K^{Gal(K/F)} = F$. Step 1 with $Gal(K/F)$ instead of G yields :

$$[K:F] = |Gal(K/F)|$$

and comparing with step 1, it follows that

$$|G| = |Gal(K/F)|.$$

STEP 3 : $Gal(K/K^H) = H$ and $(G:H) = [K^H:F]$ for any subgroup H of G.

The first equality readily follows from step 2, with H instead of G. To obtain the second equality, we compare the following equalities which are derived from step 1 :

$$[K:F] = |G| \quad \text{and} \quad [K:K^H] = |H|.$$

Since the ranks of field extensions are multiplicative (lemma 3),

$$[K:F] = [K:K^H][K^H:F]$$

hence the preceding equalities yield :

$$[K^H:F] = \frac{|G|}{|H|} = (G:H).$$

STEP 4 : $K^{Gal(K/L)} = L$ and $[L:F] = (G:Gal(K/L))$ for any subfield L of K containing F.

By restriction to L, each $\sigma \in G$ induces a homomorphism of fields :

$$\sigma|_L : L \to K.$$

If two such homomorphisms coincide, say :

$$\sigma|_L = \tau|_L \qquad \text{for some } \sigma, \tau \in G,$$

then $\sigma(\ell) = \tau(\ell)$ for all $\ell \in L$

hence $\tau^{-1}\sigma(\ell) = \ell$ for all $\ell \in L$

and therefore

$$\tau^{-1}\sigma \in \mathrm{Gal}(K/L),$$

which amounts, by lemma 14.27, to :

$$\sigma\, \mathrm{Gal}(K/L) = \tau\, \mathrm{Gal}(K/L).$$

Therefore, if $(G:\mathrm{Gal}(K/L)) = r$ and if $\sigma_1, \ldots, \sigma_r$ are elements of G in pairwise different cosets of $\mathrm{Gal}(K/L)$, then the homomorphisms $\sigma_i|_L$ are pairwise different. Since $\sigma_1(x) = \ldots = \sigma_r(x)$ for any $x \in F$, corollary 5 implies that

$$[L:F] \geq r \; (= (G:\mathrm{Gal}(K/L))).$$

On the other hand, the inclusion $L \subset K^{\mathrm{Gal}(K/L)}$ and step 3 yield :

$$[L:F] \leq [K^{\mathrm{Gal}(K/L)}:F] = (G:\mathrm{Gal}(K/L)),$$

hence by the preceding inequality :

$$[L:F] = (G:\mathrm{Gal}(K/L)).$$

Moreover, since $L \subset K^{\text{Gal}(K/L)}$ and since these fields both have the same (finite) dimension over F,

$$L = K^{\text{Gal}(K/L)}.$$

Steps 3 and 4 show that the maps $H \to K^H$ and $L \to \text{Gal}(K/L)$ are reciprocal bijections between the set of subgroups of G and the set of subfields of K containing F. These maps clearly reverse the inclusions :

$$H \subset J \;\Rightarrow\; K^H \supset K^J \quad \text{and} \quad L \subset M \;\Rightarrow\; \text{Gal}(K/L) \supset \text{Gal}(K/M)$$

Moreover, from step 4 it also follows that each subfield L of K containing F is the field of invariants of some finite group of automorphisms (viz. of Gal(K/L)), hence K is Galois over L.

To complete the proof of the fundamental theorem, it now suffices to show that normal subgroups correspond to fields which are Galois over F.

<u>STEP 5</u> : $\text{Gal}(K/\sigma(L)) = \sigma\, \text{Gal}(K/L)\, \sigma^{-1}$ for any $\sigma \in G$ and any subfield L of K containing F.

This readily follows from the inclusions :

$$\sigma\, \text{Gal}(K/L)\, \sigma^{-1} \subset \text{Gal}(K/\sigma(L)) \quad \text{and} \quad \sigma^{-1}\, \text{Gal}(K/\sigma(L))\, \sigma \subset \text{Gal}(K/L),$$

which are both obvious.

<u>STEP 6</u> : If L is a subfield of K containing F such that Gal(K/L) is normal in G, then L is Galois over F, and there is an isomorphism $G/\text{Gal}(K/L) \to \text{Gal}(L/F)$ obtained by restricting to L the automorphisms in G.

By step 5, the hypothesis that Gal(K/L) is normal in G implies that any $\sigma \in G$ induces by restriction to L an automorphism :

$$\sigma|_L : L \to L$$

which leaves F elementwise invariant. We thus have a restriction map :

$$\text{res} : G \to \text{Gal}(L/F).$$

Since the kernel of this map is Gal(K/L), there is an induced injective map :

$$\overline{\text{res}} : G/\text{Gal}(K/L) \to \text{Gal}(L/F).$$

Now, steps 4 and 1 yield :

$$(G : \text{Gal}(K/L)) = [L:F] = |\text{Gal}(L/F)|,$$

hence the groups $G/\text{Gal}(K/L)$ and $\text{Gal}(L/F)$ have the same finite order, and the injective map $\overline{\text{res}}$ is therefore an isomorphism.

STEP 7 : If L is a Galois extension of F contained in K, then Gal(K/L) is a normal subgroup of G.

Let $(G:\text{Gal}(K/L)) = r$ $(= [L:F]$, by step 4), and let $\sigma_1,\ldots,\sigma_r$ be elements of G in the r different cosets of Gal(K/L). We have shown in the proof of step 4 that the restrictions $\sigma_i|_L : L \to K$ yield pairwise different homomorphisms of L in K which leave F elementwise invariant. On the other hand, it follows from corollary 5 that there are at most r homomorphisms of L in K which leave F elementwise invariant. Therefore, any such homomorphism has the form $\sigma_i|_L$ for some $i=1,\ldots,r$.
Now, since L is assumed to be Galois over F, we can find r automorphisms of L leaving F elementwise invariant. These

automorphisms can be regarded as homomorphisms from L into K, since L ⊂ K, hence they have the form $\sigma_i\big|_L$. Thus,

$$\text{Gal}(L/F) = \{\sigma_1\big|_L, \ldots, \sigma_r\big|_L\},$$

and consequently $\sigma_1, \ldots, \sigma_r$ map L into L :

$$\sigma_i(L) = L \qquad \text{for } i=1, \ldots, r.$$

By step 5, it follows that

$$\sigma_i \text{Gal}(K/L)\sigma_i^{-1} = \text{Gal}(K/L) \text{ for } i=1,\ldots,r.$$

Since every element $\sigma \in G$ has the form $\sigma_i\tau$ for some $i=1,\ldots,r$ and some $\tau \in \text{Gal}(K/L)$ we also have :

$$\sigma \text{Gal}(K/L)\sigma^{-1} = \text{Gal}(K/L) \quad \text{for all } \sigma \in G,$$

hence Gal(K/L) is normal in G.

Exercises for the Appendix

1) Let $G = \{\sigma_1, \ldots, \sigma_n\}$ be a group of automorphisms of a field K and let $e_1, \ldots, e_n$ be a basis of K over K^G. Show that the matrix $(\sigma_i(e_j))_{1 \leq i,j \leq n}$ is invertible in the ring of $n \times n$ matrices over K.

The aim of the following exercise is to provide another approach to the proof of the fundamental correspondence (theorem 2) :

2) Let K/F be a Galois extension of fields with Galois group G and let $\mathcal{F}(G,K)$ be the K-vector space of all maps from G to K.
a) Show that there is a well-defined F-linear map

$$\varphi : K \otimes_F K \to \mathcal{F}(G,K)$$

such that

$$\varphi(a \otimes b)(\sigma) = \sigma(a)b.$$

b) Consider $K \otimes_F K$ as a vector space over K by : $(a \otimes b).k = a \otimes bk$. Show that φ is K-linear and bijective. [Hint : show that the matrix of φ with respect to suitable bases of $K \otimes K$ and of $\mathcal{F}(G,K)$ over K is that of exercise 1].

c) Let G act on $K \otimes K$ by : $\sigma(a \otimes b) = \sigma(a) \otimes b$. Show that the corresponding action on $\mathcal{F}(G,K)$ is : $^\sigma f(\tau) = f(\tau\sigma)$.

d) Let G act on $K \otimes K$ by : $\sigma(a \otimes b) = a \otimes \sigma(b)$. Show that the corresponding action on $\mathcal{F}(G,K)$ is : $^\sigma f(\tau) = \sigma(f(\sigma^{-1}\tau))$.

e) Show that a K-subalgebra of $K \otimes K$ is globally invariant under the action of G defined in (d) if and only if it has the form $L \otimes_F K$ for some subfield L of K containing F. [Hint : for any K-subalgebra A of $K \otimes K$, define $L = \{x \in K \mid x \otimes 1 \in A\}$. Show that every $\Sigma_{i=1}^r a_i \otimes b_i \in A$ is in $L \otimes K$ by induction on the number r of terms].

f) Show that there is a 1-1 correspondence between K-subalgebras of $\mathcal{F}(G,K)$ and partitions of G, which associates to any partition $G = \cup_{i \in I} G_i$ the subalgebra $\{f : G \to K \mid f(\sigma) = f(\tau)$ if σ and τ belong to the same $G_i\}$. Show that a subalgebra of $\mathcal{F}(G,K)$ is globally invariant under the action of G in (d) if and only if the corresponding partition is a decomposition of G into left cosets of some subgroup.

g) Use the bijection φ and exercises (e) and (f) to set up a 1-1 correspondence between subfields of K containing F and subgroups of G. Use the action of G defined in (c) to show that this correspondence is the same as that in theorem 2.

Selected solutions

10.1 For i=1,2,3, the permutations of $X_1,\ldots,X_4$ which leave u_i invariant are the same as those which leave v_i invariant. Therefore, v_i is a rational fraction in u_i with symmetric coefficients. Denoting by s_1, s_2 the first two elementary symmetric polynomials (see equation (2) in chapter 8), it is easily checked that

$$v_i = s_2 - u_i \qquad \text{for } i=1,2,3$$

and

$$w_1 = s_1^2 - 4u_2 \qquad w_2 = s_1^2 - 4u_1 \qquad w_3 = s_1^2 - 4u_3.$$

Therefore, if $P(X)$ (resp. $Q(X)$, resp. $R(X)$) is the monic cubic polynomial with roots u_1, u_2, u_3 (resp. v_1, v_2, v_3, resp. w_1, w_2, w_3), then

$$P(X) = -Q(s_2 - X) = -R(s_1^2 - 4X)/64$$

$$Q(X) = -P(s_2 - X) \qquad R(X) = (-64)P((s_1^2 - X)/4).$$

10.2 In order to reproduce the notations of theorem 10.5, let

$$g_1 = X_1 + X_2 \qquad g_2 = X_1 + X_3 \qquad g_3 = X_2 + X_3$$

$$f_1 = X_1 X_2 \qquad f_2 = X_1 X_3 \qquad f_3 = X_2 X_3.$$

Then $a_0 = s_2$, $a_1 = s_1 s_2 - 3s_3$, $a_2 = s_1^2 s_2 - 5s_1 s_3$,

$$\theta(Y) = Y^3 - 2s_1 Y^2 + (s_1^2 + s_2) Y - (s_1 s_2 - s_3)$$

$$\psi(Y) = Y^2 + (g_1 - 2s_1) Y + (g_1^2 - 2s_1 g_1 + s_1^2 + s_2),$$

and

$$f_1 = \frac{s_2 g_1^2 - s_1 s_2 g_1 - 3 s_3 g_1 + s_2^2 + s_1 s_3}{3 g_1^2 - 4 s_1 g_1 + s_1^2 + s_2}.$$

This expression is not unique; indeed, it is clear that $f_1 = s_3/X_3$ and $g_1 = s_1 - X_3$, hence $f_1 = s_3/(s_1 - g_1)$. This non-uniqueness stems from the fact that $\theta(g_1) = 0$; indeed, it is easy to check that

$$\frac{s_2 Y^2 - s_1 s_2 Y - 3 s_3 Y + s_2^2 + s_1 s_3}{3 Y^2 - 4 s_1 Y + s_1^2 + s_2} = \frac{s_3}{s_1 - Y} - \frac{s_2 \theta(Y)}{(3 Y^2 - 4 s_1 Y + s_1^2 + s_2)(s_1 - Y)}.$$

10.3 Let $\sigma : X_1 \to X_2 \to X_3 \to X_1$. If $\sigma(f^3) \neq f^3$, then f^3, $\sigma(f^3)$ and $\sigma^2(f^3)$ are pairwise distinct. This contradicts the hypothesis that f^3 takes only two values; therefore $\sigma(f^3) = f^3$ and it follows that $\sigma(f) = \omega f$ for some cube root of unity ω. Comparing coefficients in $\sigma(f)$ and f, we obtain $A = \omega B = \omega^2 C$, whence

$$f = A(X_1 + \omega^2 X_2 + \omega X_3).$$

Moreover, $\omega \neq 1$ since X_1, X_2 and X_3 can be rationally expressed from f.

10.4 By theorem 10.5, it suffices to prove that $t(\omega^k) t(\omega)^{-k}$ is invariant by the permutations which leave $t(\omega)^n$ invariant, i.e. by $\tau : X_1 \to X_2 \to \ldots \to X_n \to X_1$ (and its powers) (compare proposition 10.9). This is clear, since $\tau(t(\omega^k)) = \omega^{-k} t(\omega^k)$.

10.5 Identifying $\{0,1,\ldots,n-1\}$ with $\mathbb{F}_n$, we can represent σ_i and τ as follows :

$$\sigma_i(x) = ix \qquad \tau(x) = x+1 \qquad \text{for } x \in \mathbb{F}_p.$$

Then $\tau\sigma_i(x) = ix+1$ and $\sigma_i\tau^k(x) = i(x+k)$, hence $\tau\sigma_i = \sigma_i\tau^k$ if $ik \equiv 1 \bmod n$. One easily checks that $\sigma_i\tau^j = \sigma_k\tau^\ell \Rightarrow i = k$ and $j = \ell$ (for $i=1,\ldots,n-1$ and $j=0,\ldots,n-1$), and it follows that $|GA(n)| = n(n-1)$.

11.1
$$[\alpha \ \beta \ \gamma \ \delta \ \varepsilon] = a^\alpha b^\beta c^\gamma d^\delta e^\varepsilon + a^\delta b^\varepsilon c^\beta d^\gamma e^\alpha + a^\gamma b^\alpha c^\varepsilon d^\beta e^\delta$$
v iii iv i ii
$$+ a^\beta b^\delta c^\alpha d^\varepsilon e^\gamma + a^\varepsilon b^\gamma c^\delta d^\alpha e^\beta$$

$$[\alpha \ \varepsilon \ \delta \ \beta \ \gamma] = a^\alpha b^\varepsilon c^\delta d^\beta e^\gamma + a^\beta b^\gamma c^\varepsilon d^\delta e^\alpha + a^\delta b^\alpha c^\gamma d^\varepsilon e^\beta$$
v iii iv i ii
$$+ a^\varepsilon b^\beta c^\alpha d^\gamma e^\delta + a^\gamma b^\delta c^\beta d^\alpha e^\varepsilon$$

$$[\alpha \ \gamma \ \beta \ \varepsilon \ \delta] = a^\alpha b^\gamma c^\beta d^\varepsilon e^\delta + a^\varepsilon b^\delta c^\gamma d^\beta e^\alpha + a^\beta b^\alpha c^\delta d^\gamma e^\varepsilon$$
v iii iv i ii
$$+ a^\gamma b^\varepsilon c^\alpha d^\delta e^\beta + a^\delta b^\beta c^\varepsilon d^\alpha e^\gamma$$

$$[\alpha \ \delta \ \varepsilon \ \gamma \ \beta] = a^\alpha b^\delta c^\varepsilon d^\gamma e^\beta + a^\gamma b^\beta c^\delta d^\varepsilon e^\alpha + a^\varepsilon b^\alpha c^\beta d^\delta e^\gamma$$
v iii iv i ii
$$+ a^\delta b^\gamma c^\alpha d^\beta e^\varepsilon + a^\beta b^\varepsilon c^\gamma d^\alpha e^\delta.$$

The sum of these four partial types is a partial type corresponding to the subgroup generated by the permutations : $\tau : a \to e \to b \to c \to d \to a$ and $\sigma : a \to a; b \to d \to c \to e \to b$. If a,b,c,d,e are numbered as : $a = 0$, $b = 2$, $c = 3$, $d = 4$, $e = 1$, then the subgroup is $GA(5) \subset S_5$ (see n° 10.7).

11.2 Applying $a \to b \to c \to d \to e \to a$ to both sides of $a^2 = b + 2$ (for instance), one gets : $b^2 = c + 2$, a relation which does not hold.

11.3 The relations between a,b,c are the following :

$$a^2 = b + 2 \qquad b^2 = c + 2 \qquad c^2 = a + 2$$
$$ab = a + c \qquad bc = b + a \qquad ca = c + b.$$

It is readily verified that these relations are preserved under $a \to b \to c \to a$.

11.4 Using the fact that 2 is a primitive root of 13, it follows from proposition 12.22 (see also proposition 12.29) that $2 \cos \frac{2\pi}{13} \to 2 \cos \frac{4\pi}{13} \to 2 \cos \frac{8\pi}{13} \to 2 \cos \frac{10\pi}{13} \to 2 \cos \frac{6\pi}{13} \to 2 \cos \frac{12\pi}{13} \to 2 \cos \frac{2\pi}{13}$ preserves the relations among $2 \cos \frac{2\pi}{13}, \ldots, 2 \cos \frac{12\pi}{13}$.

12.3 Periods with an even number of terms are sums of periods of two terms, which are real numbers : see n° 12.24.

12.4 Let $\zeta = \zeta_k'$ ($= \zeta'^{g'^k}$) for some $k=0,\ldots,p-2$ and let $g = g'^\ell$ for some integer ℓ, which is prime to $p-1$ since g is a primitive root of p (see proposition 7.15). A straightforward computation yields : $\zeta_i = \zeta'_{\sigma(i)}$, where $\sigma : \{0,\ldots,p-2\} \to \{0,\ldots,p-2\}$ is defined by : $\sigma(\alpha) \equiv \ell\alpha + k$ (mod $p-1$), for $\alpha = 0,\ldots,p-2$. The same arguments as in proposition 10.7 show that σ is a permutation. Moreover, $\alpha \equiv i$ (mod $p-1$) implies $\sigma(\alpha) \equiv \sigma(i)$ (mod $p-1$), hence

$$\Sigma_{\alpha \equiv i \bmod e} \zeta_\alpha = \Sigma_{\beta \equiv \sigma(i) \bmod e} \zeta'_\beta.$$

12.5 Let $ef = gh = p-1$. If $K_g \subset K_f$, then

$$\sigma^e(\zeta_0 + \zeta_h + \ldots + \zeta_{h(g-1)}) = \zeta_0 + \zeta_h + \ldots + \zeta_{h(g-1)}.$$

In particular, $\sigma^e(\zeta_0) = \zeta_e$ is of the form $\zeta_{h\ell}$ for some ℓ. It follows that $e = h\ell$, hence h divides e and f divides g.

Let η be a period of f terms. From proposition 12.33, it follows that $K_f = K_g(\eta)$. Now, proposition 12.35 shows that η is a root of a polynomial of degree $k = g/f$ over K_g. It is not a root of a polynomial of smaller degree, since if $P(\eta) = 0$ with $P \in K_g[X]$, then $P(\sigma^h(\eta)) = P(\sigma^{2h}(\eta)) = \ldots = P(\sigma^{h(k-1)}(\eta)) = 0$. Therefore, g/f is the degree of the minimum polynomial of η over K_g (see remark 12.19), and it follows from proposition 12.18 that $\dim_{K_g} K_f = g/f$.

13.1 All the roots of any polynomial equation $P(X) = 0$, with $P \in \mathbb{R}[X]$ (resp. $\mathbb{C}[X]$) are in $\mathbb{C}$, which is a radical extension of $\mathbb{R}$ (resp. $\mathbb{C}$).

13.2 From Lagrange's result 10.5, it is known that X_1, X_2 and X_3 can be rationally expressed from $t = X_1 + \omega X_2 + \omega^2 X_3$, where $\omega = (-1 + \sqrt{-3})/2$. Explicitly,

$$X_1 = (1/3)(s_1 + t + (s_1^2 - 3s_2)/t).$$

A straightforward computation yields :

$$t^3 = s_1^3 - (9/2)(s_1 s_2 - 3s_3) + (3\sqrt{-3}/2)\Delta,$$

where $\Delta = (X_1 - X_2)(X_1 - X_3)(X_2 - X_3) = \sqrt{D(s_1, s_2, s_3)}$, where $D(s_1, s_2, s_3)$ is the discriminant (see n° 8.15). Therefore, a radical extension of $\mathbb{Q}(s_1, s_2, s_3)$ containing X_1 can be constructed as follows :

$$R_0 = \mathbb{Q}(s_1, s_2, s_3)$$

$$R_1 = R_0(\sqrt{-3D(s_1, s_2, s_3)})$$

$$R_2 = R_1(\sqrt[3]{s_1^3 - (9/2)(s_1 s_2 - 3s_3) + (3/2)\sqrt{-3D(s_1, s_2, s_3)}}).$$

The field R_2 is not contained in $\mathbb{Q}(X_1, X_2, X_3)$ since $\sqrt{-3D(s_1, s_2, s_3)} \notin \mathbb{Q}(X_1, X_2, X_3)$. In order to show that

$\mathbb{Q}(X_1,X_2,X_3)$ does not contain any radical extension of $\mathbb{Q}(s_1,s_2,s_3)$ containing X_1 (or X_2 or X_3), one can argue as in section 13.4 : if $u \in \mathbb{Q}(X_1,X_2,X_3)$ has the property that some power u^p (with p prime) is invariant under
$\sigma : X_1 \to X_2 \to X_3 \to X_1$, then u is invariant under σ; indeed, from $\sigma(u^p) = u^p$, it follows that $\sigma(u) = \omega u$ for some p-th root of unity ω. Since $\sigma^3 = I$, one has $u = \sigma^3(u) = \omega^3 u$, hence p = 3. But 1 is the only cube root of unity in $\mathbb{Q}(X_1,X_2,X_3)$, so $\sigma(u) = u$.

In order to obtain a solution of the general equation of degree 4, one first solves a resolvent cubic equation, for instance the equation with root $u = (X_1+X_2)(X_3+X_4)$, i.e.

$$X^3 - a_1 X^2 + a_2 X - a_3 = 0$$

with $a_1 = 2s_2$

$a_2 = s_2^2 + s_1 s_3 - 4 s_4$

$a_3 = s_1 s_2 s_3 - s_1^2 s_4 - s_3^2$ (see exercise 3 of chapter 8).

Then, $v = X_1 + X_2$ is obtained as a root of the quadratic equation

$$X^2 - s_1 X + u = 0$$

(the other root is $X_3 + X_4$) and then X_1 and X_2 as roots of

$$X^2 - vX + \frac{(s_1-2v)s_4}{(s_1-v)(s_2-u)-s_3} = 0.$$

(Observe that the independent term is $X_1 X_2$, expressed as a function of v,u and the symmetric polynomials). Therefore, a radical extension of $\mathbb{Q}(s_1,s_2,s_3,s_4)$ containing X_1 can be constructed as follows :

$$R_0 = \mathbb{Q}(s_1, s_2, s_3, s_4)$$

$$R_1 = R_0(\sqrt{-3D(a_1,a_2,a_3)}) \quad (= R_0(\sqrt{-3D(s_1,s_2,s_3,s_4)}); \text{ see exercise 3 of chapter 8}),$$

$$R_2 = R_1(t)$$

where $t = \sqrt[3]{(a_1^3 - (9/2)(a_1 a_2 - 3a_3) + (3/2)\sqrt{-3D(a_1,a_2,a_3)})}$;

then $u = (1/3)(a_1 + t + (a_1^2 - 3a_2)/t) \in R_2$.

Let then

$$R_3 = R_2(\sqrt{s_1^2 - 4u});$$

then $v = (1/2)(s_1 + \sqrt{s_1^2 - 4u}) \in R_3$

and $x_1 \in R_4 = R_3(\sqrt{(v^2 - 4(s_1 - 2v)s_4[(s_1-v)\cdot(s_2-u)-s_3]^{-1})})$.

13.3 Since 3 is a primitive root of 7, proposition 12.29 shows that there is an automorphism σ of $\mathbb{Q}(\zeta_7)$ defined by $\sigma(\zeta_7) = \zeta_7^3$. Suppose $\mathbb{Q}(\zeta_7)$ is radical over $\mathbb{Q}$; then there is a tower of extensions

$$\mathbb{Q}(\zeta_7) = R_0 \supset R_1 \supset \ldots \supset R_h = \mathbb{Q}$$

where $R_i = R_{i+1}(u_i)$ with $u_i^{p_i} = a_i$ for some prime number p_i and some element $a_i \in R_{i+1}$ which is not a p_i-th power in R_{i+1}.

We shall prove that if a_i is invariant under σ^2, then u_i is invariant under σ^2 too; therefore every element in R_i is invariant under σ^2. By induction, it follows that every element in $R_0 = \mathbb{Q}(\zeta_7)$ is invariant under σ^2, a contradiction.

From $\sigma^2(a_i) = a_i$, it follows that $\sigma^2(u_i)^{p_i} = u_i^{p_i}$, whence $\sigma^2(u_i) = \omega u_i$ for some p_i-th root of unity $\omega \in \mathbb{Q}(\zeta_7)$. By theorem 12.42, the cyclotomic polynomial

Φ_{p_i} is irreducible over $\mathbb{Q}(\zeta_7)$ if $p_i \neq 7$. Therefore, the only prime numbers p_i such that $\mathbb{Q}(\zeta_7)$ contains a p_i-th root of unity other than 1 are $p_i = 2$ or 7.

Now, corollary 13.10 shows that $\dim_{R_{i+1}} R_i = p_i$; therefore, it is impossible that $p_i = 7$, since $\dim_{\mathbb{Q}} \mathbb{Q}(\zeta_7) = 6$, by theorem 12.15. If $p_i = 2$, then $\omega = 1$ or -1. However, it is impossible that $\sigma^2(u_i) = -u_i$, since by applying σ^2 twice to both sides of this equation one gets: $\sigma^6(u_i) = -u_i$, a contradiction since $\sigma^6 = I$. Therefore, the only possibility is that $\sigma^2(u_i) = u_i$, and the claim is proved.

On the other hand, $\mathbb{Q}(\zeta_7, \zeta_3)$ is radical over $\mathbb{Q}$. Indeed, let

$$\eta_0 = \zeta_7^{3^0} + \zeta_7^{3^2} + \zeta_7^{3^4} = \zeta_7 + \zeta_7^2 + \zeta_7^{-3}$$

$$\eta_1 = \zeta_7^{3^1} + \zeta_7^{3^3} + \zeta_7^{3^5} = \zeta_7^3 + \zeta_7^{-1} + \zeta_7^{-2}$$

be the periods of 3 terms of $\Phi_7 = 0$. It is readily checked that $\eta_0 + \eta_1 = -1$ and $\eta_0 \eta_1 = 2$, hence $\eta_0, \eta_1 = (1 \pm \sqrt{-7})/2$. Now, let $t = \zeta_7 + \zeta_3 \zeta_7^2 + \zeta_3^2 \zeta_7^{-3}$; then

$$\zeta_7 = (1/3)(\eta_0 + t + (2\eta_0 + 1)/t)$$

and $t^3 = 8 + 2\eta_0 + 3\zeta_3 + 6\zeta_3 \eta_0 \in \mathbb{Q}(\zeta_3, \eta_0) = \mathbb{Q}(\sqrt{-3}, \sqrt{-7})$, hence $\mathbb{Q}(\zeta_3, \zeta_7) = \mathbb{Q}(\zeta_3, \eta_0, t) = \mathbb{Q}(\sqrt{-3}, \sqrt{-7}, t)$ is a radical extension of $\mathbb{Q}$.

13.4 If $R = F(u)$ with $u^p = a$, an isomorphism $f: F[X]/(X^p - a) \xrightarrow{\sim} R$ is given by: $f(P(X) + (X^p - a)) = P(u)$.

13.5 The cube roots of 2 in $\mathbb{C}$ are $\sqrt[3]{2} \in \mathbb{R}$, $\sqrt[3]{2}(-1 + i\sqrt{3})/2$ and $\sqrt[3]{2}(-1 - i\sqrt{3})/2$. Since the latter two are not in $\mathbb{R}$, the field $\mathbb{Q}(\sqrt[3]{2}) \subset \mathbb{R}$ contains only one cube root of 2. Since, by the preceding exercise, all the fields obtained from $\mathbb{Q}$ by adjoining a cube root of 2 are isomorphic, they all contain only one cube root of 2. Therefore,

$\mathbb{Q}(\sqrt[3]{2})$, $\mathbb{Q}(\sqrt[3]{2}(-1+i\sqrt{3})/2)$ and $\mathbb{Q}(\sqrt[3]{2}(-1-i\sqrt{3})/2)$

are pairwise distinct subfields of $\mathbb{C}$.

14.1 By proposition 14.36, $\mathrm{Gal}(P/F)$ acts transitively on the roots of P. By theorem 14.39, it follows that the number of roots of P divides $|\mathrm{Gal}(P/F)|$.

14.2 Applying σ to the expressions $r_i = f_i(V)$, we get: $\sigma(r_i) = f_i(\sigma(V))$, and since $\{\sigma(r_1),\ldots,\sigma(r_n)\} = \{r_1,\ldots,r_n\}$ it follows that $r_1,\ldots,r_n$ have a rational expression in $\sigma(V)$.

14.3 This follows from proposition 14.37 by induction on the number of irreducible factors of P.

14.4 Irreducibility of $(X-u_1)\ldots(X-u_r)$ over F readily follows from proposition 14.36.

14.5 If T is a set of representatives of the right cosets of H in G, then

$$G(\alpha) = \bigcup_{\tau \in T} H\tau(\alpha)$$

is a decomposition of $G(\alpha)$ into subgroups which have H as group of substitutions. If this decomposition is the same as $G(\alpha) = \bigcup_{\sigma \in R} \sigma H(\alpha)$, then for each $\sigma \in R$, there is a $\tau \in T$ such that $\sigma H = H\tau$; in particular, $\sigma = \sigma.1 = \eta\tau$ for some $\eta \in H$, and for all $\xi \in H$:

$$\sigma\xi\tau^{-1} = \sigma\xi\sigma^{-1}\eta \in H,$$

hence $\sigma\xi\sigma^{-1} \in H$. This proves that H is normal in G. The converse is clear, since if H is normal, then $\sigma H = H\sigma$ for all $\sigma \in G$.

References

[A]   ABEL, N.-H. "Oeuvres complètes" (2 vol.) (L. Sylow and S. Lie, ed.), Grøndahl & Søn, Christiania, 1881.
[Ar1]   ARTIN, E. "Galois Theory", Notre Dame Math. Lectures, Notre Dame Univ. Press, Ind., 1948.
[Ar2]   ARTIN, E. "Collected Papers" (S. Lang and J. Tate, ed.), Addison-Wesley, Reading, Mass., 1965.
[Ay]   AYOUB, R.G. "Paolo Ruffini's contributions to the quintic", Arch. Hist. Exact Sci. 23 (1980/81) 253-277.
[Bos]   BOSMANS, H. "Sur le 'Libro de algebra' de Pedro Nuñez", Biblioth. Mathem. (Sér. 3) 8 (1908) 154-169.
[Bou1] BOURBAKI, N. "Algèbre, chapitres 4 et 5", Hermann, Paris, 1967.
[Bou2] BOURBAKI, N. "Algèbre, chapitres 4 à 7", Masson, Paris, 1981.
[Boy]   BOYER, C.B. "A History of Mathematics", J. Wiley & sons, New York, N.Y., 1968.
[Bu]   BÜHLER, W.K. "Gauss. A biographical study", Springer, Berlin, 1981.
[Caj]   CAJORI, F. "A History of Mathematical Notations, vol. 1 : Notations in Elementary Mathematics", Open Court, La Salle, Ill., 1974.
[C]   CARDANO, G. "The Great Art, or the Rules of Algebra", translated and edited by T.R. Witmer, MIT Press, Cambridge, Mass., 1968.
[Car]   CARREGA, J.-C. "Théorie des corps. La règle et le compas", Coll. formation des enseignants et formation continue, Hermann, Paris, 1981.
[CHR]   CHASE, S.U., HARRISON, D.K. and ROSENBERG, A. "Galois theory and Galois cohomology of commutative rings", Mem. Amer. Math. Soc. 52 (1968), 1-19.
[Coh]   COHN, P.M. "Algebra, vol. 2", J. Wiley & sons, London, 1977.
[Co]   COTES, R. "Theoremata tum Logometrica tum Trigonometrica Datarum Fluxionum Fluentes exhibentia, per Methodum Mensurarum ulterius extensam", pp 111-249 in : "Harmonia Mensurarum, sive Analysis & Synthesis per Rationum & Angulorum Mensuras promotae : Accedunt Alia Opuscula Mathematica per Rogerum Cotesium" (R. Smith, ed.) Cantabrigiae, 1722.
[D]   DESCARTES, R. "The Geometry", translated from the French and Latin by D.E. Smith and M.L. Latham, Dover, New York, N.Y., 1954.

[D*]   DESCARTES, R. "Geometria" (2 vol.), trad. F. Van Schooten, Ex typographia Blaviana, Amstelodami, 1683 (ed. tertia).

[Di]   DIEUDONNE, J. "Abrégé d'histoire des mathématiques 1700-1900" (2 vol.), Hermann, Paris, 1978.

[Dou]   DOUADY, R. et DOUADY, A. "Algèbre et théories galoisiennes" (2 vol.) Cedic/Fernand Nathan, Paris, 1977 et 1979.

[E]   EDWARDS, H.M. "Galois Theory", Graduate Texts in Math. 101, Springer, New York, N.Y., 1984.

[G]   GALOIS, E. "Ecrits et mémoires mathématiques d'Evariste Galois", (R. Bourgne et J.-P. Azra, éd.), Gauthier-Villars, Paris, 1962.

[Gz]   GANDZ, S. "The origin and development of the quadratic equations in Babylonian, Greek and early Arabic algebra", Osiris 3 (1937) 405-557.

[Gau1] GAUSS, C.F. "Demonstratio nova theorematis omnem functionem algebraicam rationalem integram unius variabilis in factores reales primi vel secundi gradus resolvi posse", Apud C.G. Fleckeisen, Helmstadii, 1799. (Werke Bd III, Georg Olms, Hildesheim, 1981, pp 1-30).

[Gau2] GAUSS, C.F. "Disquisitiones Arithmeticae", Apud Gerh. Fleischer Iun; Lipsiae, 1801. (Werke Bd I, Herausg. König. Ges. Wiss. Göttingen, 1870).

[Gau3] GAUSS, C.F. "Demonstratio nova altera theorematis omnem functionem algebraicam rationalem integram unius variabilis in factores reales primi vel secundi gradus resolvi posse", Comm. soc. regiae scient. Gottingensis recentiores 3 (1816). (Werke Bd III, Georg Olms, Hildesheim, 1981, pp 31-56).

[Gi]   GIRARD, A. "Invention Nouvelle en l'Algèbre", réimpression par D. Bierens De Haan, Muré Frères, Leiden, 1884.

[Go]   GOLDSTINE, H.H. "A History of Numerical Analysis from the 16th through the 19th century", Studies in the History of Math. and Phys. sciences 2, Springer, New York, N.Y., 1977.

[Ha]   HANKEL, H. "Zur Geschichte der Mathematik in Alterthum und Mittelalter", Teubner, Leipzig, 1874.

[HW]   HARDY, G.H. and WRIGHT, E.M. "An Introduction to the Theory of Numbers", Clarendon Press, Oxford, 1979.

[He]   HEATH, T.L. "The thirteen books of Euclid's Elements", Cambridge Univ. Press, Dover, New York, N.Y., 1956.

[H1]   HUDDE, J. "Epistola prima, de Reductione Aequationum", in [D*, vol. 1], pp 406-506.

[H2]   HUDDE, J. "Epistola secunda, de Maximis et Minimis", in [D*, vol. 1], pp 507-516.

[J1]   JACOBSON, N. "Lectures in Abstract Algebra, vol. III", Van Nostrand, New York, N.Y., 1964.

[J2]   JACOBSON, N. "Basic Algebra I", Freeman, San Francisco, Ca., 1974.

[K]   KAPLANSKY, I. "Fields and Rings", Chicago Lectures in Math., Univ. Chicago Press, Chicago, Ill., 1972.

[Ka]   KARPINSKI, L.C. "Robert of Chester's Latin translation of the Algebra of al-Khowarizmi", Univ. of Michigan Studies, Humanistic series 11, MacMillan, New York, N.Y., 1915.
[Ki]   KIERNAN, B.M. "The Development of Galois Theory from Lagrange to Artin", Arch. Hist. Exact Sci. 8 (1971) 40-154.
[Kl]   KLINE, M. "Mathematical thought from ancient to modern times", Oxford Univ. Press, New York, N.Y., 1972.
[Kn]   KNORR, W.R. "The evolution of the Euclidean Elements", Synthese Historical Lib. 15, D. Reidel, Dordrecht, 1975.
[L]   LAGRANGE, J.-L. "Réflexions sur la résolution algébrique des équations", Nouveaux Mémoires de l'Acad. Royale des sciences et belles-lettres, avec l'histoire pour la même année, 1 (1770) 134-215; 2 (1771) 138-253. ("Oeuvres de Lagrange, vol. 3" (J.-A. Serret, éd.) Gauthier-Villars, Paris, 1869, pp 203-421).
[Le]   LEBESGUE, H. "L'oeuvre mathématique de Vandermonde", Enseignement Math. Sér. II, 1 (1955) 201-223.
[Lz1] LEIBNIZ, G.W. "Mathematische Schriften, Bd V", herausg. von C.I. Gerhardt, Georg Olms, Hildesheim, 1962.
[Lz2] LEIBNIZ, G.W. "Der Briefwechsel von Gottfried Wilhelm Leibniz mit Mathematikern", herausg. von C.I. Gerhardt, Georg Olms, Hildesheim, 1962.
[N1]   NEWTON, I. "The mathematical papers of Isaac Newton", ed. by D.T. Whiteside, vol. I : 1664-1666, Cambridge Univ. Press, Cambridge, 1967; vol. IV : 1674-1684, Cambridge Univ. Press, Cambridge, 1971; vol. V : 1683-1684, Cambridge Univ. Press, Cambridge, 1972.
[N2]   NEWTON, I. "The mathematical works of Isaac Newton, vol. 2", assembled with an introduction by D.T. Whiteside, The sources of science, Johnson Reprint Corp., New York, London, 1967.
[No]   NOVY, L. "Origins of modern algebra", Noordhoff, Leyden, 1973.
[Ro]   ROMANUS, A. "Ideae Mathematicae Pars Prima, sive Methodus Polygonorum", apud Ioannem Masium, Lovanii, 1593.
[R]   RUFFINI, P. "Opere Matematiche" (3 vol.), E. Bortolotti, ed., Ed. Cremonese della Casa Editrice Perrella, Roma, 1953-1954.
[Sam] SAMUEL, P. "Théorie algébrique des nombres", Coll. Méthodes, Hermann, Paris, 1971.
[Se]   SERRET, J.-A. "Cours d'algèbre supérieure" (2 vol.), Gauthier-Villars, Paris, 1866 (3ème éd.).
[Sm]   SMITH, D.E. "A Source Book in Mathematics" (2 vol.), Dover, New York, N.Y., 1959.
[S]   STEVIN, S. "The principal works of Simon Stevin. Volume II B : Mathematics". Ed. by D.J. Struik, C.V. Swets en Zeitlinger, Amsterdam, 1958.
[St]   STEWART, I. "Galois Theory", Chapman and Hall, London, 1973.

[Ta]   TATON, R. "Les relations d'Evariste Galois avec les mathématiciens de son temps", Rev. Hist. Sc. 1 (1948), 114-130.

[T]   TSCHIRNHAUS, E.W. "Methodus Anferendi Omnes Terminos intermedios ex data aequatione", Acta Eruditorum (Leipzig) (1683), 204-207.

[VM]   VANDERMONDE, A.T. "Mémoire sur la résolution des équations", Histoire de l'Acad. Royale des Sciences (avec les mémoires de Math. & de Phys. pour la même année, tirés des registres de cette Acad.) (1771), 365-416.

[VP]   VAN DER POORTEN, A. "A Proof that Euler missed ... Apéry's Proof of the Irrationality of $\zeta(3)$", Math. Intel. 1 (1979) 195-203.

[VW1]   VAN DER WAERDEN, B.L. "Algebra" (2 vol.) (7th ed. of "Moderne Algebra"), Heidelberger Taschenbücher 12 & 23, Springer, Berlin, 1966 & 1967.

[VW2]   VAN DER WAERDEN, B.L. "Science Awakening I", Noordhoff, Leyden, 1975.

[VW3]   VAN DER WAERDEN, B.L. "Die Galoissche Theorie von Heinrich Weber bis Emil Artin", Arch. Hist. Exact Sci. 9 (1972) 240-248.

[V1]   VIETE, F. "Ad Problema quod omnibus Mathematicis totius orbis construendum proposuit Adrianus Romanus Francisci Vietae Responsum", apud Iametium Mettayer, Parisiis, 1595.

[V2]   VIETE, F. "The Analytic Art", translated by T.R. Witmer, Kent State Univ. Press, Kent, Ohio, 1983.

[Web]   WEBER, H. "Lehrbuch der Algebra, Bd I", F. Vieweg u. Sohn, Braunschweig, 1898.

[W1]   WEIL, A. "De la métaphysique aux mathématiques", Sciences (1960), 52-56. ("Oeuvres Scientifiques - Collected Papers, vol. 2", Springer, New York, N.Y., 1979, pp 408-412).

[W2]   WEIL, A. "Two lectures on number theory, past and present", Enseignement Math. 20 (1974) 87-110. ("Oeuvres Scientifiques - Collected Papers, vol. 3", Springer, New York, N.Y., 1979, pp 279-302).

Index

Abel, Niels Henrik, 2-4, 273, 275-78, 280, 286-87, 295, 299, 301-03, 316-17, 368, 383-84
abelian group, 316
Alembert, Jean Le Rond d', 153
algebra :
   Arabic, 16-20, 32
   Babylonian, 7-11, 13-15, 17, 32
   Greek, 11-16, 31
al-Khowarizmi, Mohammed ibn Musa, 16-19, 32
alternating group A_n, 299, 361
Apéry, Roger, 149
Artin, Emil, 3, 399-401, 404

Bachet de Méziriac, Claude, 115
Bernoulli, Jacques, 146
Bernoulli, Nicholas, 99
Bezout, Etienne, 94, 115, 163-64, 167, 169, 177, 179, 182, 274
Bolzano, Bernhard, 160
Bombelli, Rafaele, 30, 38
Bourbaki, Nicolas, 3, 400-01

Cardano, Girolamo, 2-3, 21, 23-24, 26, 29-32, 36-37, 42, 55, 163, 169-70, 175
Cardano's formula, 24-26, 28-30, 86, 96, 102, 106, 363
Cauchy, Augustin-Louis, 155, 275-76, 299
Chase, Stephen, 400
Cohn, Paul Moritz, 3
completely solvable (by radicals), 345
congruence modulo an integer, 221
constructible point, 263
coset, 189, 347
Cotes, Roger, 3, 97, 100, 102, 105, 107, 109, 124-25
Crelle, August Leopold, 275
cyclic group, 118
cyclotomic polynomial, 120-22, 204, 209 ff, 215-16, 219-21, 230, 233 ff, 239 ff, 301-02, 314-16
cyclotomy, 109

Dedekind, Richard, 230, 257, 398-99, 404
Delambre, Jean-Baptiste, 274
derivative, 73
Descartes, René, 3, 32, 37, 40-42, 51-52, 86
Diophantus of Alexandria, 16, 41
discriminant, 142

Edwards, Harold, 3
Eisenstein, Ferdinand Gotthold Max, 230, 232, 234, 383
elementary symmetric polynomial, 132
elimination theory, 77, 92-93, 103, 116, 164-65
equation :
   cubic, 21 ff, 83-86, 94-96, 106, 146, 165-66, 169-74, 207, 362-64
   cyclotomic : see cyclotomic polynomial
   general : 130 ff, 193 ff, 204 ff, 273 ff, 286 ff, 295 ff, 301, 314, 362, 383-84
   quadratic : 6 ff, 204-05
   quartic : 31-35, 86-87, 166-67, 177-79, 207-08, 341-44, 363-64
   resolvent cubic : 35, 341-44
   of squared differences : 150
Euclid, 12-14, 19, 32
Euclid's algorithm, 38, 61, 63-65, 76, 95
Euclidean division property, 60, 81, 115, 121, 134
Euler, Leonhard, 77, 97, 106, 127, 146, 148-49, 153-54, 163-64, 167, 169, 177, 179, 182, 203, 224, 269, 271, 274
exponent :

of an element in a group : 229
of an integer modulo a prime
number : 226
of a root of unity : 114

Fermat, Pierre de, 224, 227, 229, 259, 269-71, 315, 373
Ferrari, Ludovico, 31-32, 86, 177, 195, 342, 363-64
Ferro, Scipione del, 21
Fior, Antonio Maria, 21
Foncenex, Daviet François de, 153
Fontana, Niccolo : see Tartaglia
Fourier, Joseph, 303
fundamental theorem :
of algebra : 50, 98, 105, 139, 141, 144, 152 ff, 219, 274
of symmetric fractions or polynomials : 132
of Galois theory : 190, 402

Galois, Evariste, 1-4, 301, 303-09, 313, 317, 320, 322, 325, 345-46, 351, 368, 371-72, 381, 383, 387-88, 392, 396-99
Galois extension of fields : 399-402, 410-12
Galois group : 304-06, 308-10, 312-17, 326-27, 329-35, 337-38, 340-46, 353-55, 357-66, 368, 381-84, 395, 399, 401-02, 407-12
Galois resolvent : 308-11, 313, 315-16, 320, 324-26, 330-31, 333, 335, 337-38, 355, 395
Galois theory : 2-4, 11, 183, 190, 216, 252, 303-04, 368, 396-402
Gandz, Solomon, 11
Gauss, Carl Friedrich, 2-4, 46, 139, 141, 153, 160, 216, 219-22, 230, 233, 242-46, 252, 255-58, 268, 271, 273-74, 283, 301-02
Girard, Albert, 3, 48-53, 88, 129, 152-55, 285, 288, 332
greatest common divisor (G.C.D.), 61
group, 183 ff, 208, 304, 387 ff.
(See also : Galois group)

Harrison, D.K., 400
height of a radical extension, 278
Hero, 16

Hippasus of Metapontum, 12
Hudde, Johann, 73, 75-77, 306, 384

ideal, 154
index (of a subgroup), 189, 333, 347
irreducible polynomial, 65
isotropy subgroup, 184, 369

Jacobson, Nathan, 3, 400

Kaplansky, Irving, 3
Khayyam, Omar, 19-20
Kiernan, Melvin, 398
Knorr, Wilbur Richard, 12
Kronecker, Leopold, 66, 155, 230, 257, 287, 398-99

Lacroix, Sylvestre-François, 273
Lagrange, Joseph-Louis, 2, 3, 11, 46, 132, 153, 163, 168-70, 173-75, 177, 179-81, 183, 185-88, 190, 192-93, 203-04, 206, 215, 219, 229, 243, 253, 273-75, 283-84, 333, 342-43, 347, 356, 361, 369, 373
Lagrange resolvent : 182-83, 195-200, 203, 206-07, 213, 215, 253, 257, 355
Landau, Edmund, 230
leading coefficient, 59
Lebesgue, Henri, 203, 217
Legendre, Adrien-Marie, 273
Leibniz, Gottfried Wilhelm, 91, 98-100, 105, 122-23, 146, 152
Liouville, Joseph, 303, 397

Mertens, Franz, 230
minimum polynomial, 238, 308
Moivre, Abraham de, 2, 4, 97-98, 102-07, 110, 112-13, 120, 124, 152, 163, 210, 216
monic polynomial, 59

Newton, Isaac, 3, 53-55, 98, 100, 122-24, 126, 147
normal subgroup, 334
Nový, Lubos, 398
Nunes, Pedro, 38, 51

orbit, 365, 368
order :
of an element in a group, 229
of a group, 184

428

Pacioli, Luca, 20
period (of a cyclotomic equation), 242
primitive :
   root of unity : 114
   root of a prime number : 221
Pythagoras, 12, 101

quotient ring (by an ideal), 155

radical :
   expression (solution) by radicals : 110 ff, 209 ff, 252 ff, 279, 295 ff, 302, 345 ff, 381 ff, 396
   radical extension of fields : 278
rank of a field extension, 401
Recorde, Robert, 37
regular polygon, 109, 219, 245, 263 ff, 302
resolvent : see equation (cubic resolvent), Galois resolvent or Lagrange resolvent
resultant, 77-78, 93, 116, 127, 151
Riemann, Bernhard, 149
root :
   of a complex number : 105, 107
   of a polynomial : 70
   of unity : 108
   common roots of two polynomials : 77 ff
   multiple root of a polynomial : 71, 72 ff, 144-46
   rational roots of a rational polynomial : 88-90
Romanus, Adrianus : see Van Roomen
Rosenberg, Alex, 400
Ruffini, Paolo, 3, 273-78, 280, 286, 295, 299, 301, 383
ruler and compass constructions, 219, 245, 263-71, 302

Schur, Issai, 230
solvable group, 346, 386
Stevin, Simon, 6-7, 37-40, 55
Stewart, Ian, 3
symmetric group S_n, 183
symmetric polynomial (or rational fraction), 131

Tartaglia, Niccolo, 21-23, 27

transitive group of permutations, 365
Tschirnhaus, Ehrenfried Walter, 2-3, 83, 90, 92, 97, 164, 167, 169, 175, 177, 179

Van Collen, Ludolf, 43
Vandermonde, Alexandre-Théophile, 2-3, 106, 113, 132, 150, 168, 203-13, 216-17, 219, 240, 242
Van der Waerden, Bartel Leendert, 3, 398-99
Van Roomen, Adriaan, 43, 45-47
Viète, François, 2-3, 7, 37, 41-42, 44-47, 56, 83-86, 96, 106, 129

Wantzel, Pierre Laurent, 268, 276, 295
Waring, Edward, 132-34, 138, 141, 143, 168
Weil, André, 11, 398

Date Due

DEC 05 '89			
AP·5-31-0(ORBIS			

BRODART, INC.   Cat. No. 23 233   Printed in U.S.A.